Mathematical Optimization and Modeling: Concepts, Theories and Applications

Mathematical Optimization and Modeling: Concepts, Theories and Applications

Edited by **Lucas Lincoln**

Willford Press
New York

Published by Willford Press,
118-35 Queens Blvd., Suite 400,
Forest Hills, NY 11375, USA
www.willfordpress.com

Mathematical Optimization and Modeling: Concepts, Theories and Applications
Edited by Lucas Lincoln

International Standard Book Number: 978-1-68285-084-8 (Hardback)

Printed in the United States of America.

Contents

Preface

Optimization theory and techniques are a primary area of study under applied mathematics. This book on mathematical optimization and modeling seeks to delve into its various concepts, theories and models. Different approaches, evaluations, methodologies and advanced studies on mathematical optimization have been included in this text which will provide useful insights to the readers. Mathematical theory of uncertainty, optimization under uncertainty, applications of uncertainty, etc. are some of the topics discussed extensively in this book. As this field is emerging at a rapid pace, the contents of the book will help the readers understand the upcoming concepts and prospective applications of mathematical modeling and mathematical optimization.

After months of intensive research and writing, this book is the end result of all who devoted their time and efforts in the initiation and progress of this book. It will surely be a source of reference in enhancing the required knowledge of the new developments in the area. During the course of developing this book, certain measures such as accuracy, authenticity and research focused analytical studies were given preference in order to produce a comprehensive book in the area of study.

This book would not have been possible without the efforts of the authors and the publisher. I extend my sincere thanks to them. Secondly, I express my gratitude to my family and well-wishers. And most importantly, I thank my students for constantly expressing their willingness and curiosity in enhancing their knowledge in the field, which encourages me to take up further research projects for the advancement of the area.

Editor

1

Stochasticity and noise-induced transition of genetic toggle switch

Wei-Yin Chen

Correspondence: cmchengs@olemiss.edu
Department of Chemical Engineering, University of Mississippi, Anderson Hall, University, MS 38677-9740, USA

Abstract

The ability to predict and analyze the function of genetic circuits will enhance the design of autonomous, programmable, complex regulatory genetic structures. An abundance of modeling techniques has recently been developed to delineate simple genetic structures in terms of their constituents. Simple systems with characteristics of feedback inhibition, multi-stability, switching, and oscillatory expression have often been the focus. The present work is an attempt to improve existing deterministic models that fail to oblige to the crucial aspect of noise in genetic modeling.
The objective of this work is to analyze, model, and simulate the protein populations in gene expression mechanisms by resorting to stochastic algorithms. The system involves two types of genes; the protein produced from the expression of one gene is capable of turning off the expression of the other gene. Rates of degradation of these proteins are assumed to be proportional to their concentrations. The master equation of this 'genetic toggle switch' is formulated using the probabilistic population balance around a particular state and by considering five mutually exclusive events. The efficacy of the present methodology is mainly attributable to the ability to derive the governing equations for the means, variances, and covariance of the random variables by the method of system-size expansion of the nonlinear master Equation. A less laborious approach based on Kurtz's limit theorems for the derivation of the stochastic characteristics is also presented for comparison. Solving the resultant ordinary differential equations governing the means, variances, and covariance of the master equations simultaneously using the published data yield information concerning not only the means of the two populations of proteins but also the minimal uncertainties of the populations inherent in the expressions. It is demonstrated that systems with small populations are susceptible to large internal fluctuations (or uncertainties) in their population evolution. Large uncertainties are observed after the populations enter the proximity of the saddle node, which is likely to cause transition of system's steady state from one to another. Independent Monte-Carlo simulation runs clearly demonstrates that the occurrence of such internal noise-induced transition.

Keywords: Stochastic; Nonlinear; Model; Genetic networks; Genetic circuits; Metastable; Noise-induced transitions

Introduction

One of the earliest examples of a bistable genetic switch is represented in the rightward operator of bacteriophage lambda [1,2]. The essential elements of this type of genetic switch, are a pair of promoters that each produces a repressor protein capable of inhibiting the production of the opposing repressor. Overlayed on these essential elements are several layers of regulatory nuance. To elucidate the impacts of these essential elements of a simplified regulatory circuit, a series of synthetic toggle switches were created.

Figure 1 shows the two-state genetic toggle switch consisting of two protein repressor genes and two promoters, which was investigated by Gardner et al. [3]. Each promoter enables the production of one repressor and is inhibited by the other. They elegantly designed experiments that demonstrated switching of a toggle circuit from one steady state to another by switching system's parameters across the bifurcation curve to a bistable region through either thermal inactivation of Repressor A or ligand binding-induced dissociation of the Repressor B-DNA complex. In the proximity of the bifurcation point, the final steady-state protein population possesses a bimodal distribution in their green fluorescent protein (GFP) fluorescence. It does not have a sharp jump from one fluorescence level to another, as the deterministic model predicts. The authors surmise that the stochastic nature of the dynamics blurs the bifurcation point.

McAdams and Arkin's [4,5] Monte-Carlo simulations of gene expression revealed the importance of fluctuations, or noises or uncertainties, of small systems. In such small

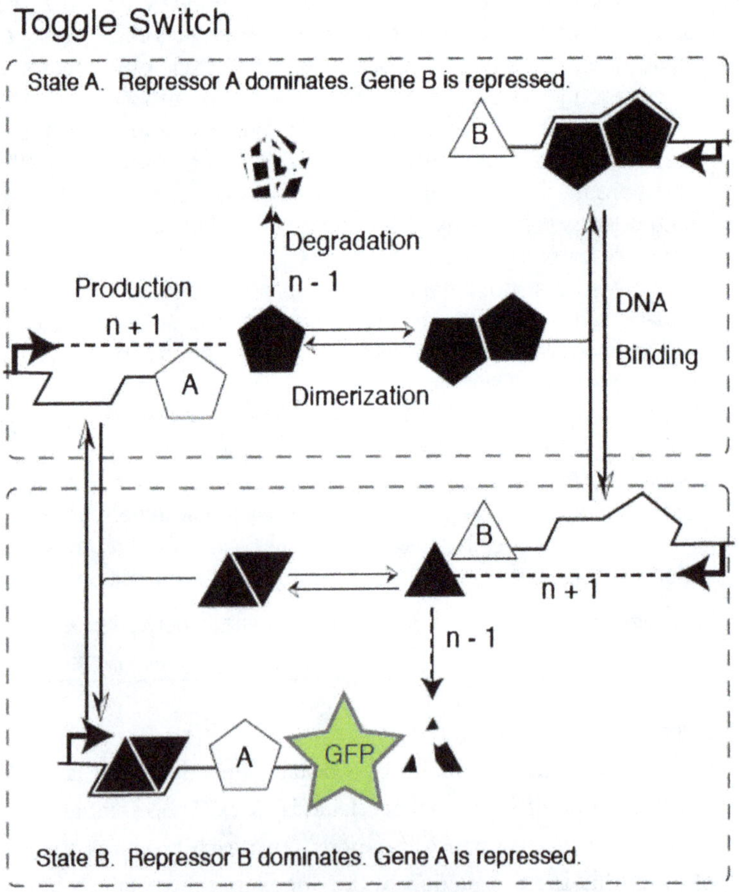

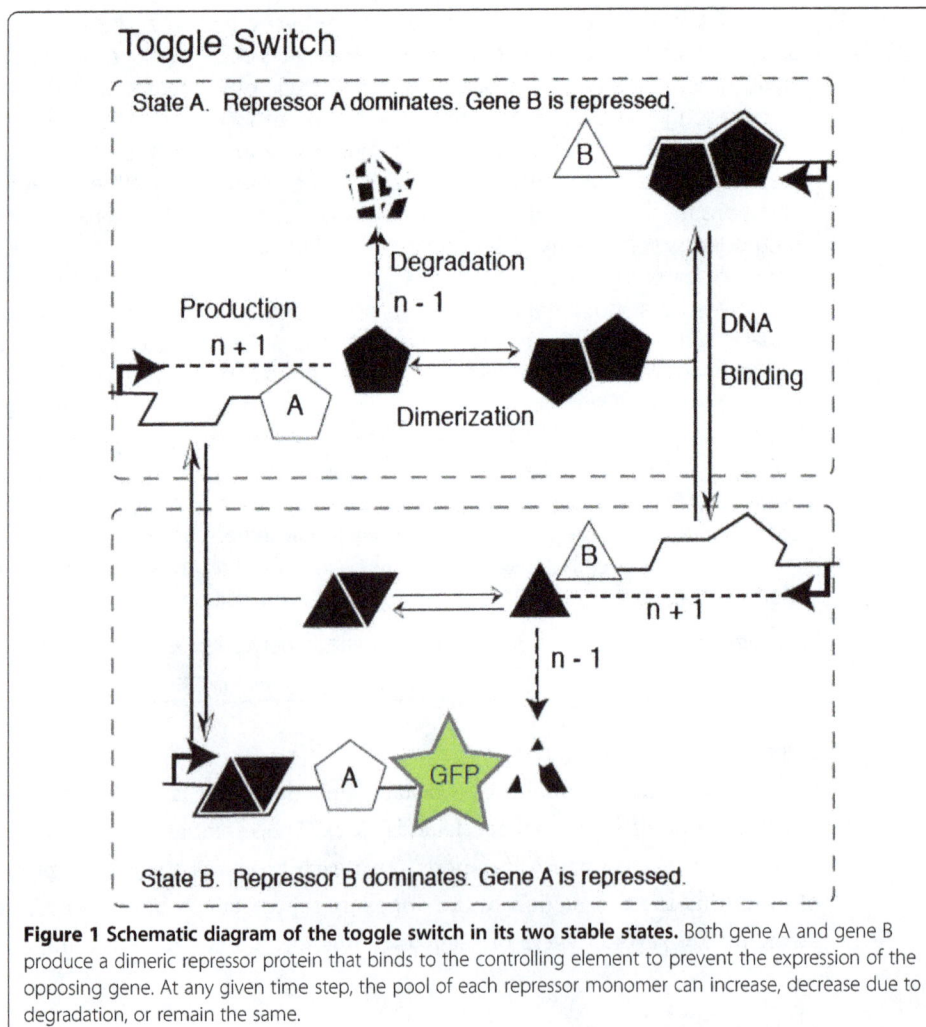

Figure 1 Schematic diagram of the toggle switch in its two stable states. Both gene A and gene B produce a dimeric repressor protein that binds to the controlling element to prevent the expression of the opposing gene. At any given time step, the pool of each repressor monomer can increase, decrease due to degradation, or remain the same.

systems, proteins are produced from an activated promoter in short bursts of variable numbers of proteins that occur at random time intervals. As a result, there can be large differences in the time between successive events in regulatory cascades across a cell population, which, in turn, creates both special and temporal heterogeneity of cell populations in biological systems. Soon after the discovery of the potential impacts of the stochasticity of genetic regulatory system, stochastic algorithms developed by chemical physicists have been introduced in analyzing gene expression (e.g., [6,7]). The stochastic nature of a competitive expression mechanism can produce probabilistic outcomes in switching mechanisms that select between alternative regulatory paths, such as toggle switch.

Stochastic algorithms have been developed for analyzing noise of different origins and internal and external noises (e.g., [8-10]). External noises are the fluctuations created in an otherwise deterministic system by the application of an external random force, whose stochastic properties are supposed to be known. A Langevine equation is commonly adopted in the analysis of dynamics caused by external noises. Internal noise arises from discrete systems where only a limited number of variables affecting the populations of the discrete entities can be included in the analysis. Small discrete systems, such as genes of small populations, often exhibit notable internal fluctuations. A master equation, derived from probabilistic population balance around a particular state of the system by taking into account all mutually exclusive events, has been adopted this type of discrete state, continuous-time stochastic processes.

The stochasticity of gene expression is complicated by its nonlinearity. Multiple steady states, stability, and bifurcation in gene expressions (e.g., [11]) could mingle with the analysis of noise, or fluctuations. The efficacy of the master equation algorithm in gene expression is mainly attributable to its powerful ability to solve the nonlinear master equations through the system size expansion [9,12]. In this approach, a suitable expansion parameter must be identified in the master equation. The expansion parameter represents the size of the fluctuations, and therefore, the magnitude of the jumps, or transitions, of system's state. Since the internal noises are expected to be low when the system size is large, the system size has been proposed as an expansion parameter. Master equation formulation along with the system-size expansion has indeed applied to the analysis of noise in gene expression. It should be mentioned that the limit theorems of Kurtz [13-15] have rendered the complex procedure of system size expansion simple and highly accessible. Kurtz's proof demonstrated the solution of a Langevine equation approaches to van Kampen's system size expansion as the system size approaches to infinite.

Kepler and Elston [16] examined the stochastic dynamics of the single-gene system with and without feedback and a switching system composed of two mutually repressed genes. Several assumptions were made in their simplified model: the two genes share the same operator and same degradation rate, proteins bind to the operator as dimers, and rate of dimerization is fast. Both master equation and Monte-Carlo simulation were adopted in their study. Scott et al. [17] adopted the master equation along with the system size expansion algorithm in the estimation of internal noise of the single-gene system that involves the mRNA formation and degradation and protein formation and degradation.

The system size expansion has several limitations in modeling the gene regulatory process. It is a good approximation to the master equation for small internal noise and large system size. Moreover, the noise should be well within the boundary of attraction [9]. Thus, noises in oscillatory process and those away from the steady states have been a focus of several studies. Tao et al. [18] studied the noise far from the steady states and revealed that during the approach to equilibrium, the noise is not always reduced by the strength of the feedback. This is contrary to results seen in the equilibrium limit which show decreased noise with feedback strength. Ito and Uchida [19] found that the internal noise of a regulatory single-gene system grows without bound in oscillatory networks and developed an alternative method for estimating the evolution of internal noise in such systems.

Kepler and Elston's simulation work [16] demonstrated that simple noisy genetic switch have rich bifurcation structures. Among them, bifurcations driven solely by changing the rate of operator fluctuations even as the underlying deterministic system remains unchanged. They find stochastic bistability where the deterministic equations predict monostability and vice versa. Ochab-Marcinek [20] investigated the stationary behavior of a nonlinear system, a reduced, deterministic Yildirim and Mackey [21] model of the gene regulatory system, and discovered the transition of a steady state induced by noise. A perturbed Gaussian white noise term was introduced in the deterministic model followed by numerical simulations. Turcotte et al. [22] studied noise-induced stabilization of an unstable state of a genetic switch that undergoes a variety of bifurcations in response to parameter changes. Their Monte Carlo simulations showed that near one such bifurcation, noise induces oscillations around an unstable spiral point and thus effectively stabilizes this unstable fixed point.

In addition to the master equation algorithm, Monte Carlo simulation has been adopted in simulating the dynamic behaviors in genetic regulatory systems under the influences of internal noise (e.g., [11,23,24]). The Monte Carlo simulation shares the same assumption, the Markov property, as the master equation, and the noise can be obtained directly from master equation's deterministic counterpart. Moreover, the Monte Carlo simulation is capable of revealing the various characteristics of nonlinear dynamic system, such as the number of steady states, bifurcation, and internal noises.

In this expositional work, the master equations are formulated by stochastic population balance. Van Kampen's system size expansion of the resultant nonlinear master equation gives rise to the variances of the processes. We demonstrate the implementation of Kurtz's limit theorems can efficiently achieve the same goal. Simulations are conducted based on both the master equations and the Monte Carlo procedure for three systems: bistable, monostable, and on the bifurcation curve. Finally, we demonstrate the possibility of transition induced by internal noises for a bistable system.

Model formulation

A genetic toggle switch with negative feedback to the genes consists of two mutually coupled genes. The transcription products of these genes are two inhibitory repressor proteins competing to shut off the production of two constitutive promoters [1,3,25]; the protein transcribed by a gene of one type is capable of deactivating the transcription of the other gene. A toggle switch typically has more than one possible stable steady state depending on the reaction parameters under consideration [3]. There are a number of instances in nature where this switch-like behavior is utilized. The lysogeny/lysis switch of the bacteriophage λ virus infecting the bacterium *Escherichia coli* is a representative example and has been discussed in detail by Ptashne [1] and Ptashne and Gann [25].

Gardner [26] discussed results generated from their deterministic model of a negative feedback toggle switch. Each type of the repressor protein is involved in two types of processes. The first process corresponds to the production of the protein. The rate of protein production is proportional to the concentration of mRNA, which, in turn, is proportional to the concentration of the un-repressed gene, G. The repressor binding on un-repressed gene is commonly assumed to be in a quasi-steady state with the repressor, R, and the repressed gene, GR_m, i.e.,

$$G + m\ R \leftrightarrows GR_m$$

Moreover, by assuming the total number of un-repressed genes is much larger than that of R so that G remains constant during the process, it can be shown that the rate of production of protein is proportional to $\frac{1}{1+KR^m}$ where K is the equilibrium constant of the above reaction and R the concentration of the repressor monomer [27-29]. The second process in the model of Gardner et al. is degradation of the protein that is assumed to be first order.

Similar to the work of Gardner, we will assume that the genes are in equilibrium with their repressed genes in the current work. The stochastic nature of a competitive expression mechanism can produce probabilistic outcomes in switching mechanisms that select between alternative regulatory paths, such as toggle switch.

The master equation describing the stochastic nature of the toggle switch is developed through the probabilistic population balance. The formulation of the master equation given below follows what Oppenheim et al. [8], Gardiner [10], and van Kampen [9] established. We have previously adopted this algorithm in the analysis of disease spread [30].

Mathematical assumptions

Let the random variables, $N_1(t)$ and $N_2(t)$ represent the populations of the repressor protein R_1 and repressor protein R_2 at time t, respectively. The random vector of the system is $\boldsymbol{N}(t)$ such that $\boldsymbol{N}(t) = [N_1(t), N_2(t)]$ and the realization of this random vector representing the state of the system at time t is given by $\boldsymbol{n}(t)$, where $\boldsymbol{n}(t) = [n_1(t), n_2(t)]$. Moreover, the probability of the system to be in state $\boldsymbol{n}$ at time t is denoted by $P_{n1,n2}(t)$ or $P[n_1(t),n_2(t);t]$. The following assumptions are imposed in driving the master equation governing the transition of the system among various states.

1. The random vector, $N(t)$, is Markovian, i.e., for any set of successive times, $t_1 < t_2 < \ldots < t_q$, we have $P\ [N(t_q) * N(t_1), N(t_2), \rfloor, N(t_{q-1})] = P\ [N(t_q) * N(t_{q-1})]$.
2. The number of increments or decrements in population numbers of the classes depends only on the time interval, Δt, but not time, i.e., it is temporally homogeneous, signifying that $N(\Delta t)$and $[N(t + \Delta t) - N(t)]$ are identically distributed.
3. The probability of an individual to produce or degrade is proportional to the duration of time interval, $(t, t + \Delta t)$, if the value of Δt is sufficiently small.
4. The probabilities of two or more transitions to take place are negligible during the time interval, $(t, t + \Delta t)$, so that at most, one transition occurs during this period.
5. Individual proteins in the same class have the same probability of contacting the genes, and therefore, have the same probability of repressing the genes. Similarly, the individual proteins in the same class have the same probability of being degraded.

Transition intensity functions

On the basis of the assumptions given in the proceeding subsection, the transition probability of each event can be written in terms of the transition intensity functions, k_1, k_2, α_2, and α_4, as follows:

The first transition-intensity function, k_1, is the production probability of a type-1 repressor protein from a particular active (not repressed) gene of type-1 per unit time. Based on the assumption of temporal homogeneity, we have

$$\begin{aligned}&\Pr\begin{pmatrix}\text{a particular active gene of type}-1\text{ will produce a repressor protein of}\\ \text{type}-1\text{ during the time interval, }(t, t+\Delta t)\end{pmatrix}\\ &\quad = k_1\ \Delta t\ +\ o(\Delta t),\end{aligned}$$

where $\lim\limits_{\Delta t\to 0}\frac{o(\Delta t)}{\Delta t}=0$. By considering all active type-1 genes in the system, the probability that the population of the type-1 protein will increase by one is $k_1 G_{a1}$, where G_{a1} denotes the number of active gene of type-1, i.e., the genes that are not repressed. Mathematically,

$$\begin{aligned}&\Pr\left(\text{a type}-1\text{ protein will be produced during the time interval, }(t, t+\Delta t)\right)\\ &\quad = k_1\ G_{a1}\ \Delta t\ +\ o(\Delta t)\\ &\quad = \alpha_1\ f_1\ \Delta t\ +\ o(\Delta t)\end{aligned}$$

where f_1 is the ratio of populations of active gene to total, active and repressed, genes of type-1. In writing the last line of the above statement, we assume that the total number of gene remains constant during the process of interest. Thus, the parameter, α_1, is the probability that a particular active gene will transcribe and produce a type-1 protein per unit time multiplied by the total number of genes.

For a negative feedback genetic circuit, Goodwin [27,28] and Griffith [29] showed that

$$f_1=\frac{1}{1+K_a {n_2}^m},$$

where K_a is the equilibrium constant of the combination reaction of the active gene of type-1 and repressor, a m-mer, and m is the number of protein monomers of type-2 in the repressor. Combining the last two equations yields

$$\begin{aligned}&\Pr\left(\text{a type}-1\text{ protein will be produced in time interval, }(t, t+\Delta t)\right)\\ &\quad = \frac{\alpha_1}{1+K_a {n_2}^m}\ \Delta t\ +\ o(\Delta t)\end{aligned}\tag{1}$$

The second transition intensity function, α_2, is the overall consumption probability of a particular active protein of type-1 in time interval, $(t,\ t+\Delta t)$, including its function in repressing protein type-2. Mathematically,

$$\begin{aligned}&\Pr\left(\text{a particular repressor protein of type}-1\text{ will be consumed in time interval, }(t, t+\Delta t)\right)\\ &\quad = \alpha_2\Delta t+o(\Delta t)\end{aligned}$$

By considering all repressor protein of type-1 genes in the system, the probability that the population of the type-1 protein will decrease by one is $\alpha_2 n_1$, or,

$$\begin{aligned}&\Pr\left(\text{a type}-1\text{ repressor protein will be consumed in time interval, }(t, t+\Delta t)\right)\\ &\quad = \alpha_2\ n_1\ \Delta t+o(\Delta t)\end{aligned}\tag{2}$$

By analogy, the third transition intensity function, k_2, is the production probability of a type-2 repressor protein from a particular active (not repressed) gene of type-2 per unit time, or,

$$\Pr\left(\begin{array}{l}\text{a particular active gene of type–2 will produce a repressor protein of type–2 during} \\ \text{time interval, } (t, t+\Delta t)\end{array}\right)$$
$$= k_2 \ \Delta t \ + \ o(\Delta t)$$

This definition will lead to the following transition probability:

$$\begin{aligned}
&\Pr\,(\text{a type–2 protein will be produced during the time interval, } (t, t+\Delta t)) \\
&\quad = k_2 \ G_{a2} \ \Delta t \ + \ o(\Delta t) \\
&\quad = \alpha_3 \ f_2 \ \Delta t \ + \ o(\Delta t) \\
&\quad = \frac{\alpha_3}{1+K_b{n_1}^M}\Delta t \ + \ o(\Delta t)
\end{aligned} \tag{3}$$

where G_{a2} denotes the number of active gene of type-2, f_2 the ratio of populations of active gene to total, active and repressed, genes of type-2, or,

$$f_2 = \frac{1}{1+K_b{n_1}^M},$$

K_b the equilibrium constant of the combination reaction of the active gene of type-2 and repressor of type-1, a M-mer, and M is the number of protein monomers of type-1 in the repressor.

Also by analogy, the fourth transition-intensity function, α_4, is the consumption probability of a particular active protein of type-2 during the time interval, $(t, t + \Delta t)$, or,

$$\Pr\,(\text{a particular repressor protein of type–2 will be consumed in time interval, } (t, t+\Delta t))$$
$$= \alpha_4 \Delta t + o(\Delta t)$$

By considering all repressor protein of type-2 genes in the system, we have,

$$\begin{aligned}
&\Pr\,(\text{a type–2 repressor protein will be consumed in time interval, } (t, t+\Delta t)) \\
&\quad = \alpha_4 n_2 \Delta t + o(\Delta t)
\end{aligned} \tag{4}$$

It should be noted that the rates adopted in deterministic models and discussed earlier in the outset of the 'Model Formulation' section are used in defining the transition intensity functions below. The transition intensity functions have pivotal importance in master equation models and Monte Carlo simulations. More importantly, the adoption of deterministic rate constants in master equation is a cornerstone in the interpretation of intrinsic (or internal) noise van Kampen [9].

Master equations

Based on the transition intensity functions defined above, the master equation can be obtained by taking probability balance of the following five mutually exclusive events leading to the evolution of the state of the system:

- a R_1 is produced while R_2 remains constant
- a R_1 is degraded while R_2 remains constant
- a R_2 is produced while R_1 remains constant
- a R_2 is degraded while R_1 remains constant
- both R_1 and R_2 remain the same.

As illustrated in Figure 2, the probabilities that these five exclusive events will lead the system to state $\boldsymbol{n}$ at arbitrary time $(t+\Delta t)$ can be written as follows:

$$\begin{aligned}
&\Pr\begin{pmatrix}\text{the system will transfer from state } \boldsymbol{n}'(t) = (n_1-1,\ n_2) \text{ into state} \\ \boldsymbol{n}(t+\Delta t) = (n_1,\ n_2) \text{ due to the production of one} \\ \text{type-1 protein in time interval, } (t, t+\Delta t)\end{pmatrix} \\
&= W_{\boldsymbol{n}(t+\Delta t),\, \boldsymbol{n}'(t)}\, P_{n_1-1,\, n_2}(t)\,\Delta t = W_{(n_1,\, n_2),\, (n_1-1,\, n_2)}\, P_{n_1-1,\, n_2}(t)\,\Delta t \\
&= \frac{\alpha_1}{1+K_a n_2^m} P_{n_1-1,\, n_2}(t)\,\Delta t + o(\Delta t)
\end{aligned} \tag{5}$$

where $W_{\boldsymbol{n},\, \boldsymbol{n}'}(t)$ is the conditional probability of the system transition from state $\boldsymbol{n}'(t)$ to state $\boldsymbol{n}(t+\Delta t)$ per unit time.

$$\begin{aligned}
&\Pr\begin{pmatrix}\text{the system will transfer from state into state } \boldsymbol{n}'(t) = (n_1+1,\ n_2) \text{ into} \\ \text{state } \boldsymbol{n}(t+\Delta t) = (n_1,\ n_2) \text{ due to the degradation of a type-1} \\ \text{protein in time interval, (t, t} + \Delta\text{t)}\end{pmatrix} \\
&= W_{\boldsymbol{n}(t+\Delta t),\, \boldsymbol{n}'}(t) P_{n_1+1,\, n_2}(t)\,\Delta t = W_{(n_1,\, n_2),\, (n_1+1,\, n_2)} P_{n_1+1,\, n_2}(t)\,\Delta t \\
&= (n_1+1)\,\alpha_2\, P_{n_1+1,\, n_2}(t)\,\Delta t + o(\Delta t)
\end{aligned} \tag{6}$$

$$\begin{aligned}
&\Pr\begin{pmatrix}\text{the system will transfer from state into state } \boldsymbol{n}'(t) = (n_1,\ n_2-1) \text{ into} \\ \text{state } \boldsymbol{n}(t+\Delta t) = (n_1,\ n_2) \text{ due to production of one type-2} \\ \text{protein in time interval, } (t, t+\Delta t)\end{pmatrix} \\
&= W_{\boldsymbol{n}(t+\Delta t),\, \boldsymbol{n}'}(t)\, P_{n_1,\, n_2-1}(t)\,\Delta t = W_{(n_1,\, n_2),\, (n_1,\, n_2-1)}\, P_{n_1,\, n_2-1}(t)\,\Delta t \\
&= \frac{\alpha_3}{1+K_b n_1^M} P_{n_1,\, n_2-1}(t)\,\Delta t + o(\Delta t)
\end{aligned} \tag{7}$$

$$\begin{aligned}
&\Pr\begin{pmatrix}\text{the system will transfer from state } \boldsymbol{n}'(t) = (n_1,\ n_2+1) \text{ into} \\ \text{state } \boldsymbol{n}(t+\Delta t) = (n_1,\ n_2) \text{ due to the degradation of a type-2} \\ \text{protein in time interval, } (t, t+\Delta t)\end{pmatrix} \\
&= W_{\boldsymbol{n}(t+\Delta t),\, \boldsymbol{n}'}(t) P_{n_1,\, n_2+1}(t)\,\Delta t = W_{(n_1,\, n_2),\, (n_1,\, n_2+1)} P_{n_1,\, n_2+1}(t)\,\Delta t \\
&= (n_2+1)\,\alpha_4\, P_{n_1,\, n_2+1}(t)\,\Delta t + o(\Delta t)
\end{aligned} \tag{8}$$

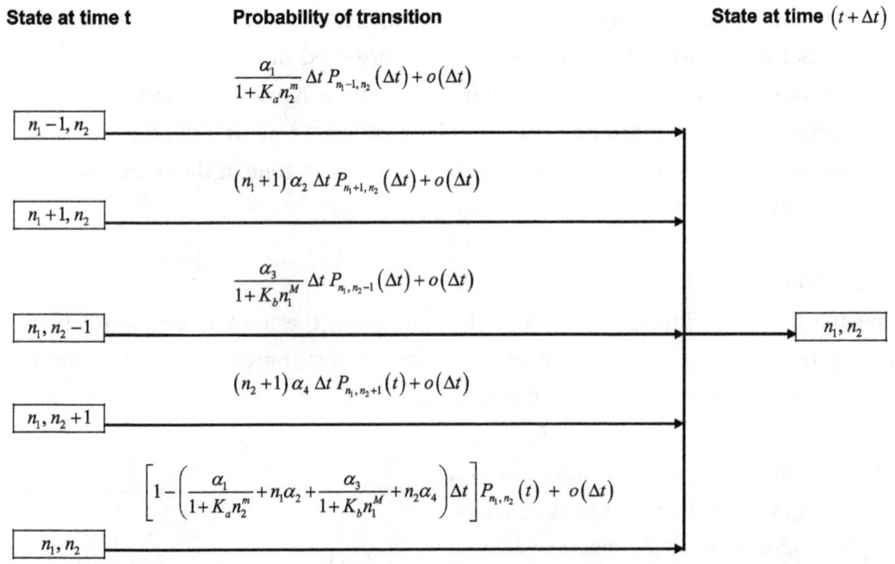

Figure 2 Probabilistic population balance involving five mutually exclusive events in time interval, $(t, t+\Delta t)$.

$$\begin{aligned}
&\Pr\text{ (the system will remain at state } \boldsymbol{n}=(n_1,n_2) \text{ during the time interval, } (t,t+\Delta t))\\
&= W_{\boldsymbol{n}(t+\Delta t),\,\boldsymbol{n}'}(t)P_{n_1,\,n_2}(t)\,\Delta t \;=\; W_{(n_1,\,n_2),\,(n_1,\,n_2)}P_{n_1,\,n_2}(t)\,\Delta t\\
&= \left[1-\left(\frac{\alpha_1}{1+K_a{n_2}^m}+n_1\alpha_2+\frac{\alpha_3}{1+K_b{n_1}^M}+n_2\alpha_4\right)\Delta t\right]P_{n_1,\,n_2}(t)\;+\;o(t)
\end{aligned} \tag{9}$$

Since these five events are mutually exclusive, we have

$$\begin{aligned}
P_{n_1,\,n_2}(t+\Delta t) \;=\;& W_{(n_1,\,n_2),\,(n_1-1,\,n_2)}P_{n_1-1,\,n_2}(t)\;+\;W_{(n_1,\,n_2),\,(n_1+1,\,n_2)}P_{n_1+1,\,n_2}(t)\\
&+W_{(n_1,\,n_2),\,(n_1,\,n_2-1)}P_{n_1,\,n_2-1}(t)+\;W_{(n_1,\,n_2),\,(n_1,\,n_2+1)}P_{n_1,\,n_2+1}(t)\\
&+W_{(n_1,\,n_2),\,(n_1,\,n_2)}P_{n_1,\,n_2}(t)
\end{aligned}$$

By substituting all the transition probabilities discussed in Equations 5 through 9 into the above expression, we obtain the probability of the system at state $\boldsymbol{n}$ at arbitrary time $(t+\Delta t)$ as follows:

$$\begin{aligned}
P_{n_1,\,n_2}(t+\Delta t)\,t \;=\;& \frac{\alpha_1}{1+K_a{n_2}^m}P_{n_1-1,\,n_2}(t)\,\Delta t\;+\;(n_1+1)\,\alpha_2P_{n_1+1,\,n_2}(t)\,\Delta\\
&+\frac{\alpha_3}{1+K_b{n_1}^M}P_{n_1,\,n_2-1}(t)\,\Delta t\;+\;(n_2+1)\,\alpha_4\,P_{n_1,\,n_2+1}(t)\,\Delta t\\
&+\left[1-\left(\frac{\alpha_1}{1+K_a{n_2}^m}+n_1\alpha_2+\frac{\alpha_3}{1+K_b{n_1}^M}+n_2\alpha_4\right)\Delta t\right]P_{n_1,\,n_2}(t)
\end{aligned}$$

By rearranging the above equation and taking the limit as $\Delta t \rightarrow 0$, we obtain the following master equation:

$$\begin{aligned}
\frac{dP_{n_1,\,n_2}(t)}{dt} \;=\;& \frac{\alpha_1}{1+K_a{n_2}^m}P_{n_1-1,\,n_2}(t)\;+\;(n_1+1)\alpha_2P_{n_1+1,\,n_2}(t)\\
&+\frac{\alpha_3}{1+K_b{n_1}^m}P_{n_1,\,n_2-1}(t)\;+\;(n_2+1)\alpha_4P_{n_1,\,n_2+1}(t)\\
&-\left[\frac{\alpha_1}{1+K_a{n_2}^m}+n_1\alpha_2+\frac{\alpha_3}{1+K_b{n_1}^M}+n_2\alpha_4\right]P_{n_1,\,n_2}(t)
\end{aligned} \tag{10}$$

For convenience, the one-step operator, $\boldsymbol{E}$, is defined through its effect on arbitrary function $f(n)$ as van Kampen [9]:

$$\boldsymbol{E}(f(n)) = f(n+1) \text{ and } \boldsymbol{E}^{-1}(f(n)) = f(n-1) \tag{11}$$

The master equation is rewritten compactly in terms of the one-step operator as follows:

$$\begin{aligned}
\frac{dP_{n_1,\,n_2}(t)}{dt} \;=\;& \frac{\alpha_1}{1+K_a{n_2}^m}\left(\boldsymbol{E}_{n_1}^{-1}-1\right)P_{n_1,\,n_2}(t)\;+\;(\boldsymbol{E}_{n_1}-1)\alpha_2\,n_1\,P_{n_1,\,n_2}(t)\\
&+\frac{\alpha_3}{1+K_b{n_1}^M}\left(\boldsymbol{E}_{n_2}^{-1}-1\right)P_{n_1,\,n_2}(t)\;+\;(\boldsymbol{E}_{n_2}-1)\alpha_4\,n_2\,P_{n_1,\,n_2}(t)
\end{aligned} \tag{12}$$

The solution to the equation with the step operator yields the time-dependent joint probability distribution of the populations of repressor proteins.

System-size expansion based on van Kampen's procedure

The approximation of the master equation, Equation 10 or 12, leads the evolution of the joint probability distribution of the populations of the two competing repressors, $P_n(t)$. Equation 10 comprises a set of ordinary differential equations with the joint probability function, $P_n(t)$, as its unknown. Each equation in the set represents a particular outcome of $\boldsymbol{n}$; thus, solving Equation 12 for the joint probability distribution of an exceedingly large number of all possible ***n***s is extremely difficult, if not impossible. In practice, however, it often suffices to determine only the expressions that govern a limited number of moments, especially the first and second moments, of the resultant population distribution. These expressions yield the means, variances, and covariances that can be correlated or compared with the experimental data.

Moreover, Equation 12 is nonlinear, which prevents the moments from being evaluated by averaging techniques or joint probability generating function techniques [9]. This difficulty is circumvented by resorting to the system-size expansion, a rational approximation technique based on the power series expansion [9,12,31]. The technique gives rise to the deterministic macroscopic equations as well as the equations of fluctuations for the master equation.

To apply the system-size expansion, a suitable expansion parameter must be identified in the master equation, specifically in the transition intensity functions. The expansion parameter must govern the size of the fluctuations, and therefore, the magnitude of the jumps, or transitions. The macroscopic features are determined by the average behavior of all particles, while internal fluctuations are caused by the discrete nature of matter. Hence, we expect the fluctuations to be relatively small when the system size is large. The system size, Ω, has been proposed as an expansion parameter because it measures the relative importance of the fluctuations [9,12,31]. In the current genetic regulatory network, the total initial number of promoter population, or the total number of initial reactants, is chosen as Ω so that the noises estimated based on both the master equation and Monte Carlo simulations discussed below represent the standard deviations from the means.

For a linear system, fluctuations are of the order of $\Omega^{1/2}$ in a collection of Ω entities. As a result, their effect on the macroscopic properties is of the order of $\Omega^{-1/2}$ [9,12]. In the system under consideration, therefore, we expect that the joint probability, $P_n(t)$, will have a sharp maximum around the macroscopic value, $n(t) = \Omega\Theta(t)$, with a width of the order of $\Omega^{1/2}$. Here, $\Theta(t)$ is a vector where elements are the mean numbers of the two protein populations, $\varnothing(t)$ and $\theta(t)$ obtained through the solution of the macroscopic equations as will be elaborated later. To exploit these characteristics of the system, a new random vector $\mathbf{Y}(t)$ is defined as follows:

$$N_1(t) = \Omega\,\phi(t) + \Omega^{1/2} Y_1(t) \tag{13}$$

$$N_2(t) = \Omega\,\theta(t) + \Omega^{1/2} Y_2(t) \tag{14}$$

The equations of realizations of these expressions are given, respectively, by

$$n_1(t) = \Omega\phi(t) + \Omega^{1/2} y_1(t) \tag{15}$$

$$n_2(t) = \Omega\theta(t) + \Omega^{1/2} y_2(t) \tag{16}$$

Accordingly, the joint probability of n_1 and n_2 i.e., $P_n(t)$, is now transformed into that of y_1 and y_2, i.e., $\Psi_y(t)$. Subsequently, the new random vector, ***Y***, the new joint probability distribution, $\Psi_y(t)$, and the definition of the one-step operator, ***E***, Equation 11, are substituted into Equation 12. By expanding the right-hand side of the resultant expression into a Taylor's series, the master equation in terms of the new variables is obtained, see Appendix 1. All appendices to this paper can be found in the supporting materials for this Journal.

Collecting the terms of order $\Omega^{1/2}$ in the right-hand side of the expanded equation gives rise to the following expressions governing the evolution of the macroscopic equation of the system:

$$\frac{d\phi}{dt} = \frac{\alpha_1'}{1 + K_a' \, \theta(t)^m} - \alpha_2 \, \phi(t) \tag{17}$$

$$\frac{d\theta}{dt} = \frac{\alpha_3'}{1 + K_b' \, \phi(t)^M} - \alpha_4 \, \theta(t) \tag{18}$$

where the constants, α_1', K_a', α_3', and K_b' correspond respectively to the parameters α_1, K_a, α_3, and K_b, normalized with Ω or a specific power of Ω so that collected terms in system size expansion have the same order of magnitude, i.e.,

$$\alpha_1 = \alpha_1' \, \Omega, \qquad K_a = K_a' \, \Omega^{-m} \tag{19}$$

$$\alpha_3 = \alpha_3' \, \Omega, \qquad K_b = K_b' \, \Omega^{-M} \tag{20}$$

Equations 17 and 18 are of the same forms as the macroscopic equations of Gardner [26].

Similarly, by collecting the terms of order Ω^0 gives rise to the following linear Fokker-Plank equation [9], see Appendix 1, that governs the first and the second moments associated with the fluctuations of the system:

$$\frac{\partial \Psi}{\partial t} = -\sum_{i,j} A_{ij} \frac{\partial}{\partial y_i} \left(y_j \Psi \right) + \frac{1}{2} \sum_{i,j} \left(B_{ij} \frac{\partial^2 \Psi}{\partial y_i \partial y_j} \right) \tag{21}$$

where the two matrices A and B are

$$A = \begin{bmatrix} A_{11} & A_{12} \\ A_{21} & A_{22} \end{bmatrix} = \begin{bmatrix} -\alpha_2 & \dfrac{-\alpha_1' K_a' m (\theta(t))^{m-1}}{\left(1 + K_a' (\theta(t))^m\right)^2} \\ \dfrac{-\alpha_3' K_b' M (\phi(t))^{M-1}}{\left(1 + K_b' (\phi(t))^M\right)^2} & -\alpha_4 \end{bmatrix} \tag{22}$$

$$B = \begin{bmatrix} B_{11} & B_{12} \\ B_{21} & B_{22} \end{bmatrix} = \begin{bmatrix} \dfrac{\alpha_1'}{1 + K_a' (\theta(t))^m} + \alpha_2 \phi(t) & 0 \\ 0 & \dfrac{\alpha_3'}{1 + K_b' (\phi(t))^M} + \alpha_4 \theta(t) \end{bmatrix} \tag{23}$$

A Fokker-Planck equation is considered linear if the coefficient matrix A, the drift term, is a linear function of Y and the coefficient matrix B, the diffusion term, is constant [9]. Note that the macroscopic trajectories, N and 2, are functions of t only and they can be obtained by integrating Equations 17 and 18. Thus, the coefficients of the equation governing the fluctuations, A and B in Equations 22 and 23, are independent of the fluctuations, Y. For a linear Fokker-Planck equation, the ordinary differential equations governing the means and variances of the fluctuations, Y, can be derived by taking the first and second moments of Equation 21.

Taking the first moment of Equation 21 yields the expression governing the mean of the fluctuations, *Y*:

$$\frac{d}{dt}\boldsymbol{E}[Y_k] = \sum_{j=1}^{2} A_{kj}\boldsymbol{E}[Y_j], \qquad k = 1, 2$$

By substituting Equations 22 and 23 into the above expression gives rise to

$$\frac{d}{dt}\boldsymbol{E}[Y_1] = A_{11}\boldsymbol{E}[Y_1] + A_{12}\boldsymbol{E}[Y_2] = -\alpha_2\boldsymbol{E}[Y_1] - \frac{\alpha_1' K_a' m(\theta(t))^{m-1}}{(1 + K_a'(\theta(t))^m)^2}\boldsymbol{E}[Y_2] \tag{24}$$

$$\frac{d}{dt}\boldsymbol{E}[Y_2] = A_{21}\boldsymbol{E}[Y_1] + A_{22}\boldsymbol{E}[Y_2] = \frac{-\alpha_3' K_b' M(\phi(t))^{M-1}}{\left(1 + K_b'(\phi(t))^M\right)^2}\boldsymbol{E}[Y_1] - \alpha_4\boldsymbol{E}[Y_2] \tag{25}$$

Similarly, taking the second moment of Equation 21 yields the expression governing the second moment of the fluctuations, *Y*:

$$\frac{d}{dt}\boldsymbol{E}[Y_i Y_j] = \sum_{k=1}^{2} A_{ik}\boldsymbol{E}[Y_k Y_j] + \sum_{k=1}^{2} A_{jk}\boldsymbol{E}[Y_i Y_k] + B_{ij}$$

By substituting Equations 22 and 23 into the above expression gives rise to

$$\begin{aligned}\frac{d}{dt}\boldsymbol{E}[Y_1^2] &= 2A_{11}\boldsymbol{E}[Y_1^2] + 2A_{12}\boldsymbol{E}[Y_1 Y_2] + B_{11} \\ &= -2\alpha_2\boldsymbol{E}[Y_1^2] - \frac{\alpha_1' K_a' m(\theta(t))^{m-1}}{(1 + K_a'(\theta(t))^m)^2}\boldsymbol{E}[Y_1 Y_2] \\ &\quad + \frac{\alpha_1'}{1 + K_a'(\theta(t))^m} + \alpha_2\phi(t)\end{aligned} \tag{26}$$

$$\begin{aligned}\frac{d}{dt}\boldsymbol{E}[Y_2^2] &= 2A_{22}\boldsymbol{E}[Y_2^2] + 2A_{21}\boldsymbol{E}[Y_1 Y_2] + B_{22} \\ &= -2\alpha_4\boldsymbol{E}[Y_2^2] - \frac{\alpha_3' K_b' M(\phi(t))^{M-1}}{\left(1 + K_b'(\phi(t))^M\right)^2}\boldsymbol{E}[Y_1 Y_2] \\ &\quad + \frac{\alpha_3'}{1 + K_b'(\phi(t))^M} + \alpha_4\theta(t)\end{aligned} \tag{27}$$

$$\begin{aligned}\frac{d}{dt}\boldsymbol{E}[Y_1 Y_2] &= A_{11}\boldsymbol{E}[Y_1 Y_2] + A_{12}\boldsymbol{E}[Y_2^2] + A_{21}\boldsymbol{E}[Y_1^2] + A_{22}\boldsymbol{E}[Y_1 Y_2] \\ &= -\alpha_2\boldsymbol{E}[Y_1 Y_2] - \frac{\alpha_1' K_a' m(\theta(t))^{m-1}}{(1 + K_a'(\theta(t))^m)^2}\boldsymbol{E}[Y_2^2] \\ &\quad - \frac{\alpha_3' K_b' M(\phi(t))^{M-1}}{\left(1 + K_b'(\phi(t))^M\right)^2}\boldsymbol{E}[Y_1^2] - \alpha_4\boldsymbol{E}[Y_1 Y_2]\end{aligned} \tag{28}$$

System size expansion based on Kurtz's limit theorems

The approximation of the master equation discussed in the preceding section, i.e., system size expansion method, can be derived and stated compactly in a general form based on

Kurtz's limit theorems [13-15] under the condition $\Omega \to \infty$. First, the master equation, Equation 11, can be written in the following continuous state, gain-loss form [9]:

$$\frac{\partial P(\boldsymbol{n},\, t)}{\partial t} = \int [W(\boldsymbol{n};\boldsymbol{n}+\boldsymbol{r})P(\boldsymbol{n},\, t) - W(\boldsymbol{n}-\boldsymbol{r};\boldsymbol{n})P(\boldsymbol{n}-\boldsymbol{r},\, t)]\, d\boldsymbol{r} \tag{29}$$

where $W(n;n+\boldsymbol{r})$ is the transition probability from state $\boldsymbol{n}$ to state $\boldsymbol{n}+\boldsymbol{r}$ per unit time. Both $\boldsymbol{n}$ and $\boldsymbol{r}$ in Equation 29 are now treated as continuous varying vectors. The convergence of the system size expansion procedure relies on two criteria for transition probability rate: small jump and slow varying [9]. Mathematically, the small-jump criterion implies that there is a small δ so that

$$W(\boldsymbol{n};\ \boldsymbol{n}+\boldsymbol{r}) \approx 0 \quad \mathit{for}\ |\boldsymbol{r}| > \delta \tag{30}$$

and the slow varying assumption means that there is a small δ so that

$$W(\boldsymbol{n}+\Delta\boldsymbol{n};\ \boldsymbol{r}) \approx W(\boldsymbol{n};\ \boldsymbol{r}) \quad \mathit{for}\ |\Delta\boldsymbol{n}| < \delta \tag{31}$$

To satisfy these criteria, the unit jumps associated with the mutually exclusive events in the formulation of the master equation are replaced by jumps of size Ω^{-1}, the system size or the largeness parameter. Thus, the random vector $\boldsymbol{N}(t) = (n_1(t),n_2(t))$ is replaced by $\tilde{\boldsymbol{N}}(t) = {}^{N}(t)/_{\Omega}$ and time is replaced by $\tilde{\boldsymbol{t}} = {}^{t}/_{\Omega}$. The resultant master equation of Equation 29 becomes

$$\frac{\partial P(\tilde{\boldsymbol{n}},\, t)}{\partial \tilde{t}} = \int \left[\tilde{W}(\tilde{\boldsymbol{n}};\tilde{\boldsymbol{n}}+\boldsymbol{r})P(\tilde{\boldsymbol{n}},\, t) - \tilde{W}(\tilde{\boldsymbol{n}}-\boldsymbol{r};\tilde{\boldsymbol{n}})P(\tilde{\boldsymbol{n}}-\boldsymbol{r},\, t)\right] d\boldsymbol{r} \tag{32}$$

Comparing Equations 10 and 29 yields the transition probability per unit time for the current problem can be stated in the following form:

$$\begin{aligned}\tilde{W}\left(\tilde{\boldsymbol{n}};\ \tilde{\boldsymbol{n}}+\frac{\boldsymbol{r}}{\Omega}\right) &= \Omega\, W\left(\boldsymbol{n};\ \boldsymbol{n}+\frac{\boldsymbol{r}}{\Omega}\right)\\ &= \Omega \left[\frac{\alpha_1'}{1+K_a'\tilde{n}_2{}^m}\delta_{\tilde{n}_1,\,{}^1/_\Omega}\delta(\tilde{n}_2) + \alpha_2\tilde{n}_1\delta_{\tilde{n}_1,\,-{}^1/_\Omega}\delta(\tilde{n}_2)\right.\\ &\qquad \left. + \frac{\alpha_3'}{1+K_b'\tilde{n}_1{}^M}\delta(\tilde{n}_1)\delta_{\tilde{n}_2,\,{}^1/_\Omega} + \alpha_4\tilde{n}_2\delta(\tilde{n}_1)\delta_{\tilde{n}_2,\,-{}^1/_\Omega}\right]\end{aligned} \tag{33}$$

where $\delta(\boldsymbol{n})$ and $\delta_{i,j}$ are Dirac and Kronecker delta functions, respectively. The four parameters on the right-hand side of Equation 33 are obtained from the definitions of transition intensity functions.

Kurtz's limit theorems state that, as $\Omega \to \infty$ with an error of $O(\ln\Omega/\Omega)$, the statistical properties of the master equation, Equation 32, can be approximated by the following Fokker-Planck equation:

$$\frac{\partial P(\tilde{\boldsymbol{n}},\, \tilde{t})}{\partial \tilde{t}} = -\sum_i \frac{\partial}{\partial \tilde{n}_i}\left[K_i^{(1)\infty}(\tilde{\boldsymbol{n}})P(\tilde{\boldsymbol{n}},\, \tilde{t})\right] + \frac{1}{2}\sum_{i,j}\frac{\partial^2}{\partial \tilde{n}_i\, \partial \tilde{n}_j}\left[K_{ij}^{(2)\infty}(\tilde{\boldsymbol{n}})P(\tilde{\boldsymbol{n}},\, \tilde{t})\right], \tag{34}$$

where the deterministic drift, $K_i^{(1)\infty}(\tilde{\boldsymbol{n}})$, and diffusion coefficients, $K_{ij}^{(2)\infty}(\tilde{\boldsymbol{n}})$, are

$$K_i^{(1)\infty}(\tilde{\boldsymbol{n}}) = \int r_i\, \tilde{W}(\tilde{\boldsymbol{n}};\ \tilde{\boldsymbol{n}}+\boldsymbol{r})dr_i \tag{35}$$

$$K_{ij}^{(2)\infty}(\tilde{\boldsymbol{n}}) = \iint r_i r_j \tilde{W}(\tilde{\boldsymbol{n}};\, \tilde{\boldsymbol{n}} + \boldsymbol{r}) dr_i\, dr_j \tag{36}$$

Substituting Equation 33 into Equation 35 yields the first moments of $\tilde{W}$:

$$K_1^{(1)\infty}(\tilde{\boldsymbol{n}}) = \Omega\left(\frac{\alpha_1'}{1+K_a'\tilde{n}_2{}^m} - \alpha_2\tilde{n}_1\right) = \frac{\alpha_1}{1+K_a n_2{}^m} - \alpha_2 n_1 \tag{37}$$

$$K_2^{(1)\infty}(\tilde{\boldsymbol{n}}) = \Omega\left(\frac{\alpha_3'}{1+K_b'\tilde{n}_1^M} - \alpha_4\tilde{n}_2\right) = \frac{\alpha_1}{1+K_a n_2{}^m} - \alpha_2 n_1 \tag{38}$$

Similarly, substituting Equation 33 into Equation 36 yields the following second moments of $\tilde{W}$:

$$K_{11}^{(2)\infty}(\tilde{\boldsymbol{n}}) = \Omega\left(\frac{\alpha_1'}{1+K_a'\tilde{n}_2^m} + \alpha_2\tilde{n}_1\right) = \frac{\alpha_1}{1+K_a n_2^m} + \alpha_2 n_1 \tag{39}$$

$$K_{22}^{(2)\infty}(\tilde{\boldsymbol{n}}) = \Omega\left(\frac{\alpha_3'}{1+K_b{}'\tilde{n}_1^M} + \alpha_4\tilde{n}_2\right) = \frac{\alpha_3}{1+K_b n_1^M} + \alpha_4 n_2 \tag{40}$$

$$K_{12}^{(2)\infty}(\tilde{\boldsymbol{n}}) = 0 \tag{41}$$

$$K_{21}^{(2)\infty}(\tilde{\boldsymbol{n}}) = 0 \tag{42}$$

The approximation of the master equation, Equation 12, can be found base on the fact that the Fokker-Planck equation, Equation 34, can be obtained by integrating the following nonlinear Langevin equation in Ito's interpretation [9]

$$\frac{d\tilde{n}_i}{d\tilde{t}} = K_i^{(1)\infty}(\tilde{n}) + \eta_i(\tilde{\boldsymbol{n}},\, \tilde{t}) = K_i^{(1)\infty}(\tilde{\boldsymbol{n}}) + C_i(\tilde{\boldsymbol{n}})L(\tilde{t}), \tag{43}$$

where the first term on the right-hand side of the above equation represents the deterministic, or macroscopic characteristic of the process, $\eta_i(\tilde{\boldsymbol{n}},\, \tilde{t})$ denotes a Gaussian white noise having the following means and covariance matrix

$$\boldsymbol{E}\left[\eta_i(\tilde{\boldsymbol{n}},\, \tilde{t})\right] = 0 \tag{44}$$

$$\boldsymbol{E}\left[\eta_i(\tilde{\boldsymbol{n}},\, \tilde{t})\eta_j(\tilde{\boldsymbol{n}},\, \tilde{t})\right] = K_{ij}^{(2)\infty}(\tilde{\boldsymbol{n}})\delta(\tilde{t}-\tilde{t}') \tag{45}$$

$L(\tilde{t})$ denotes a Gaussian white noise with a unit strength, and $C_i(\tilde{\boldsymbol{n}})$ denotes the effects of interactions of the noise and the system on the random variable. The discontinuity of Gaussian white noise has been the source of evolution of several algorithms in interpreting $C_i(\tilde{\boldsymbol{n}})$ during the process, and thus the conversion of a Langevin equation to its Fokker-Plank counterpart. In Ito's algorithm, the value of $C_i(\tilde{\boldsymbol{n}})$ before the arrival of white noise is used in averaging. In Stratonovich's algorithm, the averaged value of $C_i(\tilde{\boldsymbol{n}})$ during the time of noise is used in averaging, which yields an extra term in the macroscopic part of the Fokker-Plank equation. Since $L(\tilde{t})$ is never infinitely sharp and it lasts a finite time, the Ito and Stratonovich's calculus are more appropriate in modeling internal and external noises, respectively [9].

With this Langevin representation in hand, the equations derived in the last section, i.e., Equations 17, 18, 22, and 23, can be readily obtained. Specifically, substituting Equations 37 and 38 into Equation 43 and ignoring the noise term yields Equations 17 and 18. Since the drift coefficient in a Fokker-Planck equation, matrix A in Equation 21, is the Jacobian matrix of the functions on the right-hand side Equations 17 and 18 [9],

Equation 22 can be obtained by taking derivatives. Finally, it is obvious that the elements of the covariance matrix, Equations 39 through 42, are identical to those shown in Equation 23.

System size expansion based on Kurtz's theorems is substantially simpler than the original procedure proposed by van Kampen [9]. This efficiency was previously utilized by Aparicio and Solari [32] and Chua et al. [33] in their studies of stochastic population dynamics of disease transmission and chemical vapor deposition, respectively.

It should be mentioned that the system size expansion method discussed in this and last sections suffers several limitations. Simulation with the system-size expansion converges to the steady state within its boundary of attraction just like its deterministic counterpart, and it cannot be generate noise-induced transition, as it will be discussed later in the simulation section [9]. The system size expansion near the steady-state boundary of attraction (i.e., away from the steady state) yields noises that are not compatible to those generated from near the steady states [18].

Simulations

The genetic toggle switch model presented in the preceding section has been simulated by two approaches. The first approach relies on the solution of the governing equations for the first and second moments of the random variables derived from the master equations. The second approach resorts to the event-driven Monte Carlo algorithm.

Simulation based on the master equations

To effectively analyze the impact of system parameters, the equations governing the first and second moments are converted to dimensionless forms. Following Gardner's procedure [3], we introduce the following variables, with the assumption $\alpha_2 = \alpha_4$:

$$u = \phi(t)\left(K_b'\right)^{1/M} \tag{46}$$

$$v = \theta(t)\left(K_a'\right)^{1/m} \tag{47}$$

$$\bar{t} = t\alpha_2 = t\alpha_4 \tag{48}$$

Substituting these three variables into Equations 17 and 18 yields the following compact set of equations:

$$\frac{du}{d\bar{t}} = \frac{\alpha_1''}{1+v^m} - u \tag{49}$$

$$\frac{dv}{d\bar{t}} = \frac{\alpha_2''}{1+u^M} - v \tag{50}$$

where

$$\alpha_1'' = \frac{\alpha_1'\left(K_b'\right)^{1/M}}{\alpha_2} \tag{51}$$

$$\alpha_2'' = \frac{\alpha_3'\left(K_a'\right)^{1/m}}{\alpha_4} \tag{52}$$

When the effective rates of synthesis of the two proteins are comparable, we have

$$\left(K'_a\right)^{1/m} \cong \left(K'_b\right)^{1/M} \tag{53}$$

Then Equations 24 through 28 can be transformed into the following compact forms

$$\frac{d}{dt}\boldsymbol{E}[Y_1] = -\boldsymbol{E}[Y_1] - \frac{\alpha_1'' \, m \, v^m}{v(1+v^m)^2}\boldsymbol{E}[Y_2] \tag{54}$$

$$\frac{d}{dt}\boldsymbol{E}[Y_2] = -\frac{\alpha_2'' \, M \, u^M}{u(1+u^M)^2}\boldsymbol{E}[Y_1] - \boldsymbol{E}[Y_2] \tag{55}$$

$$\frac{d}{dt}\boldsymbol{E}\left[Y_1^2\right] = -2\boldsymbol{E}\left[Y_1^2\right] - \frac{\alpha_1'' \, m \, v^m}{v(1+v^m)^2}\boldsymbol{E}[Y_1Y_2] + \frac{\alpha_1''}{1+v^m} + u \tag{56}$$

$$\frac{d}{dt}\boldsymbol{E}\left[Y_2^2\right] = -2\boldsymbol{E}\left[Y_2^2\right] - \frac{\alpha_2'' \, M \, u^M}{u(1+u^M)^2}\boldsymbol{E}[Y_1Y_2] + \frac{\alpha_2''}{1+u^M} + v \tag{57}$$

$$\frac{d}{dt}\boldsymbol{E}[Y_1Y_2] = -\boldsymbol{E}[Y_1Y_2] - \frac{\alpha_1'' \, m \, v^m}{v(1+v^m)^2}\boldsymbol{E}\left[Y_2^2\right] - \frac{\alpha_2'' \, M \, u^M}{u(1+u^M)^2}\boldsymbol{E}\left[Y_1^2\right] - \boldsymbol{E}[Y_1Y_2] \tag{58}$$

Equations 49, 50, and 54 through 58 can be integrated simultaneously to obtain the statistical characteristics of the dynamical processes. Equations 49 and 50 yield the means of the populations while Equations 54 and 55 yield the means of the fluctuations, which are essentially zero due to the assumption of symmetric noises around the means, i.e., Equations 13 and 14. Equations 56 through 58 generate the variance and covariance of the two constituent populations. The integration was conducted in Matlab by ode45, a subroutine based on Gear's method for stiff sets of ordinary differential equations.

As we will demonstrate later, some of the simulation results, including noise-induced transitions, depend on the parameter values and initial conditions, which, in turn, are closely related to the properties of the deterministic system, i.e., Equations 49 and 50. For a nonlinear system governed by Equations 49 and 50, the location of the parameters α_1'' and α_2'' in the bifurcation diagram and the initial population in the phase diagram have significant effects on the evolution of system's state. In order to analyze the process under selected conditions, the values of the four parameters used for simulation, α_1'', α_2'', m, and M, are taken from published experimental results [3,5,34,35] as well as the inference that can be drawn from the phase and bifurcation diagrams. A thorough review of the protein and mRNA reaction rates involved in the control mechanism can be found in Santallin and Mackey [36]. The values of several of these variables can also be found in other regulatory modeling literature [7,37-39]. As shown in Figure 3, for $m = M = 2$ and $\alpha_1'' = \alpha_2'' = 15.6$ the traces of (u,v) by setting the right-hand sides of Equations 49 and 50 being zero yield with three interceptions. Liapunov stability analysis reveals that two of these steady states are stable, and the one in the middle is unstable, i.e., a saddle node. The bifurcation analysis, for $m = M = 2$, illustrates that the system has one or two stable steady states depending on the values of α_1'' and α_2'' (see Figure 4 and [3]).

As marked in Figure 4, three possible sets of α_1'' and α_2'' are sufficient to characterize the different cases of population dynamics: monostable, bistable, and bifurcation. Thus,

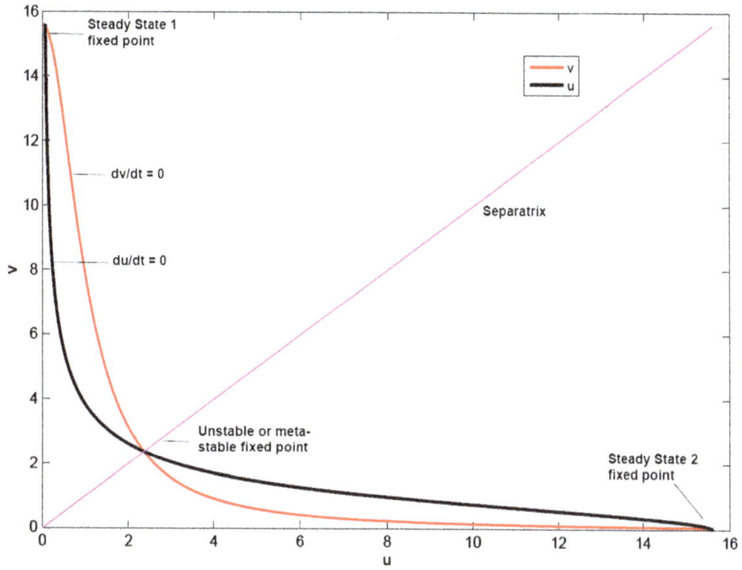

Figure 3 Phase plane drawn by setting right-hand side of Equations 49 and 50 be zero for $m = M = 2$, $\alpha_1'' = 15.6$, and $\alpha_2'' = 15.6$. The diagram shows three steady states. Stability analysis reveals that two of them are stable and one unstable.

the following three sets of parameter values are chosen in our simulations for characterizing the dynamics in different regions:

Case A, in bistable region: $\alpha_1'' = 15.6$ and $\alpha_2'' = 15.6$,
Case B, on bifurcation curve: $\alpha_1'' = 15.6$ and $\alpha_2'' = 4.0$,
Case C, in monostable region: $\alpha_1'' = 15.6$ and $\alpha_2'' = 1.2$.

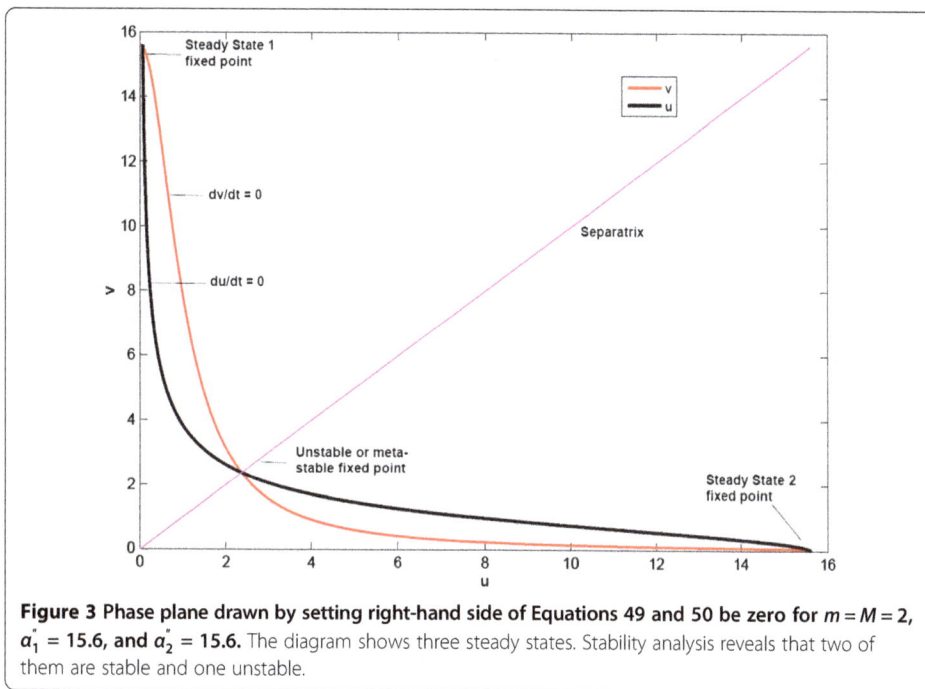

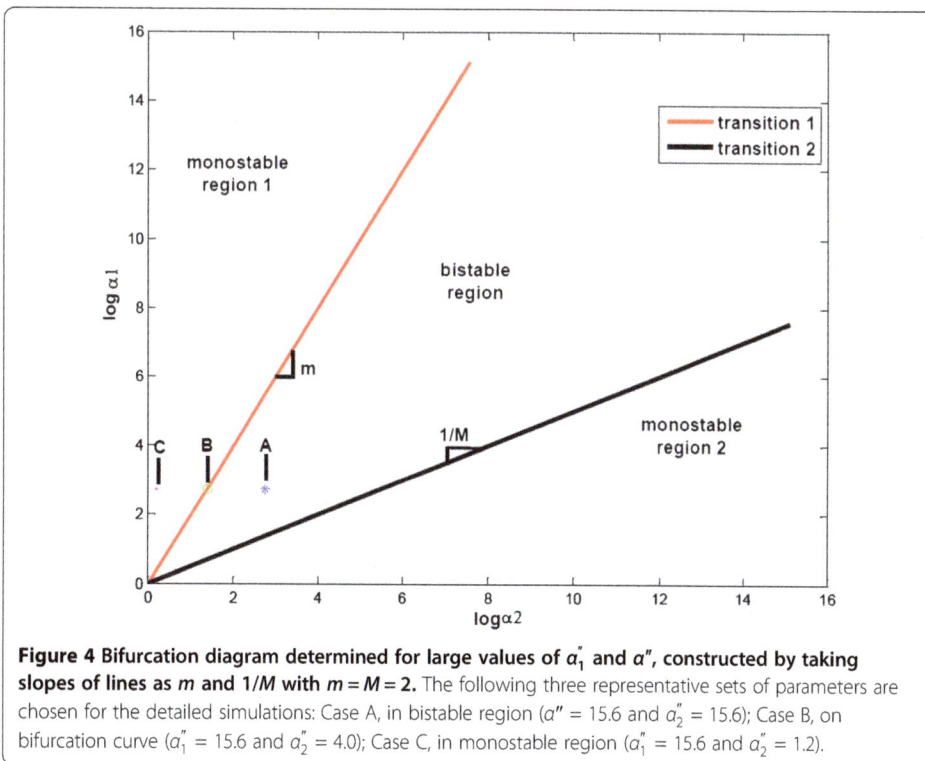

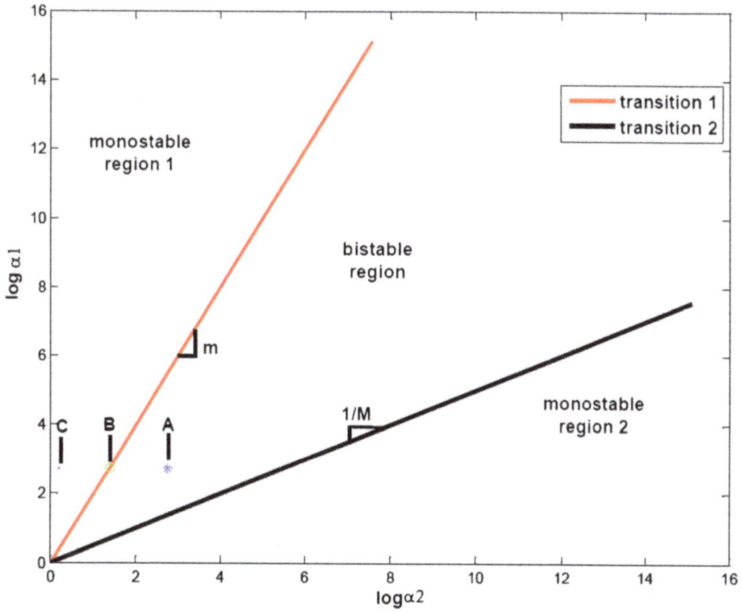

Figure 4 Bifurcation diagram determined for large values of α_1'' and α'', constructed by taking slopes of lines as m and $1/M$ with $m = M = 2$. The following three representative sets of parameters are chosen for the detailed simulations: Case A, in bistable region ($\alpha'' = 15.6$ and $\alpha_2'' = 15.6$); Case B, on bifurcation curve ($\alpha_1'' = 15.6$ and $\alpha_2'' = 4.0$); Case C, in monostable region ($\alpha_1'' = 15.6$ and $\alpha_2'' = 1.2$).

We assume $m = 2$ and $M = 2$ for all the simulations presented herein.

Initial protein populations are also important to the evolution of the dynamics in several aspects. It is established in nonlinear dynamics that different initial conditions could lead to different steady states, and the evolution of the dynamics may be altered significantly by small variation of initial conditions. In this work, we will demonstrate that noise could induce system transition from one steady state to another when the populations pass through the neighborhood of an unstable steady state (or the saddle node) of a bistable system. Moreover, for very small initial populations, the numerical equations become invalid as the protein values tend to become so small that they drive the bifurcation lines beyond the domain of application. Thus, the choice of initial population should be such that it is between 10s and 100 s. In the present work, the initial populations are chosen $u(0) = 155$ and $v(0) = 154$ for the three cases discussed above. To further illustrate the effects of the initial populations, a simulation is conducted with $u(0) = 15$, $v(0) = 155$, $\alpha_1'' = 15.6$, and $\alpha_2'' = 15.6$ for the bistable system, or Case A. The population trajectories do not pass through the neighborhood of the saddle node in this simulation, and possess no risk of noise-induced transition.

It should be mentioned that the process of interest is characterized by the transition intensity functions, k_1, k_2, α_2, and α_4, defining the probabilities of transitions of each type of population per unit time. If the fraction of population converted per unit time is taken to represent the intensity function, its significance is equivalent to the deterministic rate constant of the specific rate. In other words, from the change in the population of a particular protein type i due to the conversion of type i during the time interval, $(t, t + \Delta t)$, we have

$$-R_i = \lim_{\Delta t \to 0} \frac{n_i - [n_i - n_i \lambda_i \Delta t + o(\Delta t)]}{\Omega \, \Delta t} = \lambda_i \frac{n_i}{\Omega}, \tag{59}$$

where Ω stands for the system size, i.e., the total initial population; and $-R_i$, the population converted attributable to transition type i protein per unit time. A detailed discussion of the relationship between the deterministic rate constant and the intensity function can be found in [9].

Simulation based on Monte Carlo simulation

Linear or nonlinear dynamic processes have been simulated either deterministically or stochastically by Monte Carlo procedures. It is worth noting that a well-developed class of Monte Carlo simulation procedures essentially shares identical computational bases with the master equation algorithm presented in the preceding sections. Specifically, the assumptions of Markov property and temporal homogeneity of the random variables lead to the definitions of transition intensity functions [33,40,41]. As discussed in the "Model Formulation" section, probability balances of various events on the basis of these intensity functions give rise to the master equations. In the Monte Carlo simulation, the system's state is simulated by a step-wise, random-walk scheme based on the same intensity functions.

Process systems or phenomena can be simulated by time-driven and event-driven Monte Carlo procedures [42]. The difference between these two procedures is in the manner of updating the time clock of the evolution of the system. The time-driven procedure advances the simulation clock by a pre-specified time increment, t, which is sufficiently small so that at most, one event will occur in this interval. The probability of an event occurring is determined by the nature and magnitudes of the transition intensity functions. In contrast, the event-driven procedure updates the simulation clock by randomly generating the waiting time, τ_w, which has an exponential distribution [43,44]; this distribution signifies that a population transition takes place completely randomly. At the end of each waiting interval, one event will occur, and the state to which the system will transfer is also determined by the nature of the transition intensity functions.

The process of interest here, i.e., genetic toggle switch, has been simulated by the event-driven procedure; it is usually computationally faster than the time-driven procedure. The simulation starts with a given initial distribution of population; the essential task is to obtain the probability distributions of the protein numbers at any subsequent times. To determine the system transition in each time step, two random numbers are generated for two different purposes. The first random number in (0, 1), i.e., r_1, is for estimating the waiting time during which a possible transition of the system's state will take place. The second random number in (0, 1), i.e., r_2, is for identifying the transition type.

Waiting time

Let T_n be the random variable representing the waiting time of the population of the system of interest at state n prior to its transition due to the transformation of a protein production or consumption. τ_w is the realization of T_n. Moreover, let $G_n(\tau_w)$ be the probability that no transition takes place during τ_w. Thus,

$$G_n(\tau_w) = P_r(\tau_w \le T_n). \tag{60}$$

This can be expressed as (see derivation in Appendix 2)

$$G_n(\tau_w) = \exp\left(-\left\{\frac{n_1\alpha_1}{1+K_a{n_2}^m} + n_1\alpha_2 + \frac{n_2\alpha_3}{1+K_b{n_1}^M} + n_2\alpha_4\right\}\tau_w\right) \tag{61}$$

The complement of $G_n(\tau_w)$

$$H_n(\tau_w) = 1 - G_n(\tau_w) \tag{62}$$

expresses the cumulative probability distribution of T_n up to τ_w. The probability density function of T_n, i.e.,

$$h(\tau_w) \equiv \frac{dH_n(\tau_w)}{d\tau_w}. \tag{63}$$

Therefore, $h_n(\tau_w)$ has the following exponential form (see Appendix 2)

$$\begin{aligned} h_n(\tau_w) &= \left(\frac{n_1\alpha_1}{1+K_a{n_2}^m} + n_1\alpha_2 + \frac{n_2\alpha_3}{1+K_b{n_1}^M} + n_2\alpha_4\right) \\ &\quad \exp\left(-\left(\frac{n_1\alpha_1}{1+K_a{n_2}^m} + n_1\alpha_2 + \frac{n_2\alpha_3}{1+K_b{n_1}^M} + n_2\alpha_4\right)\tau_w\right) \end{aligned} \tag{64}$$

Note that $H_n(\tau_w)$ is the probability function of T_n.

Equation 64 indicates that to estimate the waiting time of a protein-regulated gene expression, τ_w, a sequence of exponentially distributed random numbers must be generated. The sequences of the computer-generated random numbers, however, are usually uniformly distributed in interval [0, 1]. This uniform distribution, therefore, need be transformed into the exponential distribution, which can be accomplished by defining a new random variable, denoted by U, whose realization, denoted by u, assumes the value of $H_n(\tau_w)$ at τ_w [43,44], i.e.,

$$\begin{aligned} u &= H_n(\tau_w) \\ &= 1-\exp\left(-\left(\frac{n_1\alpha_1}{1+K_a{n_2}^m}+n_1\alpha_2+\frac{n_2\alpha_3}{1+K_b{n_1}^M}+n_2\alpha_4\right)\tau_w\right) \end{aligned} \tag{65}$$

or, inversely,

$$\tau_w = \frac{1}{\left(\frac{n_1\alpha_1}{1+K_a{n_2}^m}+n_1\alpha_2+\frac{n_2\alpha_3}{1+K_b{n_1}^M}+n_2\alpha_4\right)}\ln(1-u) \tag{66}$$

It can be verified that if the waiting time, T_n, whose realization is τ_w, is exponentially distributed, then the random variable, U, whose realization is u, is uniformly distributed over interval [0, 1], see Appendix 3.

Probabilities of four possible transitions

After residing in state $n = (n_1,n_2)$ for a waiting time of τ_w, the system will transfer to one of its adjacent states. During the process, the transition intensity functions governing the four possible transitions of protein populations from state (n_1,n_2) to states $(n_1 - 1,n_2)$, $(n_1 + 1,n_2)$, $(n_1,n_2 - 1)$, and $(n_1,n_2 + 1)$ are α_2, k_1, α_4, and k_2, respectively. These transitions are exact equivalents of the transitions from states $(n_1 + 1,n_2)$, $(n_1 - 1,n_2)$, $(n_1,n_2 + 1)$ and $(n_1,n_2 - 1)$ to state (n_1,n_2), as shown in Figure 2. These four possible transitions are mutually exclusive events. Moreover, as discussed in the last section, one and only one of the four possible transitions takes place during the waiting time determined by the random number r_1. Thus, the probability of the system transferring from (n_1,n_2) to $(n_1 - 1,n_2)$ is

$$Q_1 = \frac{n_1\alpha_2}{\left(\frac{n_1\alpha_1}{1+K_a{n_2}^m}+n_1\alpha_2+\frac{n_2\alpha_3}{1+K_b{n_1}^M}+n_2\alpha_4\right)} \tag{67}$$

The probability of the system transferring from state (n_1,n_2) to $(n_1 + 1,n_2)$ is

$$Q_2 = \frac{\frac{n_1\alpha_1}{1+K_a{n_2}^m}}{\left(\frac{n_1\alpha_1}{1+K_a{n_2}^m}+n_1\alpha_2+\frac{n_2\alpha_3}{1+K_b{n_1}^M}+n_2\alpha_4\right)} \tag{68}$$

The probability of the system transferring from state (n_1,n_2) to $(n_1,n_2 - 1)$ is

$$Q_3 = \frac{n_2\alpha_4}{\left(\frac{n_1\alpha_1}{1+K_a{n_2}^m}+n_1\alpha_2+\frac{n_2\alpha_3}{1+K_b{n_1}^M}+n_2\alpha_4\right)} \tag{69}$$

Similarly, the probability of the system transferring from state (n_1, n_2) to $(n_1, n_2 + 1)$ is

$$Q_4 = \frac{\frac{n_2\alpha_3}{1+K_b n_1{}^M}}{\left(\frac{n_1\alpha_1}{1+K_a n_2{}^m} + n_1\alpha_2 + \frac{n_2\alpha_3}{1+K_b n_1{}^M} + n_2\alpha_4\right)} \tag{70}$$

Since the sum of Q_1 through Q_4 is 1, the transition type can be identified by the randomly generated number, r_2. Specifically, r_2 falling within the interval,

$$\left[0, \frac{n_1\alpha_2}{\left(\frac{n_1\alpha_1}{1+K_a n_2{}^m} + n_1\alpha_2 + \frac{n_2\alpha_3}{1+K_b n_1{}^M} + n_2\alpha_4\right)}\right] \tag{71}$$

implies that the population of type-1 protein decreases by 1, see Equation 67; r_2 falling within the interval,

$$\left[\frac{n_1\alpha_2}{\left(\frac{n_1\alpha_1}{1+K_a n_2{}^m} + n_1\alpha_2 + \frac{n_2\alpha_3}{1+K_b n_1{}^M} + n_2\alpha_4\right)}, \frac{\frac{n_1\alpha_1}{1+K_a n_2{}^m} + n_1\alpha_2}{\left(\frac{n_1\alpha_1}{1+K_a n_2{}^m} + n_1\alpha_2 + \frac{n_2\alpha_3}{1+K_b n_1{}^M} + n_2\alpha_4\right)}\right] \tag{72}$$

implies that the population of type-1 protein increases by 1; r_2 falling within the interval,

$$\left[\frac{\frac{n_1\alpha_1}{1+K_a n_2{}^m} + n_1\alpha_2}{\left(\frac{n_1\alpha_1}{1+K_a n_2{}^m} + n_1\alpha_2 + \frac{n_2\alpha_3}{1+K_b n_1{}^M} + n_2\alpha_4\right)}, \frac{\frac{n_1\alpha_1}{1+K_a n_2{}^m} + n_1\alpha_2 + \frac{n_2\alpha_3}{1+K_b n_1{}^M}}{\left(\frac{n_1\alpha_1}{1+K_a n_2{}^m} + n_1\alpha_2 + \frac{n_2\alpha_3}{1+K_b n_1{}^M} + n_2\alpha_4\right)}\right] \tag{73}$$

implies that the population of type-2 protein decreases by 1; r_2 falling within the interval,

$$\left[\frac{\frac{n_1\alpha_1}{1+K_a n_2{}^m} + n_1\alpha_2 + \frac{n_2\alpha_3}{1+K_b n_1{}^M}}{\left(\frac{n_1\alpha_1}{1+K_a n_2{}^m} + n_1\alpha_2 + \frac{n_2\alpha_3}{1+K_b n_1{}^M} + n_2\alpha_4\right)}, 1\right] \tag{74}$$

implies that the population of type-2 protein increases by 1.

Simulation algorithm

The event-driven Monte Carlo procedure is conducted according to Rajamani [40]. A step-wise description of the procedure is given below.

1. Define the initial populations of the two types of proteins, and let the system size, Ω, be the sum of the two protein populations. This Ω will also be the total number of independent simulations to be conducted before taking their statistics. Start the random walk from this point.
2. Select the total length of time of each simulation, T_f has to be selected. For the current work, T_f was chosen to be either 15 or 50 s.
3. Determine the length of the waiting time, τ_w. First, generate a random number, r_1, from a uniform distribution in [0, 1]; then, calculate τ_w, for a system's transition state $n(t) = (n_1(t),n_2(t))$ according to Equation 66.
4. Update the computer clock by letting $t = t + \tau_w$.
5. Calculate the transition probabilities that the system will transfer from state n to the other states Q_i's by Equations 67 through 70. Then, generate another random number r_2, from a uniform distribution in [0, 1]. Determine the transition type by examining in which interval given by Equations 71 through 74 is r_2 located.

6. Repeat steps 3 to 5 until the total time exceeds T_f; this terminates one replication of simulation.
7. Repeat steps 2 to 6 for Ω times, and store the resultant number in proteins of type i during the jth replication at time t, $n_{ij}(t)$. This yields the mean number of proteins of type i at time t as

$$\boldsymbol{E}[N_i(t)] = \frac{\sum_{j=1}^{\Omega} n_{ij}}{\Omega} \tag{75}$$

The variance of population of type i at time t can be calculated from its definition, i.e.,

$$\mathrm{Var}[N_i(t)] = \frac{\sum_{j=1}^{\Omega} \left(n_{ij} - \boldsymbol{E}[N_i(t)]\right)^2}{\Omega - 1} \tag{76}$$

The covariance around the means between the two types of populations i and j, at time t can be calculated from its definition, i.e.,

$$\mathrm{Cov}\left[N_i(t), N_j(t)\right] = \frac{\sum_{h=1}^{\Omega}\sum_{k=1}^{\Omega} \left(n_{ih} - \boldsymbol{E}[N_i(t)]\right)\left(n_{jk} - \boldsymbol{E}\left[N_j(t)\right]\right)}{\Omega - 1} \tag{77}$$

As mentioned at the outset of this section, both the Monte Carlo simulation and the simulation based on the master equations adopted in the current work are rooted in the identical set of transition intensity functions derived from the same set of assumptions. Thus, integrating the equations for the first and second moments of the master equations, Equations 54 through 58 for the process, is expected to generate results nearly identical to those from the Monte Carlo simulations, i.e., Equations 75 through 77. Equations 75 to 77 are expected to be nearly identical to Equations 54 to 58 together with 49 to 50.

Results and discussion

The present stochastic analysis of the genetic toggle switch yields the transition probabilities of mutually exclusive events through the definitions of the transition intensity functions of protein production as well as degradation. This analysis renders it possible to formulate the nonlinear master equations of the process as well as to derive the event-driven Monte Carlo simulation. Even though each of these was simulated separately they portrayed interesting analogies.

The stochastic algorithms developed here allow us to analyze the stochastic nature of the two-state toggle switch quantitatively. The master equations governing the numbers of the two types of protein are formulated from stochastic population balance. The stochastic pathways of the two proteins, i.e., their means and the fluctuations around these means, have been numerically simulated independently by the algorithm derived from the master equations, as well as by an event-driven Monte Carlo algorithm. Both algorithms have given rise to the identical results. Moreover, these analyses render it possible to circumvent the possibility of noise-induced transitions.

Simulation based on the master equations

Figures 5, 6, and 7 represent the temporal profiles of the Cases A through C discussed earlier. The left-hand parts of these figures are the exploded portion of the more completed simulations on the right. These simulations were conducted with $m = M = 2$, $\alpha_1'' = 15.6$, and the same set of initial conditions, $u(0) = 155$ and $v(0) = 154$. These initial conditions correspond to a point below the separatrix in the phase diagram, see Figure 3. The value of α_2'' varies to illustrate the characteristics of three different cases of dynamics: in the bistable region, on the bifurcation curve, and in the monostable region. The standard deviation envelopes are plotted around the macroscopic trajectories.

Case A, $\alpha_2'' = 15.6$, represents a bistable system, as marked in the bifurcation diagram in Figure 4. Figure 5 presents the simulated results of this system based on the master equations. As expected, the populations eventually reaches the stable steady state #2 marked in Figure 3 since the initial conditions consist a point below the separatrix, and our analysis of the vector field depicting the flow of dynamics [45] suggests this outcome. The protein populations decrease rapidly and stay in the proximity of the saddle node for a while before they depart for their steady states, an observation consistent with the classical dynamics. During this period, the populations of the two proteins are very similar to each other. The fluctuations around the mean trajectories increase initially from zero and then decrease when they approach the steady states. In a stable system, the standard deviation of the number of either type-1 or type-2 proteins attains the maximum because the state of the system is usually well defined at the outset of the process and the uncertainties decline eventually until it varnishes upon stabilization

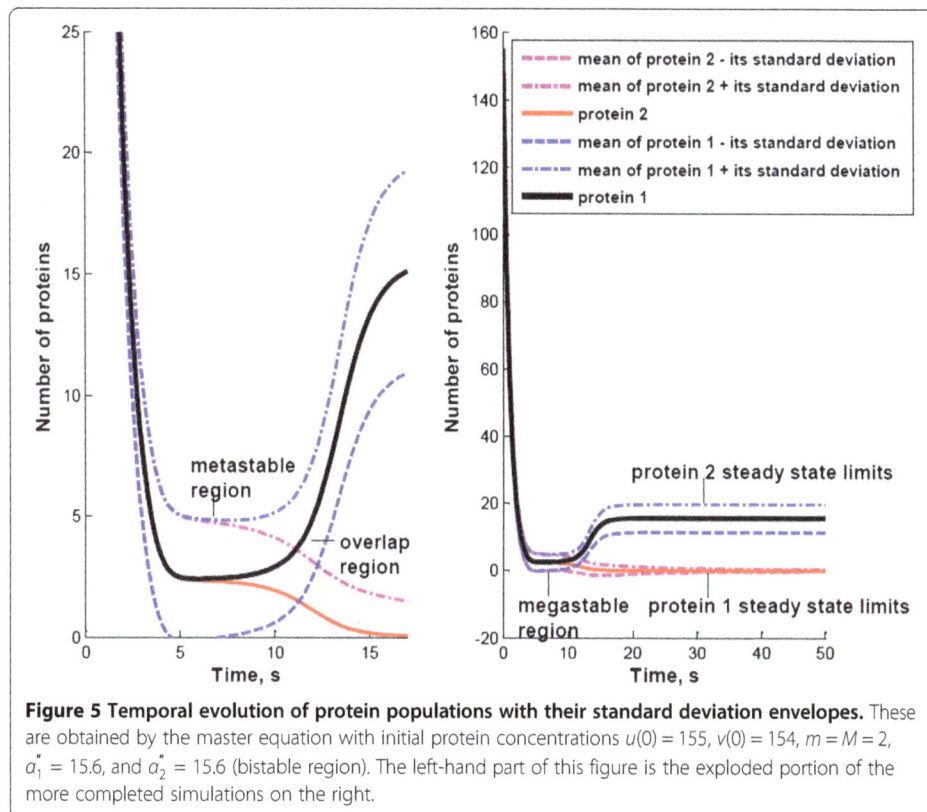

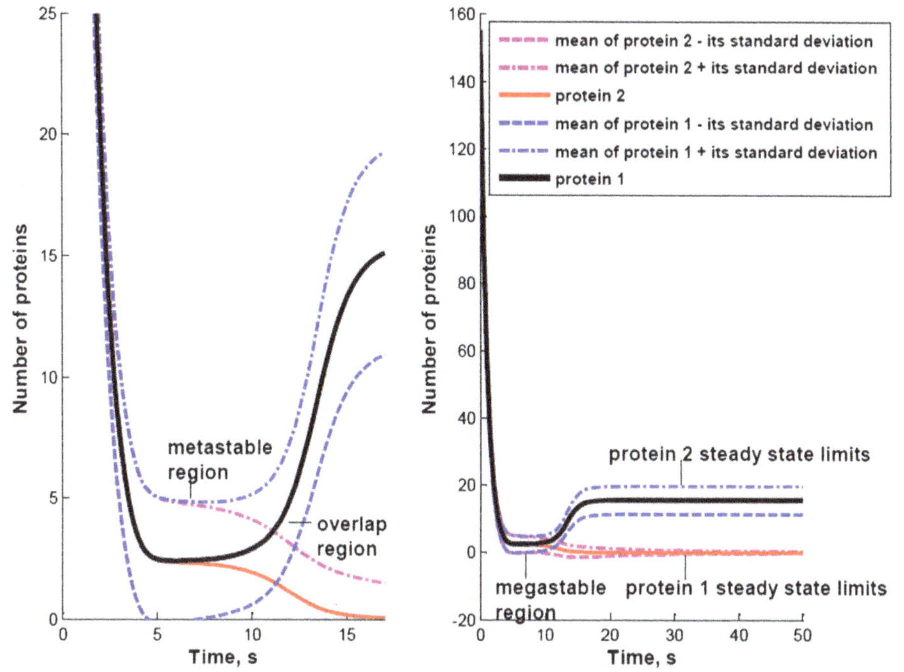

Figure 5 Temporal evolution of protein populations with their standard deviation envelopes. These are obtained by the master equation with initial protein concentrations $u(0) = 155$, $v(0) = 154$, $m = M = 2$, $\alpha_1'' = 15.6$, and $\alpha_2'' = 15.6$ (bistable region). The left-hand part of this figure is the exploded portion of the more completed simulations on the right.

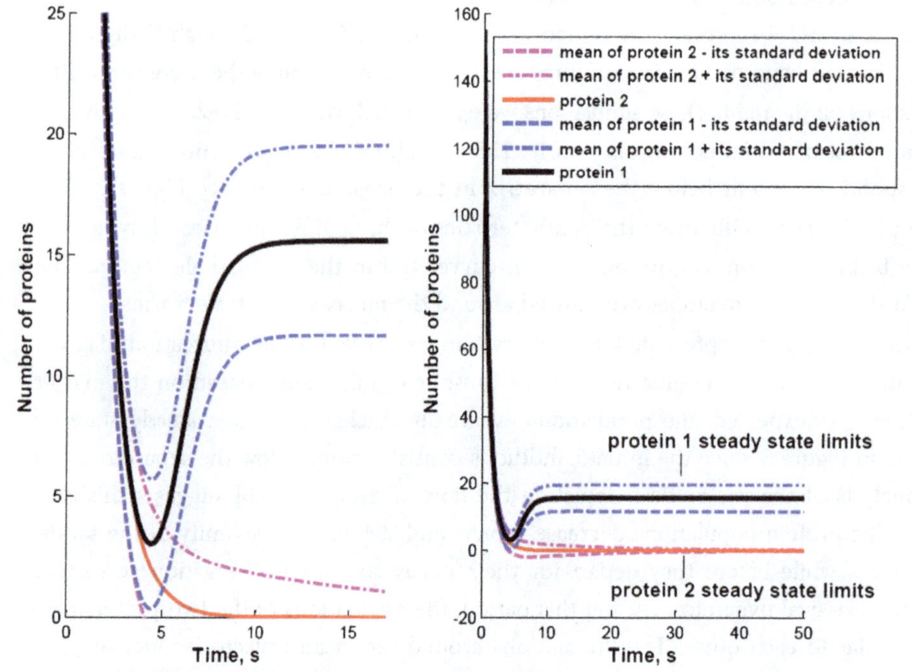

Figure 6 Temporal evolution of protein populations with their standard deviation envelopes. These are obtained by the master equation with initial protein concentrations $u(0) = 155$, $v(0) = 154$, $m = M = 2$, $\alpha_1^{''} = 15.6$, and $\alpha_2^{''} = 4.0$ (on the bifurcation line). The left-hand part of this figure is the exploded portion of the more completed simulations on the right.

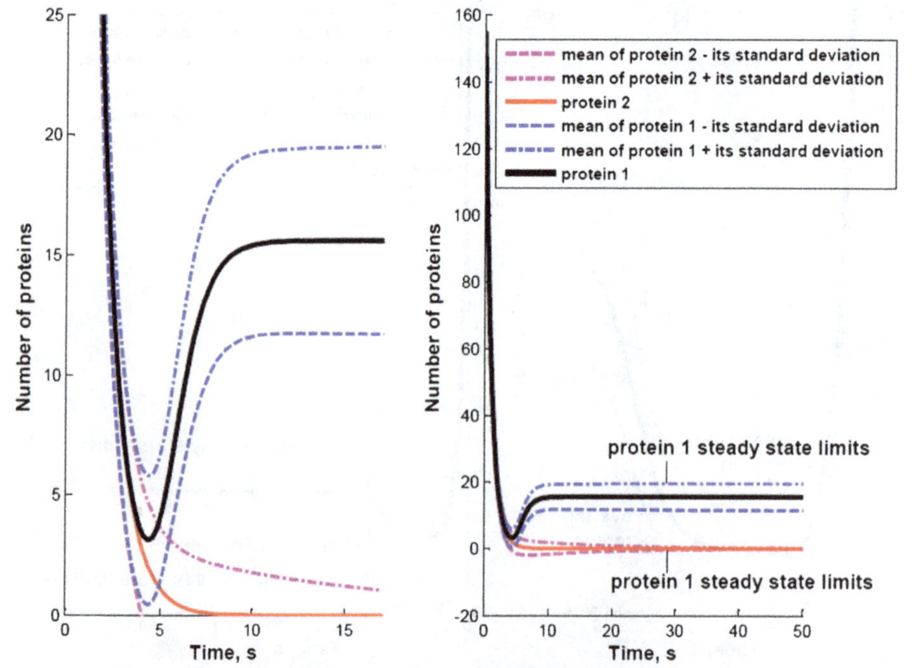

Figure 7 Temporal evolution of protein populations with their standard deviation envelopes. These are obtained by the master equation with initial protein concentrations $u(0) = 155$, $v(0) = 154$, $m = M = 2$, $\alpha'' = 15.6$, and $\alpha_2^{''} = 1.2$ (in monostable region). The left-hand part of this figure is the exploded portion of the more completed simulations on the right.

[46]. The uncertainty in the population of the type-2 protein in Figure 5 appears to remain constant; a special computer experiment was conducted with long simulation time to ensure that it indeed decreases over time.

The formulator is often confronted with a myriad of interacting factors related to a gene's expression mechanisms before settling on a strategy to assess their impact. A mathematical description of this complex process usually relies on a manageable number of system variables. This lumping procedure inevitably results in a high degree of freedom and fluctuations, or uncertainties, in the predictions of populations of discrete systems [9]. The behavior of an individual protein molecule in a discrete system with such a high degree of freedom is thus difficult to predict even when the system is monitored experimentally. The parameters in the equations, e.g., the transition intensity functions of the master equation algorithm adopted here, are presumed to depend only on the major variables of the system and to be independent of the variables of secondary importance. Neglecting these secondary variables is, in essence, the source of internal, or system, or minimal noises that should be appropriately analyzed stochastically. Thus, the internal noises caused by the discrete nature of a system are inherent in the system and they govern the minimum scattering expected of the random variable of interest. The experimentally observed scattering should always be larger than the predicted one induced by internal noises because of inevitable external noises attributable to experimental errors and imprecision of measuring devices. This implies that it is worth cautioning ourselves not to replicate the experiments excessively in an attempt to reduce the scattering far beyond what is predicted. It is interesting to note that fluctuations reported by Gardner et al. are significantly higher than what master equations will predict. The number of culture used in their fluorescence analysis was 40,000, and the actual number of culture in the sample is much larger than this number. Therefore, the noise levels reported by Gardner et al. [3] certainly involve not internal, but also external noises. External noises are the fluctuations created in an otherwise deterministic system by the application of a random force, whose stochastic properties are supposed to be known [9].

The two proteins do not have well-defined states as a deterministic model depicts when they pass the saddle node. Instead, their populations are probabilistically distributed. The two proteins have not only similar populations but also similar uncertainties in their populations. In fact, as shown in the left-hand side of Figure 5, the uncertainties in their populations are in the same order of magnitude. These characteristics imply that there is a high probability that the relative sizes of the two protein populations are switched when the system approaches the unstable steady state. This switch brings the populations to the region above the separatrix in the phase diagram in Figure 3, and the vector field in that region eventually leads the process to the steady state #1 marked in the same figure. The noise-induced phase transition has been examined in detail by Nicolis and Turner [47], Malek Mansour et al. [48], and Horsthemke and Lefever [49]. Nicolis and Turner have shown that the fluctuations enhanced at a 'critical point' (populations closest to the instable steady state); the variances are of the order of $\Omega^{-1/2}$, a result consistent with that derived by van Kampen for expanding the master equation by system size expansion. Thus, systems with low populations are more subjective to noise-induced transitions. The noise enhancement near the instability is illuminated in Figure 5. Once the system moves away from the instability, the noises decrease and noise-induced transition becomes more difficult. Internal fluctuations do not change the local stability of the system, and the position of transition points is in no way modified by the presence of these fluctuations.

It should be mentioned that the parameter values for our simulation are carefully chosen to illustrate the possibility of noise-induced transition. Gardner et al. [3] did not observe this possibility probably because the populations of their system are very large and the difference between the two protein populations at the critical point is large, as discussed earlier.

Case B, $\alpha_2'' = 4.0$, represents system on the bifurcation line, as marked in the bifurcation diagram in Figure 4. Gardner [3] has a good exposition on the dependence of bifurcation diagram and phase diagram on the parameters. There are two steady states on the phase diagram, similar to Figure 3; one is stable and the other, unstable. Figure 6 presents the simulated results of this system based on the master equations. Similar to Case A, the populations eventually reach the stable steady state. The populations do not stay in the vicinity of the saddle node for a long time as they are in Case A. Although the two protein populations are very close to each other and the fluctuations are of the same order of the populations during this time period, one steady state characteristic guarantees the system's final destination.

Case C, $\alpha_2'' = 1.2$, is a monostable system, as marked in the bifurcation diagram in Figure 4. Figure 7 presents the simulated results of this system based on the master equations. Similar to Case B, the populations eventually reach the stable steady state. Although the two protein populations have maximal fluctuations during the evolution of the dynamics, but they eventually vanish to zero.

Figure 8 presents the results from a simulation very similar to Case A. It uses the identical set of parameters for Case A and with a slightly different set of initial conditions: $u(0) = 14$ and $v(0) = 154$. It is a bistable system and the initial conditions represent a point above the separatrix in Figure 4. As expected, the process eventually reaches the steady state #1 shown in Figure 3. Unlike Case A, however, this dynamics does not pass through the proximity of the saddle node, and the fluctuations around the means do not permit easy switching between the two populations, see Figure 8.

Simulation based on Monte Carlo procedure

Monte Carlo simulations have yielded results essentially indistinguishable from those generated from the master equations. This is expected since the algorithms based on the event-driven Monte Carlo procedure and master equations derived in the present work are rooted in identical assumptions, i.e., the Markov property and temporal homogeneity of the random variables. These assumptions lead to the definitions of transition intensity functions that are the cornerstones of the formulation of the master equations and of the Monte Carlo procedure.

The fact that the two algorithms have yielded essentially the same results implies that both indeed define the evolution of dynamic process in a precisely equivalent way. The master-equation algorithm generates the equations governing the statistical moments of the process, which can be readily varied to cover a wide range of initial conditions, whereas the Monte Carlo procedure will require far more computational time and storage space under such circumstances.

Internal noise-induced transition was clearly observed during Monte Carlo simulation for Case A. Figure 9 demonstrates the two traces from two independent Monte

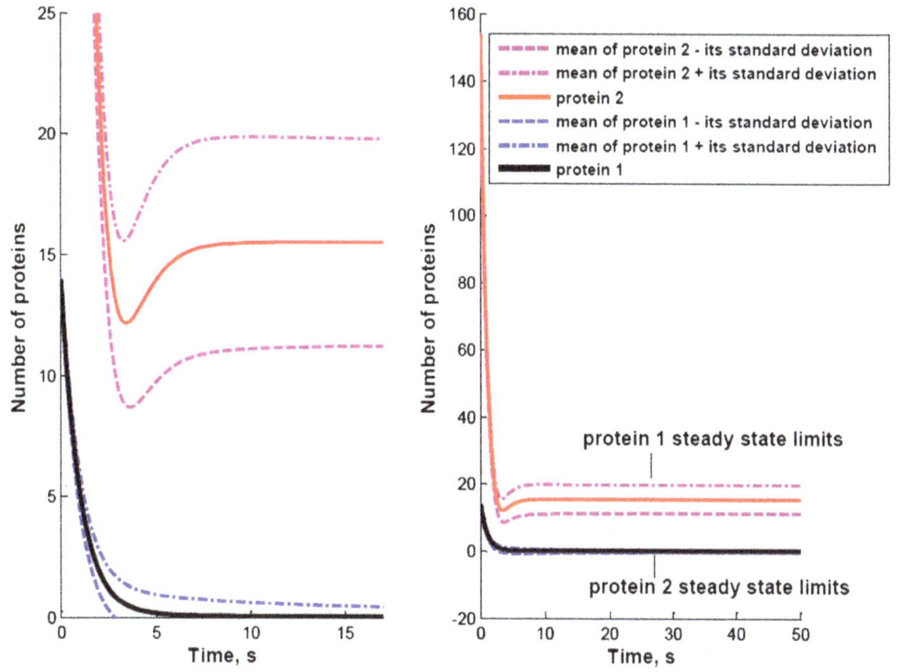

Figure 8 Temporal evolution of protein populations with their standard deviation envelopes. These are obtained by the master equation with initial protein concentrations $u(0) = 14$, $v(0) = 154$, $m = M = 2$, $\alpha_1^{*} = 15.6$, and $\alpha_2^{*} = 15.6$ (bistable region). Comparison of the results here and those from Figure 4 suggests that the initial conditions determine the ultimate steady state for a bistable system. The left-hand part of this figure is the exploded portion of the more completed simulations on the right.

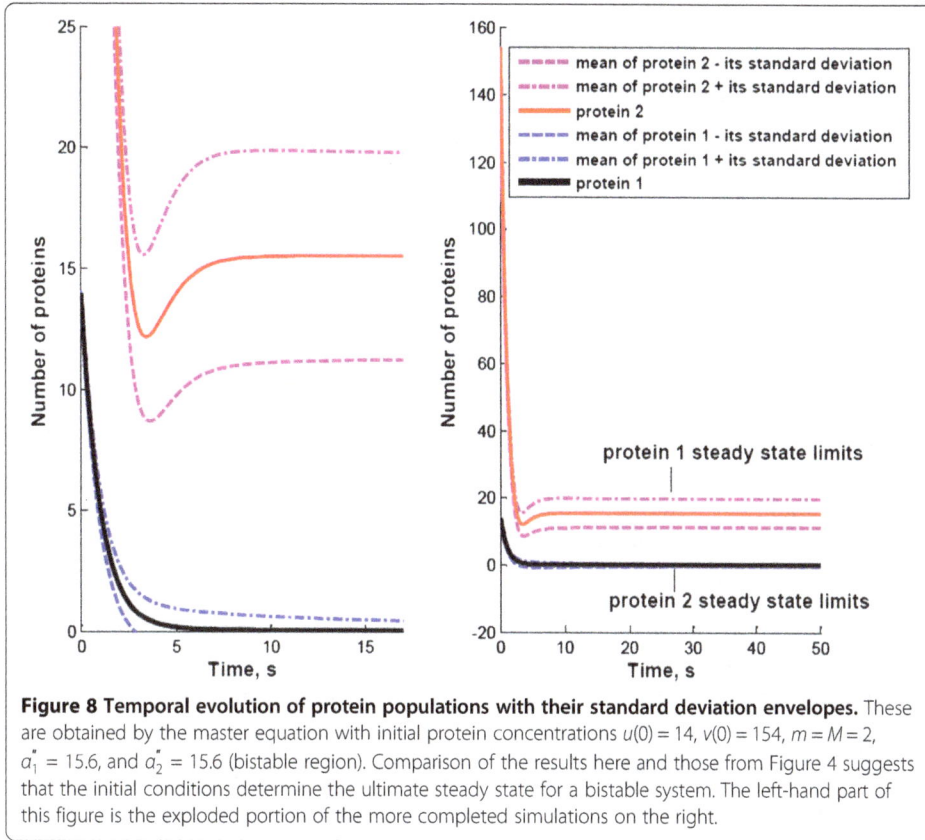

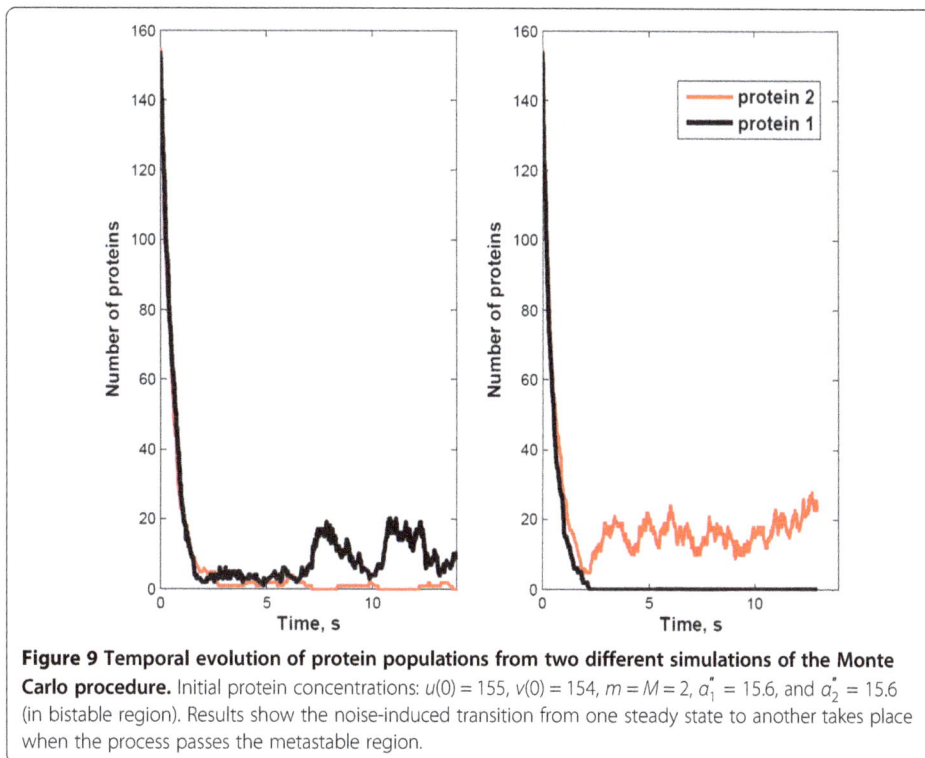

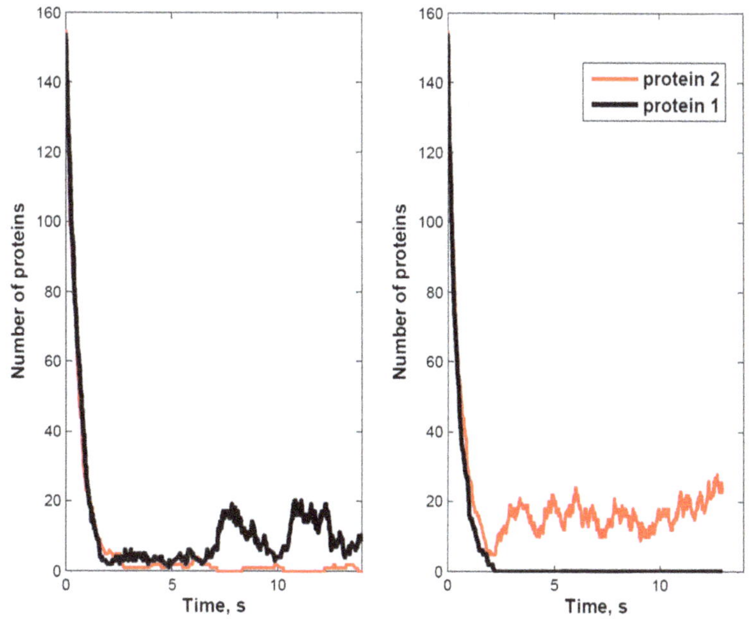

Figure 9 Temporal evolution of protein populations from two different simulations of the Monte Carlo procedure. Initial protein concentrations: $u(0) = 155$, $v(0) = 154$, $m = M = 2$, $\alpha_1^{*} = 15.6$, and $\alpha_2^{*} = 15.6$ (in bistable region). Results show the noise-induced transition from one steady state to another takes place when the process passes the metastable region.

Carlo simulations with parameters and initial conditions identical to those for Case A. These two independent Monte Carlo simulations result in two different steady states that is a consequence of internal noise-induced transition. It should be mentioned that results based on master equation, as shown in Figure 5, represent an averaged outcome of independent Monte Carlo simulations of Ω times (Ω = 155 + 154 = 309 for this case), which are indeed observed in our simulation experiments. As mentioned in the last section, systems of small populations are susceptible to large internal fluctuations (or uncertainties) in the evolution of their dynamics. The evolutions of protein statistics shown in Figure 5 also illustrate the large uncertainties after the populations enter the proximity of the saddle node. In fact, the uncertainty is of the same magnitude as the mean number of particles. Internal fluctuations are inherent characteristics of discrete systems that are beyond the regulation of external means. The results on the right-hand side of Figure 9 show a clear transition in protein numbers in a particular Monte Carlo simulation. The transition takes place soon after the populations enter the proximity of the saddle node. It is caused by the fact that the populations of both proteins are low and, therefore, there are susceptible to large internal fluctuations and noise-induced transitions. Noise-induced transition has been discussed by Nicolis and Turner [47], Malek Mansour et al. [48], and Horsthemke and Lefever [49].

As mentioned in the 'Introduction' section, the master equation and its system-size expansion suffers a few limitations. One of such limitations is that the algorithm is valid for the dynamics well within the boundary of attraction [9]. For a bistable dynamics staring in a region outside this boundary, such as Case A, the Monte Carlo simulation converges to two possible steady states. The master equation algorithm converges to only one.

Some comparisons of the three algorithms are worth mentioning. The governing equations for the system size expansion can be derived in a straightforward manner, though the detailed derivations may be cumbersome and time consuming. It requires only a minor transformation of variable for some unstable stochastic processes, such as the diffusion process, well beyond the initial transient period [9]. Unlike the Monte Carlo simulation, the derived moment equations can be repeatedly integrated for different sets of parameters and initial conditions. Consequently, system size expansion has been widely adopted in the derivation of governing equation of stochastic processes governed by internal noises.

Kurtz's algorithm is highly compact and convenient. The implementation of the rigorous Kurtz algorithm requires knowledge about the relations among master, Langevin, and Fokker-Planck equations. It allows direct derivation of the equations governing the moments. However, the algorithm merely describes the dynamics in the initial transient period of unstable systems for selected processes, such as the diffusion process [9].

The Monte Carlo method is easy to implement because it bypasses all derivations of equations. It is most efficient when the number of random variables is large and the master equation is difficult to derive. Repeated simulations have to be carried out for different sets of parameters and initial conditions. The required computational time and disk space are usually high.

Conclusions

The current model adopts the essential concepts of a nonlinear toggle switch model for analyzing a protein-regulated system. The master equation algorithm, along with its system size expansion, involves the stochastic probability balance of the two types of populations. The resultant master equation should yield not only the deterministic evolution of protein populations during gene expression, but also the fluctuations, or uncertainties inherited in the prediction or measurement. Kurtz's limit theorems significantly reduce the complex and laborious exercise of system size expansion. In fact, they will be indispensable tools for the analysis of really complex genetic networks.

The validity of the model is amply demonstrated by numerically calculating the evolution of population of both types and their fluctuations over time through two simulation algorithms, one based on the master equations and the other based on the event-driven Monte Carlo procedure. These two algorithms are implemented totally independently of each other but with the same set of system parameters, i.e., the transition intensity functions. Hence, it is indeed remarkable that the two algorithms have yielded essentially identical results.

Both simulation results demonstrate the possibility of noise-induced transition when the dynamics passes through the proximity of the saddle node. It happens when the protein populations are low and the noises are in the same order of magnitudes as the populations. This property may have practical applications in developing gene therapy, cell cycle control, and protein sensors.

Nomenclature

$\boldsymbol{E}$, one-step operator; N_1, random variable representing population of repressor 1; N_2, random variable representing population of repressor 2; n_1, realization of random variable, $N_1(t)$; n_2, realization of random variable, $N_2(t)$; $\boldsymbol{N}$, random vector, i.e., $[N_1(t),N_2(t),N_3(t)]$; $\boldsymbol{n}$, realization of random vector $\boldsymbol{N}(t)$; $P_{\boldsymbol{n}}$, probability that the system is at state $\mathbf{n}$ at time t; t, time; Y, random variable denoting the fluctuations about macroscopic behavior; y, realization of random variable Y fluctuations; Q, the transition probability; u, Gardner's concentration of repressor 1; v, Gardner's concentration of repressor 2; K_1, effective reaction rate for repressor 1 formation; K_2, effective reaction rate for repressor 2 formation.

Greek letters

α_1, the rate of production of repressor 1; α_2, the rate of production of repressor 2; α_3, the rate of degradation of repressor 1; α_4, the rate of degradation of repressor 2; α'_1, the effective rate of synthesis of repressor 1 on system size; α'_2, the effective rate of synthesis of repressor 2 on system size; $\varnothing$, macroscopic number of repressor 1; θ, macroscopic number of repressor 2; τ_w, the waiting time; λ, transition intensity function; Ψ, joint probability distribution in terms of random vector $\boldsymbol{Y}$; Ω, total number of repressors or system size; β, the multimerization constant of repressor 1; γ, the multimerization constant of repressor 2; Θ, the vector representing the two mean numbers of proteins in system-size expansion.

Subscripts

1, repressor 1; 2, repressor 2

Appendices

Appendix 1: system-size expansion

The constituent populations at any time in the genetic toggle switch system can be represented by the random vector $\boldsymbol{N}(t) = [N_1(t), N_2(t)]$, their mean values can be taken as a deterministic vector $\mathbf{\Theta}(t) = [\phi(t), \theta(t)]$ and their fluctuations can be taken as another random vector given by $\mathbf{Y}(t)$, where $\mathbf{Y}(t) = [Y_1(t), Y_2(t)]$. As stated in Equations 13 and 14 in the text:

$$N_1(t) = \Omega\,\phi(t) + \Omega^{1/2} Y_1(t) \tag{78}$$

$$N_2(t) = \Omega\,\theta(t) + \Omega^{1/2} Y_2(t) \tag{79}$$

Their realizations of these expressions are given, respectively, by

$$n_1(t) = \Omega\phi(t) + \Omega^{1/2} y_1(t) \tag{80}$$

$$n_2(t) = \Omega\theta(t) + \Omega^{1/2} y_2(t) \tag{81}$$

Accordingly, the joint probability of n_1 and n_2, $P_{\mathbf{n}}(t)$, is now transformed into that of y_1 and y_2, i.e., $\Psi_{\mathbf{y}}(t)$.

Recall that in the context of deriving the master equation, the state or dependent variable of interest is the joint probability of the population distribution, $P_{\mathbf{n}}(t)$, and the realization of random variables at time t, i.e., n_1 and n_2, are invariant with respect to time. Consequently, the time derivatives of Equations 80 and 81 are, respectively,

$$\frac{dy_1}{dt} = -\Omega^{1/2} \frac{d\phi}{dt} \tag{82}$$

$$\frac{dy_2}{dt} = -\Omega^{1/2} \frac{d\theta}{dt} \tag{83}$$

For the convenience of the subsequent expansion of the master equation, Eq. 11 in the text is restated below

$$\begin{aligned} \frac{dP_{n_1, n_2}(t)}{dt} &= \frac{\alpha_1}{1 + K_a {n_2}^m}\left(E_{n_1}^{-1} - 1\right) P_{n_1, n_2}(t) + (E_{n_1} - 1)\alpha_2\, n_1\, P_{n_1, n_2}(t) \\ &+ \frac{\alpha_3}{1 + K_b {n_1}^M}\left(E_{n_2}^{-1} - 1\right) P_{n_1, n_2}(t) + (E_{n_2} - 1)\alpha_4\, n_2\, P_{n_1, n_2}(t) \end{aligned} \tag{84}$$

Substituting Equations 82 and 83 into the right-hand side of the above expression yields

$$\begin{aligned} \frac{dP_{\mathbf{n}}}{dt} &= \frac{d\Psi_{\mathbf{y}}}{dt} = \frac{\partial \Psi_{\mathbf{y}}}{\partial t} + \frac{\partial \Psi_{\mathbf{y}}}{\partial y_1}\frac{dy_1}{dt} + \frac{\partial \Psi_{\mathbf{y}}}{\partial y_2}\frac{dy_2}{dt} \\ &= \frac{\partial \Psi_{\mathbf{y}}}{\partial t} - \frac{\partial \Psi_{\mathbf{y}}}{\partial y_1}\Omega^{1/2}\frac{d\phi}{dt} - \frac{\partial \Psi_{\mathbf{y}}}{\partial y_2}\Omega^{1/2}\frac{d\theta}{dt} \end{aligned} \tag{85}$$

Without causing confusion, the subscript $\mathbf{y}$ of $\Psi_{\mathbf{y}}(t)$ is eliminated in the subsequent discussion. The step operators, E_{n_1} and $E_{n_1}^{-1}$, convert n_1 to $n_1 + 1$ and $n_1 - 1$, respectively. Similarly, Equation 80 suggests that E_{n_1} shifts y_1 to $y_1 + \Omega^{-1/2}$. Therefore, the operations of step operators in Equation 84 are equivalent to evaluating the values of target functions at shifted points through the following Taylor series expansions, i.e.,

$$E_{n_1} = 1 + \Omega^{-1/2}\frac{\partial}{\partial y_1} + \frac{\Omega^{-1}}{2}\frac{\partial^2}{\partial y_1^2} + \ldots \tag{86}$$

$$E_{n_1}^{-1} = 1 - \Omega^{-1/2}\frac{\partial}{\partial y_1} + \frac{\Omega^{-1}}{2}\frac{\partial^2}{\partial y_1^2} - \ldots \tag{87}$$

$$E_{n_2} = 1 + \Omega^{-1/2}\frac{\partial}{\partial y_2} + \frac{\Omega^{-1}}{2}\frac{\partial^2}{\partial y_2^2} + \ldots \tag{88}$$

$$E_{n_2}^{-1} = 1 - \Omega^{-1/2}\frac{\partial}{\partial y_2} + \frac{\Omega^{-1}}{2}\frac{\partial^2}{\partial y_2^2} - \ldots \tag{89}$$

Substituting Equations 85 through 89 into 84 yields

$$\begin{aligned}
&\frac{\partial \Psi}{\partial t} - \Omega^{1/2}\frac{d\phi}{dt}\frac{\partial \Psi}{\partial y_1} - \Omega^{1/2}\frac{d\theta}{dt}\frac{\partial \Psi}{\partial y_2} \\
&= \frac{\alpha_1}{1 + K_a\left[\Omega\theta(t) + \Omega^{1/2}y_2(t)\right]^m}\left[1 - \Omega^{-1/2}\frac{\partial}{\partial y_1} + \frac{\Omega^{-1}}{2}\frac{\partial^2}{\partial y_1{}^2} - \ldots - 1\right]\Psi \\
&+ \alpha_2\left[1 + \Omega^{-1/2}\frac{\partial}{\partial y_1} + \frac{\Omega^{-1}}{2}\frac{\partial^2}{\partial y_1{}^2} + \ldots - 1\right]\left(\Omega\phi(t) + \Omega^{1/2}y_1(t)\right)\Psi \\
&+ \frac{\alpha_3}{1 + K_b\left[\Omega\phi(t) + \Omega^{1/2}y_1(t)\right]^M}\left[1 - \Omega^{-1/2}\frac{\partial}{\partial y_2} + \frac{\Omega^{-1}}{2}\frac{\partial^2}{\partial y_2{}^2} - \ldots - 1\right]\Psi \\
&+ \alpha_4\left[1 + \Omega^{-1/2}\frac{\partial}{\partial y_2} + \frac{\Omega^{-1}}{2}\frac{\partial^2}{\partial y_2{}^2} + \ldots - 1\right]\left(\Omega\theta(t) + \Omega^{1/2}y_2(t)\right)\Psi
\end{aligned} \tag{90}$$

In order to collect the terms of same power of Ω in the subsequent expansion, the Ω dependence of the parameters in the above equation have to be examined and converted to their Ω -independent counterparts. The definitions of α_1 and α_3 in the 'Model formulation' section suggest that they are proportional to the system size, Ω, i.e.,

$$\alpha_1 = \alpha_1'\Omega, \qquad \alpha_3 = \alpha_3'\Omega, \tag{91}$$

where α_1' and α_3' are independent of the system size, Ω. Moreover, the definitions of equilibrium constants, K_a and K_b, for the gene repression, $G + m\,R \leftrightarrows GR_m$, in the beginning of the 'Model formulation' section suggest

$$K_a = K_a'\Omega^{-m}, \qquad K_b = K_b'\Omega^{-M}, \tag{92}$$

where K_a' and K_b' are independent of the system size, Ω.

Substituting Equations 91 and 92 into Equation 90 gives

$$\begin{aligned}
&\frac{\partial \Psi}{\partial t} - \Omega^{1/2}\frac{d\phi}{dt}\frac{\partial \Psi}{\partial y_1} - \Omega^{1/2}\frac{d\theta}{dt}\frac{\partial \Psi}{\partial y_2} \\
&= \frac{\alpha_1'\Omega}{1 + K_a'\Omega^{-m}\left[\Omega\theta(t) + \Omega^{1/2}y_2(t)\right]^m}\left[-\Omega^{-1/2}\frac{\partial}{\partial y_1} + \frac{\Omega^{-1}}{2}\frac{\partial^2}{\partial y_1{}^2} - \ldots\right]\Psi \\
&+ \alpha_2\left[\Omega^{-1/2}\frac{\partial}{\partial y_1} + \frac{\Omega^{-1}}{2}\frac{\partial^2}{\partial y_1{}^2} + \ldots\right]\left(\Omega\phi(t) + \Omega^{1/2}y_1(t)\right)\Psi \\
&+ \frac{\alpha_3'\Omega}{1 + K_b'\Omega^{-M}\left[\Omega\phi(t) + \Omega^{1/2}y_1(t)\right]^M}\left[-\Omega^{-1/2}\frac{\partial}{\partial y_2} + \frac{\Omega^{-1}}{2}\frac{\partial^2}{\partial y_2{}^2} - \ldots\right]\Psi \\
&+ \alpha_4\left[\Omega^{-1/2}\frac{\partial}{\partial y_2} + \frac{\Omega^{-1}}{2}\frac{\partial^2}{\partial y_2{}^2} + \ldots\right]\left(\Omega\theta(t) + \Omega^{1/2}y_2(t)\right)\Psi
\end{aligned} \tag{93}$$

The first and third terms on the right-hand side of the above expression can be expanded in power of Ω through known power and binomial expansions. Specifically, for small $K'_a\Omega^{-m}\left[\Omega\theta(t)+\Omega^{1/2}y_2(t)\right]^m$, we have

$$\begin{aligned}
&\frac{1}{1+K'_a\Omega^{-m}\left[\Omega\theta(t)+\Omega^{1/2}y_2(t)\right]^m} \\
&= 1-K'_a\Omega^{-m}\left[\Omega\theta(t)+\Omega^{1/2}y_2(t)\right]^m+\left\{K'_a\Omega^{-m}\left[\Omega\theta(t)+\Omega^{1/2}y_2(t)\right]^m\right\}^2-+\ldots \\
&= 1-K'_a\Omega^{-m}\sum_{i=0}^{m}\binom{m}{i}\left[(\Omega\theta(t))^{m-i}\left(\Omega^{1/2}y_2(t)\right)^i\right] \\
&+ (K'_a\Omega^{-m})^2\sum_{i=0}^{2m}\binom{2m}{i}\left[(\Omega\theta(t))^{2m-i}\left(\Omega^{1/2}y_2(t)\right)^i\right] \\
&-(K'_a\Omega^{-m})^3\sum_{i=0}^{3m}\binom{3m}{i}\left[(\Omega\theta(t))^{3m-i}\left(\Omega^{1/2}y_2(t)\right)^i\right]+\ldots
\end{aligned} \tag{94}$$

where $\binom{m}{i}$ denotes a binomial coefficient. Lumping the terms of the same power of Ω in the above expansion gives

$$\begin{aligned}
&\frac{1}{1+K'_a\Omega^{-m}\left[\Omega\theta(t)+\Omega^{1/2}y_2(t)\right]^m} \\
&= 1-K'_a\Omega^{-m}\left(\Omega^m\theta^m+m\Omega^{m-1}\theta^{m-1}y_2\Omega^{-1/2}+\ldots\right) \\
&+ (K'_a)^2\Omega^{-2m}\left(\Omega^{2m}\theta^{2m}+2m\Omega^{2m-1}\theta^{2m-1}y_2\Omega^{-1/2}+\ldots\right) \\
&+ (K'_a)^3\Omega^{-3m}\left(\Omega^{3m}\theta^{3m}+3m\Omega^{3m-1}\theta^{3m-1}y_2\Omega^{-1/2}+\ldots\right)+\ldots \\
&= \left(1-K'_a\theta^m+(K'_a)^2\theta^{2m}-(K'_a)^3\theta^{3m}+-\ldots\right)\Omega^0 \\
&-\left(K'_a\theta^{m-1}-2(K'_a)^2\theta^{2m-1}+3(K'_a)^3\theta^{3m-1}-+\ldots\right)my_2\Omega^{-1/2}+\ldots \\
&= \frac{1}{1+K'_a\theta^m}\Omega^0-\frac{K'_a\theta^{m-1}}{(1+K'_a\theta^m)^2}m\,y_2\,\Omega^{-1/2}+\ldots
\end{aligned} \tag{95}$$

Substituting the above expression into the first term on the right-hand side of Equation 93 yields

$$\begin{aligned}
&\frac{\alpha'_1\Omega}{1+K'_a\Omega^{-m}\left[\Omega\theta(t)+\Omega^{1/2}y_2(t)\right]^m}\left[-\Omega^{-1/2}\frac{\partial}{\partial y_1}+\frac{\Omega^{-1}}{2}\frac{\partial^2}{\partial {y_1}^2}-\cdots\right]\Psi \\
&= \alpha'_1\Omega\left(\frac{1}{1+K'_a\theta^m}\Omega^0-\frac{K'_a\theta^{m-1}}{(1+K'_a\theta^m)^2}m\,y_2\,\Omega^{-1/2}+\ldots\right) \\
&\left[-\Omega^{-1/2}\frac{\partial\Psi}{\partial y_1}+\frac{\Omega^{-1}}{2}\frac{\partial^2\Psi}{\partial {y_1}^2}-\cdots\right]+\ldots \\
&= \frac{-\alpha'_1}{1+K'_a\theta^m}\frac{\partial\Psi}{\partial y_1}\Omega^{1/2}+\left(\frac{\alpha'_1K'_a\theta^{m-1}m\,y_2}{(1+K'_a\theta^m)^2}\frac{\partial\Psi}{\partial y_1}+\frac{\alpha'_1}{2}\frac{1}{1+K'_a\theta^m}\frac{\partial^2\Psi}{\partial {y_1}^2}\right)\Omega^0+\ldots
\end{aligned} \tag{96}$$

The expansion of the second term on the right-hand side of Equation 93 gives

$$\begin{aligned}&\alpha_2\left[\Omega^{-1/2}\frac{\partial}{\partial y_1}+\frac{\Omega^{-1}}{2}\frac{\partial^2}{\partial {y_1}^2}+\cdots\right]\left(\Omega\phi(t)+\Omega^{1/2}y_1(t)\right)\Psi\\&=\alpha_2\phi\frac{\partial\Psi}{\partial y_1}\Omega^{1/2}+\alpha_2\left(\Psi+\frac{\phi}{2}\frac{\partial^2\Psi}{\partial {y_1}^2}+y_1\frac{\partial\Psi}{\partial y_1}\right)\Omega^0+\ldots\end{aligned} \tag{97}$$

Following the same procedure, the third and fourth terms on the right-hand side of Equation 93 can be expanded into the following power series of $\Omega^{1/2}$, respectively:

$$\begin{aligned}&\frac{\alpha_3'\,\Omega}{1+K_b'\,\Omega^{-M}\left[\Omega\phi(t)+\Omega^{1/2}y_1(t)\right)M}\left[-\Omega^{-1/2}\frac{\partial}{\partial y_2}+\frac{\Omega^{-1}}{2}\frac{\partial^2}{\partial {y_2}^2}-\cdots\right]\Psi\\&=\frac{-\alpha_3'}{1+K_b'\,\phi^M}\frac{\partial\Psi}{\partial y_2}\Omega^{1/2}+\left(\frac{\alpha_3'K_b'\theta^{M-1}My_1}{(1+K_b'\phi^M)^2}\frac{\partial\Psi}{\partial y_2}+\frac{\alpha_3'}{2}\frac{1}{1+K_b'\phi^M}\frac{\partial^2\Psi}{\partial {y_2}^2}\right)\Omega^0+\ldots\end{aligned} \tag{98}$$

and

$$\begin{aligned}&\alpha_4\left[\Omega^{-1/2}\frac{\partial}{\partial y_2}+\frac{\Omega^{-1}}{2}\frac{\partial^2}{\partial {y_2}^2}+\cdots\right]\left(\Omega\theta(t)+\Omega^{1/2}y_2(t)\right)\Psi\\&=\alpha_4\theta\frac{\partial\Psi}{\partial y_2}\Omega^{1/2}+\alpha_4\left(\Psi+\frac{\theta}{2}\frac{\partial^2\Psi}{\partial {y_2}^2}+y_2\frac{\partial\Psi}{\partial y_2}\right)\Omega^0+\ldots\end{aligned} \tag{99}$$

Substituting Equations 96 through 99 into Equation 93 and collecting the terms of order $\Omega^{1/2}$ on both sides of the Equation 100 yields the macroscopic, or the deterministic, equations:

$$\frac{d\phi}{dt}=\frac{\alpha_1'}{1+K_a'\,\theta(t)^m}-\alpha_2\,\phi(t) \tag{100}$$

$$\frac{d\theta}{dt}=\frac{\alpha_3'}{1+K_b'\,\phi(t)^M}-\alpha_4\,\theta(t) \tag{101}$$

They are the Equations 17 and 18 in the text. By collecting terms of order Ω^0 yields

$$\begin{aligned}\frac{\partial\Psi}{\partial t}=&\left(\frac{\alpha_1'}{1+K_a'\theta(t)^m}+\alpha_2\phi(t)\right)\frac{1}{2}\frac{\partial^2}{\partial y_1^2}\Psi+\left(\frac{\alpha_3'}{1+K_b'\phi(t)^M}+\alpha_4\theta(t)\right)\frac{1}{2}\frac{\partial^2}{\partial y_2^2}\Psi\\&+\frac{\partial}{\partial y_1}\left(\alpha_2y_1\Psi+\frac{\alpha_1'K_a'm(\theta(t))^{m-1}y_2\Psi}{(1+K_a'\theta(t)^m)^2}\right)\\&+\frac{\partial}{\partial y_2}\left(\alpha_4y_2\Psi+\frac{\alpha_3'K_b'M(\phi(t))^{M-1}y_1\Psi}{\left(1+K_b'\phi(t)^M\right)^2}\right)\end{aligned} \tag{102}$$

This equation can be rearranged in a linear Fokker-Plank equation form [9] as follows:

$$\frac{\partial\Psi}{\partial t}=-\sum_{i,j}A_{ij}\frac{\partial}{\partial y_i}\left(y_j\Psi\right)+\frac{1}{2}\sum_{i,j}\left(B_{ij}\frac{\partial^2\Psi}{\partial y_i\partial y_j}\right) \tag{103}$$

or

$$\begin{aligned}\frac{\partial \Psi}{\partial t} = & -A_{11}\frac{\partial}{\partial y_1}(y_1\Psi) - A_{12}\frac{\partial}{\partial y_1}(y_2\Psi) - A_{21}\frac{\partial}{\partial y_2}(y_1\Psi) \\ & - A_{22}\frac{\partial}{\partial y_2}(y_2\Psi) + \frac{1}{2}B_{11}\frac{\partial^2 \Psi}{\partial y_1^2} + \frac{1}{2}B_{22}\frac{\partial^2 \Psi}{\partial y_2^2},\end{aligned} \tag{104}$$

where

$$A = \begin{bmatrix} A_{11} & A_{12} \\ A_{21} & A_{22} \end{bmatrix} = \begin{bmatrix} -\alpha_2 & \dfrac{-\alpha_1' K_a' m(\theta(t))^{m-1}}{\left(1 + K_a'(\theta(t))^m\right)^2} \\ \dfrac{-\alpha_3' K_b' M(\phi(t))^{M-1}}{\left(1 + K_b'(\phi(t))^M\right)^2} & -\alpha_4 \end{bmatrix} \tag{105}$$

$$B = \begin{bmatrix} B_{11} & B_{12} \\ B_{21} & B_{22} \end{bmatrix} = \begin{bmatrix} \dfrac{\alpha_1'}{1 + K_a'(\theta(t))^m} + \alpha_2\phi(t) & 0 \\ 0 & \dfrac{\alpha_3'}{1 + K_b'(\phi(t))^M} + \alpha_4\theta(t) \end{bmatrix} \tag{106}$$

Equations 103, 104, and 105 are Equations 21, 22, and 23 in the text, respectively.

Appendix 2: distribution functions of waiting time

Equation 9 in the text indicates that the probability of no transition in the small time interval, $(\tau_w,\ \tau_w + \Delta\tau_w)$, is

$$G_n(\Delta\tau_w) = \left[1 - \left\{\frac{n_1\alpha_1}{1 + K_a n_2^m} + n_1\alpha_2 + \frac{n_2\alpha_3}{1 + K_b n_1^M} + n_2\alpha_4\right\}\Delta\tau_w\right] + o(\Delta\tau_w) \tag{107}$$

The Markov property implies that succeeding time intervals, $(0, \tau_w)$ and $(\tau_w, \tau_w + \Delta\tau_w)$, are independent of each other (30), thus

$$\begin{aligned}G_n(\tau_w + \Delta\tau_w) & = G_n(\tau_w)G_n(\Delta\tau_w) + o(\Delta\tau_w) \\ & = G_n(\tau_w)\left[1 - \left\{\frac{n_1\alpha_1}{1 + K_a n_2^m} + n_1\alpha_2 + \frac{n_2\alpha_3}{1 + K_b n_1^M} + n_2\alpha_4\right\}\Delta\tau_w\right] + o(\Delta\tau_w)\end{aligned} \tag{108}$$

Rearranging the above equation yields

$$\begin{aligned}& G_n(\tau_w + \Delta\tau_w) - G_n(\tau_w) \\ = & -G_n(\tau_w)\left[\left\{\frac{n_1\alpha_1}{1 + K_a n_2^m} + n_1\alpha_2 + \frac{n_2\alpha_3}{1 + K_b n_1^M} + n_2\alpha_4\right\}\Delta\tau_w\right] + o(\Delta\tau_w)\end{aligned} \tag{109}$$

Dividing the expression by taking the $\Delta\tau_w$ and taking the limits as $\Delta\tau_w \to 0$ gives

$$\frac{G_n(\tau_w)}{d\tau_w} = -G_n(\tau_w)\left(\frac{n_1\alpha_1}{1 + K_a n_2^m} + n_1\alpha_2 + \frac{n_2\alpha_3}{1 + K_b n_1^M} + n_2\alpha_4\right) \tag{110}$$

By integrating this equation, with constant $\boldsymbol{n}$, subject to the initial condition

$$G_n(0) = 1 \tag{111}$$

we obtain

$$G_n(\tau_w) = \exp\left(-\left\{\frac{n_1\alpha_1}{1 + K_a n_2^m} + n_1\alpha_2 + \frac{n_2\alpha_3}{1 + K_b n_1^M} + n_2\alpha_4\right\}\tau_w\right) \tag{112}$$

Chapter 14 of the book by Karlin and Taylor [43,44] contains a more rigorous proof of Equation 112.

From the definition of $G_n(\tau_w)$, i.e., Equation 60 in the text, one can obtain a cumulative distribution of the form

$$\begin{aligned} G_n(\tau_w) &= P_r(T_n \geq \tau_w) \\ &= 1 - H_n(\tau_w) \\ &= 1 - \int_0^{\tau_w} h_n(\tau_w) \end{aligned} \tag{113}$$

Since $H_n(\tau_w)$ is the cumulative probability distibution form 0 to τ_w, i.e. the probability function of T_n at τ_w, $h_n(\tau_w)$, which is the derivative of $H_n(\tau_w)$ with respect to τ_w, is the probability-density function of T_n. Hence,

$$\frac{dG_n(\tau_w)}{d\tau_w} = h_n(\tau_w) = \frac{dH_n(\tau_w)}{d\tau_w} \tag{114}$$

In light of Equations 113 and 114 give rise to

$$\begin{aligned} h_n(\tau_w) = &\left(\frac{n_1\alpha_1}{1+K_a n_2{}^m} + n_1\alpha_2 + \frac{n_2\alpha_3}{1+K_b n_1{}^M} + n_2\alpha_4\right) \\ &\exp\left(-\left\{\frac{n_1\alpha_1}{1+K_a n_2{}^m} + n_1\alpha_2 + \frac{n_2\alpha_3}{1+K_b n_1{}^M} + n_2\alpha_4\right\}\tau_w\right) \end{aligned} \tag{115}$$

Equations 112 and 115 are the Equations 61 and 64 in the text, respectively. The latter signifies that the probability density function for the population to make the next transition is exponentially distributed.

Appendix 3: random number transformation

For convenience, Equation 65 in the text is reiterated below

$$\begin{aligned} u &= H_n(\tau_w) \\ &= 1 - \exp\left(-\left(\frac{n_1\alpha_1}{1+K_a n_2{}^m} + n_1\alpha_2 + \frac{n_2\alpha_3}{1+K_b n_1{}^M} + n_2\alpha_4\right)\tau_w\right) \end{aligned} \tag{116}$$

Obviously, $u = 0$ at $\tau_w = 0$ and $u = 1$ when $\tau_w \to \infty$. We are to prove that the random variable, U, whose realization u, is uniformly distributed in [0, 1].

Since u increases with τ_w monotonically according to Equation 116, there is a one-to-one correspondence between u and τ_w, thereby leading to the expression that can be visualized as signifying that the probabilities of the same event in two different domains are identical; this expression is

$$h_n(\tau_w)d\tau_w = f(u)du \tag{117}$$

where $h_n(\tau_w)$ and $f(u)$ are the probability-density functions of T_n and U respectively. For transforming a random variable, an equivalent form of Equation 117 is often given to account for a random variation in either positive or negative direction as follows [41]:

$$f(u) = h(\tau_w)\left|\frac{d\tau_w}{du}\right| \tag{118}$$

Equations 117 and 118 signifies that the probability-density functions of T_n, i.e., $h_n(\tau_w)$, is transformed into that of U, i.e., $f(u)$, such that the probability represented by $h(\tau_w)|d\tau_w|$ and that represented by $f(u)du$ are identical.

The function $h(\tau_w)$ derived from Appendix 2 and given in the text as Equation 64 is

$$\begin{aligned} h_n(\tau_w) &= \left(\frac{n_1\alpha_1}{1+K_a n_2{}^m} + n_1\alpha_2 + \frac{n_2\alpha_3}{1+K_b n_1{}^M} + n_2\alpha_4\right) \\ &\exp\left(-\left\{\frac{n_1\alpha_1}{1+K_a n_2{}^m} + n_1\alpha_2 + \frac{n_2\alpha_3}{1+K_b n_1{}^M} + n_2\alpha_4\right\}\tau_w\right) \end{aligned} \tag{119}$$

Solving Equation 116 for τ_w gives

$$\tau_w = \frac{1}{\left(\frac{n_1\alpha_1}{1+K_a n_2{}^m} + n_1\alpha_2 + \frac{n_2\alpha_3}{1+K_b n_1{}^M} + n_2\alpha_4\right)} \ln(1-u) \tag{120}$$

This is the Equation 66 in the text. By differentiating Equation 120 with respect to u we obtain

$$\frac{d\tau_w}{du} = \frac{1}{\left(\frac{n_1\alpha_1}{1+K_a n_2{}^m} + n_1\alpha_2 + \frac{n_2\alpha_3}{1+K_b n_1{}^M} + n_2\alpha_4\right)(1-u)} \tag{121}$$

Substituting Equations 119 through 121 into Equation 118 leads to

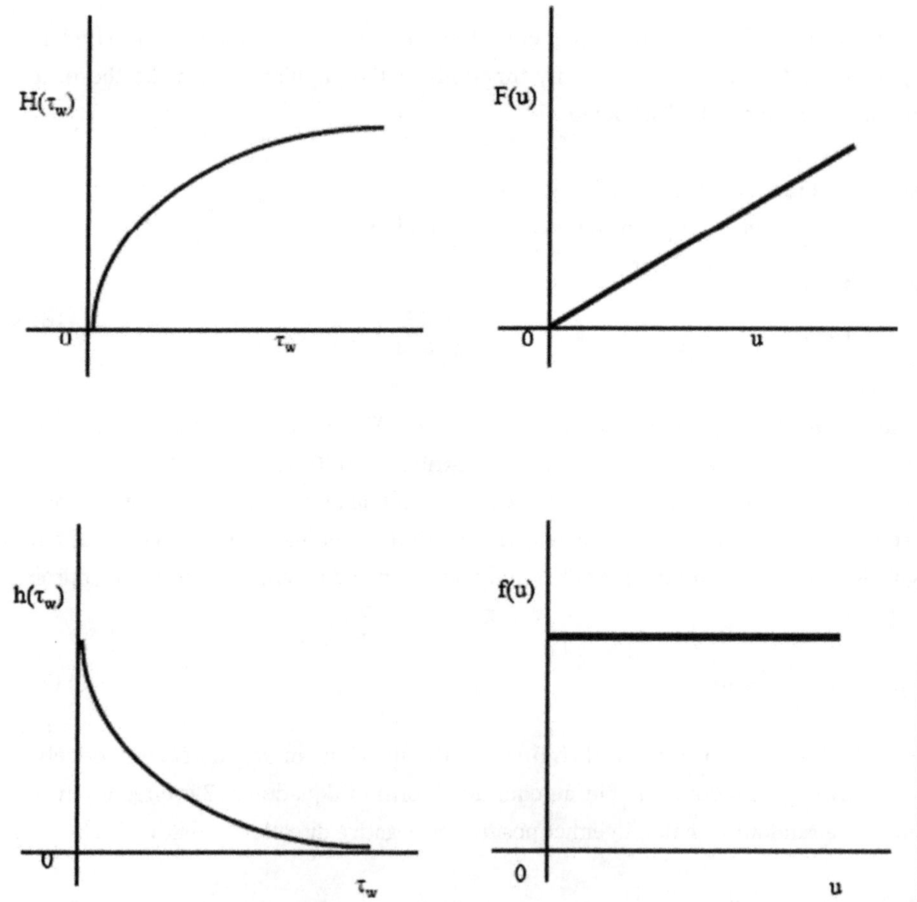

Figure 10 The relations among $H(\tau_w)$, $h(\tau_w)$, $F(u)$, and $f(u)$.

$$
\begin{aligned}
f(u) &= \exp\left(-\left\{\frac{n_1\alpha_1}{1+K_a n_2{}^m} + n_1\alpha_2 + \frac{n_2\alpha_3}{1+K_b n_1{}^M} + n_2\alpha_4\right\}\tau_w\right)\frac{1}{1-u} \\
&= \exp(\ln(1-u))\frac{1}{1-u} \\
&= 1
\end{aligned}
\tag{122}
$$

The values of u is in the interval, [0, 1], as mentioned at the outset; consequently, the probability density function of U, i.e., $f(u)$, remains 1 throughout in the interval, [0, 1] as expressed above. In other words, U, the realization of which is expressed in u, is a random variable uniformly distributed over this interval; for convenience, the probability function or cumulative probability distribution of U is $F(u)$. The relationships among $H(\tau_w)$, $h(\tau_w)$, $F(u)$, and $f(u)$ are illustrated in Figure 10.

Acknowledgements
Professor Michael Mossing of the Department of Chemistry and Biochemistry of the University of Mississippi provided valuable advices during this study. Assad Mohammed and Oluseye Adeyemi provided valuable technical supports for the completion of this work.

References

1. Ptashne, MA: A genetic switch: phage [lambda] and higher organisms. Cell Press and Blackwell Scientific Publications, Cambridge, Massachusetts (1992)
2. Oppenheim, A, Kobiler, O, Stavans, J, Adhya, S: Switches in bacteriophage lambda development. Annu. Rev. Genet. **39**, 409–429 (2005)
3. Gardner, TS, Cantor, CR, Collins, JJ: Construction of a genetic toggle switch in *Escherichia coli*. Nature **403**, 339–342 (2000)
4. McAdams, HH, Arkin, A: Stochastic mechanisms in gene expression. Proc. Natl. Acad. Sci. **94**, 814–819 (1997)
5. Arkin, A, Ross, J, McAdams, HH: Stochastic kinetic analysis of developmental pathway bifurcation in phage λ-infected *Escherichia coli* cells. Genetics **149**, 1633–1648 (1998)
6. Bower, JM, Bolouri, H: Computational modeling of genetic and biochemical network. The MIT Press, Cambridge, Massachusetts (2000)
7. Hasty, J, Pradines, J, Dolnik, M, Collins, JJ: Noise based switches and amplifiers for gene expression. Proc. Natl. Acad. Sci. **97**, 2075–2080 (2000)
8. Oppenheim, I, Shuler, KE, Weiss, GH: Stochastic processes in chemical physics: the master equation. The MIT Press, Cambridge, MA (1977)
9. Van Kampen, NG: Stochastic process in physics and chemistry, 2nd edn. Elsevier, Amsterdam, Netherlands (1992)
10. Gardiner, CW: Handbook of stochastic methods for physics, chemistry, and natural sciences, 2nd edn. Springer-Verlag, Berlin, Germany (1998)
11. Widder, S, Schicho, J, Schuster, P: Dynamic patterns of gene regulation I: simple two-gene systems. J. Theor. Biol. **246**, 395–419 (2007)
12. Van Kampen, NG: The expansion of the master equation. Adv. Chem. Phys. **34**, 245–309 (1976)
13. Kurtz, TG: Limit theorems and diffusion approximations for density dependent Markov chains. Math. Program. Study **5**, 67–78 (1976)
14. Kurtz, TG: Strong approximation theorems for density dependent Markov chains. Stoch. Process. Appl. **6**, 223–240 (1978)
15. Fox, RF, Keizer, J: Amplification of intrinsic fluctuations by chaotic dynamics in physical systems. Phys. Rev. A **43**, 1709–1720 (1991)
16. Kepler, TB, Elston, TC: Stochasticity in transcriptional regulation: origins, consequences, and mathematical representations. Biophy. J. **81**, 3116–3136 (2001)
17. Scott, M, Ingallls, B, Kærn, M: Estimations of intrinsic and extrinsic noise in models of nonlinear genetic networks. Chaos **16**(026107), 1–15 (2006)
18. Tao, Y, Jia, Y, Dewey, TG: Stochastic fluctuations in gene expression far from equilibrium: omega expansion and linear noise approximation. J. Chem. Phys. **122**, 124108–124108 (2005)
19. Ito, Y, Uchida, K: Formulas for intrinsic noise evaluation in oscillatory genetic networks. J. Theor. Biol. **267**, 223–234 (2010)
20. Ochab-Marcinek, A: Predicting the asymmetric response of a genetic switch to noise. J. Theor. Biol. **254**, 37–44 (2008)
21. Yildirim, N, Mackey, M: Feedback regulation in the lactose operon: a mathematical modeling study and comparison with experimental data. Biophys. J. **84**, 2841–2851 (2003)
22. Turcotte, M, Garcia-Ojalvo, J, Süel, GM: A genetic timer through noise-induced stabilization of an unstable state. Proc. Natl. Acad. Sci. U. S. A. **105**, 15732–15737 (2008)
23. Schultz, D, Onuchic, JN, Wolynes, PG: Understanding stochastic simulations of the smallest genetic networks. J. Chem. Phys. **126**, 245102–245102 (2007)
24. Bruggeman, FJ, Blthgen, N, Westerhoff, HV: Noise management by molecular networks. PLoS Comput. Biol. **5**, e1000506 (2009)
25. Ptashne, MA, Gann, A: Genes and signals. Cold Spring Harbor, New York (2002)
26. Gardner, TS: Design and construction of synthetic gene regulatory networks. Doctoral Dissertation, Department of Biomedical Engineering, Boston University, Boston, Massachusetts (2000)
27. Goodwin, BC: Temporal organization in cells. Academic Press, London, UK (1963)

28. Goodwin, BC: Oscillatory behavior in enzymatic control processes. Adv. Enzyme Regul. **3**, 425–438 (1965)
29. Griffith, JS: Mathematics of cellular control processes. J. Theor. Biol. **20**, 202–208 (1968)
30. Chen, WY, Bokka, S: Stochastic modeling of nonlinear epidemiology. J. Theor. Biol. **234**, 455–470 (2005)
31. Van Kampen, NG: A power series expansion of the master equations. Can. J. Phys. **39**, 551–567 (1961)
32. Aparicio, JP, Solari, HG: Population dynamics: poisson approximation and its relation to the Langevin proves. Physical Rev. Lett. **86**, 4183–4186 (2001)
33. Chua, ALS, Haselwandter, CA, Baggio, C, Vvedensky, DD: Langevin equations for fluctuating surfaces. Physical Rev. E **72**, 051103 (2005)
34. Reinitz, J, Vaisnys, JR: Theoretical and experimental analysis of the phage λ genetic switch implies missing levels of co-operativity. J. Theor. Biol. **145**, 295–318 (1990)
35. McAdams, HH, Shapiro, L: Circuit simulation of genetic networks. Science **269**, 650–656 (1995)
36. Santillan, M, Mackey, MC: Dynamic behaviour in mathematical models of the tryptophan operon. Chaos **11**, 261–268 (2001)
37. Orrell, D, Ramsey, S, Atauri, P, Bolouri, H: A method for estimating stochastic noise in large genetic regulatory networks. Bioinf **21**, 208–217 (2005)
38. Pedraza, JM, Oudenaarden, AV: Noise propagation in gene networks. Science **307**, 1965–1969 (2005)
39. Walczak, AM, Sasai, M, Wolynes, PG: Self-consistent proteomic field theory of stochastic gene switches. Biophys. J. **88**, 828–850 (2005)
40. Gillespie, DT: Exact stochastic simulation of coupled chemical reactions. J. Phys. Chem. **81**, 2340–2361 (1977)
41. Gillespie, DT: Markov processes. Academic Press, San Diego, California (1992)
42. Rajamani, K, Pate, WT, Kinneberg, DJ: Time-driven and event-driven Monte Carlo simulations of liquid-liquid dispersions: a comparison. Ind. Eng. Chem. Fundam. **25**, 746–752 (1986)
43. Karlin, S, Taylor, HM: A first course in stochastic processes, 2nd edn. Academic Press, New York (1975)
44. Karlin, S, Taylor, HM: A second course in stochastic processes. Academic Press, , New York (1981)
45. Hale, JK, Kocak, H: Dynamics and bifurcations. Springer-Verlag, New York (1991)
46. Carmichael, H: Statistical methods in quantum optics 1: master equations and Fokker-Planck equations, pp. 158–162. Springer, Berlin (1999)
47. Nicolis, G, Turner, JW: Stochastic analysis of a nonequilibrium phase transition: some exact results. Physics **A89**, 326–338 (1977)
48. Malek Mansour, M, Van den Broeck, C, Nicolis, G, Turner, JW: Asymptotic properties of Markovian master equations. Ann. Phys. **131**, 283–293 (1981)
49. Horsthemke, W, Lefever, R: Noise-induced transitions. Springer-Verlag, Berlin (1984)

2

Some moment inequalities for fuzzy martingales and their applications

Hamed Ahmadzade[1], Mohammad Amini[1*], Seyed Mahmoud Taheri[2] and Abolghasem Bozorgnia[1]

*Correspondence:
m-amini@um.ac.ir
[1] Department of Statistics, Faculty of Mathematical Sciences, Ferdowsi University of Mashhad, Mashhad 91775-1159, Iran
Full list of author information is available at the end of the article

Abstract

Martingales are a class of stochastic processes which has had profound influence on the development of probability theory and stochastic processes. Some recent developments are related to mathematical finance. In the real world, some information about these phenomena might be imprecise and represented in the form of vague quantities. In these situations, we need to generalize classical methods to vague environment. Thus, fuzzy martingales have been extended as a vague perception of real-valued martingales. In this paper, some moment inequalities are presented for fuzzy martingales. Several convergence theorems are established based on these inequalities. As an application of convergence theorems, a weak law of large numbers for fuzzy martingales is stated. Furthermore, a few examples are devoted to clarify the main results.

Keywords: Fuzzy random variable; Fuzzy martingale; Weak convergence; Strong convergence

Introduction

Over the last decades, the theory of fuzzy random variables has been extensively developed. A fuzzy random variable has been extended as a vague perception of a real-valued random variable and subsequently redefined as a particular random set, see e.g., [1-5]. We review several researches on this topic. A strong law of large numbers for fuzzy random variables was given by Miyakoshi and Shimbo [2]. Klement et al. [6] established a strong law of large numbers for fuzzy random variables, based on embedding theorem as well as certain probability techniques in the Banach spaces. Taylor et al. [7] proved a weak law of large numbers for fuzzy random variables in separable Banach spaces. Joo et al. [8] established Chung-type strong law of large numbers for fuzzy random variables based on isomorphic isometric embedding theorem. Fu and Zhang [9] obtained some strong limit theorems for fuzzy random variables with slowly varying weight. It should be mentioned that although the concept of variance has been found very convenient in studying limit theorems, but, as the authors know, the limit theorems for fuzzy random variables based on the concept of variance has not been developed, except in the works by Korner [10] and Feng [11]. Korner [10] proved the strong and weak laws of large numbers for fuzzy random variables. Based on a natural extension of the concept of variance, he extended Kolmogorov's inequality to independent fuzzy random variables and derived several limit theorems. Their methods are direct applications of classical methods in probability theory to fuzzy random variables. In this paper, some moment inequalities for fuzzy martingales are established and convergence theorems of such fuzzy random variables are studied. We survey some topics in real-valued martingale theory and fuzzy martingale theory. The term martingale originates from the gambling theory. The famous gambling strategy to double one's stake as long as one loses and leave as soon as one wins is called a martingale. Now, if a gambling value is vague with a certain membership function, the pervious strategy is called fuzzy martingale. A classic interpretation of martingales in the context of gambling is given as follows. Let X_n represent the fortune of a gambler at the end of the nth play, and let $\mathcal{F}_n$ be the information available to the gambler up to and including the nth play. Then, $\mathcal{F}_n$ contains the knowledge of all events like $\{X_j \leq r, \text{ for } r \in \mathbb{R}, j \leq n\}$ making X_n measurable with respect to $\mathcal{F}_n$. The mathematical definition of martingale $(E(X_{n+1}|\mathcal{F}_n) = X_n)$ says that the given all the information up until the end of the nth play, the expected fortune of the gambler at the end of the $(n+1)$th play remains unchanged. Thus, a martingale represents a fair game. In situation where the game puts the gambler in a favorable position, one may express sub-martingale i.e., $(E(X_{n+1}|\mathcal{F}_n) \geq X_n)$. It is mentioned that limit theorems play important roles in studying some practical instances of martingale theory such as extinction in branching process and bankruptcy in investment problems like portfolio selection.

The concept of fuzzy martingale and its properties is introduced and studied by Puri and Ralescu [12], Stojakovic [13], and Feng [14]. Puri and Ralescu [12] proved some convergence theorems of martingale. Stojakovic [13] discussed the properties of martingales. Feng [14] studied fuzzy conditional expectation and fuzzy martingale and investigated some results on convergence of fuzzy martingales. Fei et al. [15] and Fei [16] obtained several results on stopping theorem for fuzzy (super-, sub-) martingales with discrete time.

The rest of this paper is organized as follows. In Section 'Preliminaries', we briefly review some preliminary concepts of fuzzy arithmetic and fuzzy random variables. In Section 'Moment Inequalities', several useful inequalities are provided for fuzzy martingales. In Section 'Convergence Theorems', some limit theorems are established to fuzzy martingales. The conclusions are discussed in Section 'Conclusions'.

Preliminaries

In this section, we provide several definitions and elementary concepts of fuzzy arithmetic and fuzzy random variables that will be used in the next sections. For more details, the reader is referred to [14,17,18].

Define $F(\mathbb{R}) = \{\tilde{u} : \mathbb{R} \to [0,1] \mid \tilde{u}$ satisfies i to iii$\}$; where (i) $\tilde{u}$ is normal, (ii) $\tilde{u}$ is fuzzy convex, and (iii) $\tilde{u}$ is upper semicontinuous. Any $\tilde{u} \in F(\mathbb{R})$ is called a fuzzy number.

For a $\tilde{u} \in F(\mathbb{R})$, $[\tilde{u}]^r = \{x \in \mathbb{R} | \tilde{u}(x) \geq r, 0 < r \leq 1\}$ is called the r-level set of $\tilde{u}$.

We use the notations $\oplus$, $\ominus$, and $\odot$, which are defined as follows in this paper:

(i) $[\tilde{a} \oplus \tilde{b}]^r = [\tilde{a}^-(r) + \tilde{b}^-(r), \tilde{a}^+(r) + \tilde{b}^+(r)]$.

(ii) If $\lambda > 0$, then $[\lambda \odot \tilde{a}]^r = [\lambda\tilde{a}^-(r), \lambda\tilde{a}^+(r)]$.

(iii) If $\lambda < 0$, then $[\lambda \odot \tilde{a}]^r = [\lambda\tilde{a}^+(r), \lambda\tilde{a}^-(r)]$.

(iv) $[\tilde{a} \ominus \tilde{b}]^r = [\tilde{a}^-(r) - \tilde{b}^+(r), \tilde{a}^+(r) - \tilde{b}^-(r)]$.

Let $\tilde{u}, \tilde{v} \in F(\mathbb{R})$, and set $d_p(\tilde{u}, \tilde{v}) = \left(\int_0^1 h^p([\tilde{u}]^r, [\tilde{v}]^r)dr\right)^{\frac{1}{p}}, 1 \leq p < \infty$, $d_\infty(\tilde{u}, \tilde{v}) = \sup_{0<r\leq 1} h([\tilde{u}]^r, [\tilde{v}]^r)$, where h is the Hausdorff metric, i.e., $h([\tilde{u}]^r, [\tilde{v}]^r) = \max\{|u^-(r) - v^-(r)|, |u^+(r) - v^+(r)|\}$. Norm $||\tilde{u}||_p$ of a fuzzy number $\tilde{u} \in F(\mathbb{R})$ is defined

by $||\tilde{u}||_p = d_p(\tilde{u}, \tilde{0})$, where $\tilde{0}$ is the fuzzy number in $F(\mathbb{R})$ whose membership function equals 1 at 0 and zero otherwise. The norm $||.||_\infty$ of $\tilde{u}$ is defined by $||\tilde{u}||_\infty = d_\infty(\tilde{u}, \tilde{0})$. The operation $\langle .,. \rangle : F(\mathbb{R}) \times F(\mathbb{R}) \to [-\infty, \infty]$ is defined by $\langle \tilde{u}, \tilde{v} \rangle = \int_0^1 (\tilde{u}^-(r)\tilde{v}^-(r) + \tilde{u}^+(r)\tilde{v}^+(r))dr$. If the indeterminacy of the form $\infty - \infty$ arises in the Lebesgue integral, then we say that $\langle \tilde{u}, \tilde{v} \rangle$ does not exist. It is easy to see that the operation $\langle .,. \rangle$ has the following properties:

(i) $\langle \tilde{u}, \tilde{u} \rangle \geq 0$ *and* $\langle \tilde{u}, \tilde{u} \rangle = 0 \Leftrightarrow \tilde{u} = \tilde{0}$.
(ii) $\langle \tilde{u}, \tilde{v} \rangle = \langle \tilde{v}, \tilde{u} \rangle$.
(iii) $\langle \tilde{u} + \tilde{v}, \tilde{w} \rangle = \langle \tilde{u}, \tilde{w} \rangle + \langle \tilde{v}, \tilde{w} \rangle$.
(iv) $\langle \lambda \tilde{u}, \tilde{v} \rangle = \lambda \langle \tilde{u}, \tilde{v} \rangle$.
(v) $|\langle \tilde{u}, \tilde{v} \rangle| < \sqrt{\langle \tilde{u}, \tilde{u} \rangle \langle \tilde{v}, \tilde{v} \rangle}$.

For all $\tilde{u}, \tilde{v} \in F(\mathbb{R})$, if $\langle \tilde{u}, \tilde{u} \rangle < \infty$ and $\langle \tilde{v}, \tilde{v} \rangle < \infty$, then the property (v) implies that $\langle \tilde{u}, \tilde{v} \rangle < \infty$. So, we can define $d_*(\tilde{u}, \tilde{v}) = \sqrt{\langle \tilde{u}, \tilde{u} \rangle - 2\langle \tilde{u}, \tilde{v} \rangle + \langle \tilde{v}, \tilde{v} \rangle}$. In fact, d_* is a metric in $\{\tilde{u} \in F(\mathbb{R}) | \langle \tilde{u}, \tilde{u} \rangle < \infty\}$.

Moreover, the norm $||\tilde{u}||_*$ of fuzzy number $\tilde{u} \in F(\mathbb{R})$ is defined by $||\tilde{u}||_* = d_*(\tilde{u}, \tilde{0})$.

Let $(\Omega, \mathcal{A}, P)$ be a probability space. A fuzzy random variable (briefly, FRV) is a Borel measurable function $\tilde{X} : (\Omega, \mathcal{A}) \to (F(\mathbb{R}), d_\infty)$ [14]. Let $\tilde{X}$ be a FRV defined on $(\Omega, \mathcal{A}, P)$. Then $[\tilde{X}]^r = [\tilde{X}^-(r), \tilde{X}^+(r)]$, $r \in (0, 1]$, is a random closed interval, and $\tilde{X}^-(r)$ and $\tilde{X}^+(r)$ are real-valued random variables. A FRV $\tilde{X}$ is called integrably bounded if $E||\tilde{X}||_\infty < \infty$, and the expectation value $E\tilde{X}$ is defined as the unique fuzzy number which satisfies the property $[E\tilde{X}]^r = E[\tilde{X}]^r$, $0 < r \leq 1$ [3].

Definition 1. ([14]) Let $\tilde{X}$ and $\tilde{Y}$ be two FRVs in L_2 ($L_2 = \{\tilde{X} | \tilde{X}$ is FRV and $E||\tilde{X}||_2^2 < \infty\}$), then $\text{Cov}(\tilde{X}, \tilde{Y}) = \frac{1}{2}\int_0^1 (\text{Cov}(\tilde{X}^-(r), \tilde{Y}^-(r)) + \text{Cov}(\tilde{X}^+(r), \tilde{Y}^+(r)))dr$. Specially, the variance of $\tilde{X}$ is defined by $\text{Var}(\tilde{X}) = \text{Cov}(\tilde{X}, \tilde{X})$.

Theorem 1. *([14]) Let $\tilde{X}$ and $\tilde{Y}$ be two FRVs in L_2 and $\tilde{u}, \tilde{v} \in F(\mathbb{R})$ and $\lambda, k \in \mathbb{R}$. Then*

(i) $Cov(\tilde{X}, \tilde{Y}) = \frac{1}{2}(E\langle \tilde{X}, \tilde{Y} \rangle - \langle E\tilde{X}, E\tilde{Y} \rangle)$,
(ii) $Var(\tilde{X}) = \frac{1}{2}Ed_*^2(\tilde{X}, E\tilde{X})$,
(iii) $Cov(\lambda\tilde{X} \oplus \tilde{u}, k\tilde{Y} \oplus \tilde{v}) = \lambda k Cov(\tilde{X}, \tilde{Y})$,
(iv) $Var(\lambda\tilde{X} \oplus u) = \lambda^2 Var(\tilde{X})$,
(v) $Var(\tilde{X} \oplus \tilde{Y}) = Var(\tilde{X}) + Var(\tilde{Y}) + 2Cov(\tilde{X}, \tilde{Y})$.

Definition 2. ([19]) The $D_{p,q}$ distance, indexed by parameters $1 \leq p < \infty$, $0 \leq q \leq 1$, between two fuzzy numbers $\tilde{u}$ and $\tilde{v}$ is a nonnegative function on $F(\mathbb{R}) \times F(\mathbb{R})$ given as follows: $D_{p,q}(\tilde{u}, \tilde{v}) = \left[(1-q)\int_0^1 |\tilde{u}^-(r) - \tilde{v}^-(r)|^p dr + q\int_0^1 |\tilde{u}^+(r) - \tilde{v}^+(r)|^p dr\right]^{\frac{1}{p}}$.

Remark 1. If $p = 2, q = \frac{1}{2}$, then the metric $D_{p,q}$ is equal to the metric d_*; for more details, see [16].

To prove the main results, we need to apply an order relation. Thus, we use notations $\prec, \succ, \preceq$, and $\succeq$ defined as follows [18]:

(i) $\tilde{a} \prec \tilde{b}$ if and only if $a^-(r) < b^-(r)$ and $a^+(r) < b^+(r)$ $\forall r \in [0, 1]$.
(ii) $\tilde{a} \succ \tilde{b}$ if and only if $a^-(r) > b^-(r)$ and $a^+(r) > b^+(r)$ $\forall r \in [0, 1]$.

(iii) $\tilde{a} \preceq \tilde{b}$ if and only if $a^-(r) \leq b^-(r)$ and $a^+(r) \leq b^+(r)$ $\forall r \in [0,1]$.
(iv) $\tilde{a} \succeq \tilde{b}$ if and only if $a^-(r) \geq b^-(r)$ and $a^+(r) \geq b^+(r)$ $\forall r \in [0,1]$.

Definition 3. ([20]) Two fuzzy random variables $\tilde{X}$ and $\tilde{Y}$ are called independent if two σ-fields $\sigma(\tilde{X}) = \sigma(\{X^-(r), X^+(r)|r \in [0,1]\})$ and $\sigma(\tilde{Y}) = \sigma(\{Y^-(r), Y^+(r)|r \in [0,1]\})$ are independent.

Definition 4. A finite collection of FRVs $\{\tilde{X}_k,\ 1 \leq k \leq n\}$ is said to be independent if σ-fields $\sigma(\{\tilde{X}_k^-(r), \tilde{X}_k^+(r)|r \in [0,1]\}|1 \leq k \leq n)$ are independent. An infinite sequence $\{\tilde{X}_n, n \geq 1\}$ is called independent if every finite sub-collection of it is independent.

Definition 5. ([15]) A fuzzy conditional expectation of $\tilde{X}$ with respect to the sub-σ field $\mathcal{B}$ of $\mathcal{A}$, denoted as $E(\tilde{X}|\mathcal{B})$, is defined as a FRV which satisfies in the following conditions:

(i) $E(\tilde{X}|\mathcal{B})$ is $\mathcal{B}$- measurable.
(ii) $\int_B E(\tilde{X}|\mathcal{B})dP = \int_B \tilde{X}dP$ for every $B \in \mathcal{B}$. Note that $\int_B \tilde{X}dP$ is the Aumann integral of the FRV $\tilde{X}$ [21].

Proposition 1. *([15]) The fuzzy conditional expectation has the following properties:*

(i) $E(a \odot \tilde{X} \oplus b \odot \tilde{Y}|\mathcal{B}) = (a \odot E(\tilde{X}|\mathcal{B})) \oplus (b \odot E(\tilde{Y}|\mathcal{B}))$ a.s.
(ii) $\tilde{X}$ *is* $\mathcal{B}$*-measurable, then* $E(\tilde{X}|\mathcal{B}) = \tilde{X}$ a.s.
(iii) $EE(\tilde{X}|\mathcal{B}) = E\tilde{X}$.
(iv) *If* $\tilde{X} \preceq \tilde{Y}$ a.s. *then* $E(\tilde{X}|\mathcal{B}) \preceq E(\tilde{Y}|\mathcal{B})$ a.s.
(v) $d_\infty(E(\tilde{X}|\mathcal{B}), E(\tilde{Y}|\mathcal{B})) \leq E(d_\infty(\tilde{X}, \tilde{Y})|\mathcal{B})$ a.s. *and consequently*

$$||E(\tilde{X}|\mathcal{B})||_\infty \leq E(||\tilde{X}||_\infty|\mathcal{B}) \ a.s. \tag{1}$$

Definition 6. ([15]) The sequence $\{\tilde{X}_n, \mathcal{B}_n\}$ of fuzzy random variables and σ-algebras is called a fuzzy martingale if we have, for each $n \geq 1$:

(a) $\tilde{X}_n$ is $\mathcal{B}_n$-measurable and $E||\tilde{X}_n||_\infty < \infty$.
(b) $E(\tilde{X}_{n+1}|\mathcal{B}_n) = \tilde{X}_n$.

The sequence $\{\tilde{X}_n, \mathcal{B}_n\}$ is called a fuzzy sub-martingale if property (b) is replaced by

(b') $E(\tilde{X}_{n+1}|\mathcal{B}_n) \succeq \tilde{X}_n$.

For more on fuzzy martingale and related topics, see e.g. [11,15,22].

Definition 7. ([20]) Let $\tilde{X}$ and $\tilde{X}_n$ be FRVs defined on the same probability space $(\Omega, \mathcal{A}, P)$.

(i) We say that $\{\tilde{X}_n\}$ converges to $\tilde{X}$ in probability with respect to the metric d if

$$\lim_{n\to\infty} P(\omega : d(\tilde{X}_n(\omega), \tilde{X}(\omega)) > \epsilon) = 0, \ \ \forall \epsilon > 0.$$

(ii) We say that $\{\tilde{X}_n\}$ converges to $\tilde{X}$ almost surely (briefly, a.s.) with respect to the metric d, if

$$P(\omega : \lim_{n\to\infty} d(\tilde{X}_n(\omega), \tilde{X}(\omega)) = 0) = 1.$$

(iii) We say that $\{\tilde{X}_n\}$ converges to $\tilde{X}$ in L^2 with respect to the metric d, if

$$E^{\frac{1}{2}} d^2(\tilde{X}_n, \tilde{X}) \to 0, \quad as \ n \to \infty.$$

It is mentioned that in the literature of probability theory, convergence in probability and almost surely convergence are called weak and strong convergence, respectively.

Throughout the paper, $\tilde{S}_n = \oplus_{i=1}^{n} \tilde{X}_i$ and $\mathcal{F}_n = \sigma(\tilde{X}_1, \ldots, \tilde{X}_n)$.

Moment Inequalities

In this section, we prove some basic inequalities for fuzzy martingales.

The following lemma is essential to obtain our main results.

Lemma 1. *If $\{\tilde{X}_n, n \geq 1\}$ is a sequence of FRVs such that $E(\tilde{X}_n|\mathcal{F}_{n-1}) = \tilde{0}$, for all $n \geq 1$, then $\{\tilde{S}_n, \mathcal{F}_n\}$ is a fuzzy martingale.*

Proof. By using linearity property of conditional fuzzy expectation, we obtain

$$E(\tilde{S}_n|\mathcal{F}_{n-1}) = E(\tilde{S}_{n-1}|\mathcal{F}_{n-1}) \oplus E(\tilde{X}_n|\mathcal{F}_{n-1}).$$

Since $\tilde{S}_{n-1}$ is $\mathcal{F}_{n-1}$ measurable and $E(\tilde{X}_n|\mathcal{F}_{n-1}) = \tilde{0} \ \forall \ n \geq 1$, we have

$$E(\tilde{S}_n|\mathcal{F}_{n-1}) = \tilde{S}_{n-1}.$$

This completes the proof. □

Corollary 1. *If $\{\tilde{X}, n \geq 1\}$ is a sequence of independent FRVs such that $E(\tilde{X}_n) = \tilde{0}, \forall \ n \geq 1$, then $\{\tilde{S}_n, \mathcal{F}_n\}$ is a fuzzy martingale.*

The above corollary shows the importance of independence in this paper, which is used in the examples.

Remark 2. If $\{\tilde{X}_n, n \geq 1\}$ is a sequence of fuzzy random variables, then $\{\tilde{S}_n, \mathcal{F}_n\}$ is not always fuzzy martingale.

Example 1. *If $\{\tilde{X}_n, n \geq 1\}$ is a sequence of nonnegative fuzzy random variables, then $\{\tilde{S}_n, \mathcal{F}_n\}$ is not a fuzzy martingale, but it is a fuzzy sub-martingale.*

Theorem 2. *If $\{\tilde{S}_n, \mathcal{F}_n\}$ is a fuzzy martingale, then there exists a constant C depending only on p such that*

(i) $E||\tilde{S}_n||_{p,q}^p \leq C \sum_{i=1}^{n} E||\tilde{X}_i||_{p,q}^p, \ \forall \, p \in [1, 2] \ and \ q \in [0, 1]$

(ii) $E||\tilde{S}_n||_{p,q}^p \leq C n^{\frac{p}{2}} \sum_{i=1}^{n} E||\tilde{X}_i||_{p,q}^p \ \forall \, p > 2 \ and \ q \in [0, 1]$

where, $\tilde{S}_n = \oplus_{i=1}^{n} \tilde{X}_i$.

Proof. We have done the following:

(i) By invoking Fubini's theorem (FT, for short), Burkholder's inequality (BI, for short), and C_r inequality (C_r I, for short), we obtain

$$\begin{aligned}
E||\tilde{S}_n||_{p,q}^p &= E\int_0^1 q|\tilde{S}_n^+(r)|^p + (1-q)|\tilde{S}_n^-(r)|^p dr \\
&= \int_0^1 qE|\tilde{S}_n^+(r)|^p + (1-q)E|\tilde{S}_n^-(r)|^p dr \text{ (FT)} \\
&\leq C\int_0^1 qE|\sum_{i=1}^n [\tilde{X}_i^+(r)]^2|^{\frac{p}{2}} + (1-q)E|\sum_{i=1}^n [\tilde{X}_i^-(r)]^2|^{\frac{p}{2}} dr \text{ (BI)} \\
&\leq C\sum_{i=1}^n \int_0^1 qE|\tilde{X}_i^+(r)|^p + (1-q)E|\tilde{X}_i^-(r)|^p dr \; (C_r\, I) \\
&= C\sum_{i=1}^n E||\tilde{X}_i||_{p,q}^p.
\end{aligned}$$

(ii) Fubini's theorem (FT), Rosenthal's inequality (RI), C_r inequality, and conditional Jensen's inequality (CJI) imply that

$$\begin{aligned}
E||\tilde{S}_n||_{p,q}^p &= E\int_0^1 q|\tilde{S}_n^+(r)|^p + (1-q)|\tilde{S}_n^-(r)|^p dr \\
&= \int_0^1 qE|\tilde{S}_n^+(r)|^p + (1-q)E|\tilde{S}_n^-(r)|^p dr \text{ (FT)} \\
&\leq \int_0^1 q\left\{E\left(\sum_{i=1}^n E\left(\tilde{X}_i^{+2}(r)|\mathcal{F}_{i-1}\right)\right)^{\frac{p}{2}} + \sum_{i=1}^n E|\tilde{X}_i^+(r)|^p\right\} \\
&+ (1-q)\left\{E\left(\sum_{i=1}^n E\left(\tilde{X}_i^{-2}(r)|\mathcal{F}_{i-1}\right)\right)^{\frac{p}{2}} + \sum_{i=1}^n E|\tilde{X}_i^-(r)|^p\right\} dr \text{ (RI)} \\
&\leq Cn^{\frac{p}{2}}\int_0^1 q\left\{\sum_{i=1}^n E\left[E\left(\tilde{X}_i^{+^2}(r)|\mathcal{F}_{i-1}\right)\right]^{\frac{p}{2}} + E|\tilde{X}_i^+(r)|^p\right\} \\
&+ (1-q)\sum_{i=1}^n \left\{E\left[E\left(\tilde{X}_i^{+^2}(r)|\mathcal{F}_{i-1}\right)\right]^{\frac{p}{2}} + E|\tilde{X}_i^+(r)|^p\right\} dr \; (C_r\, I) \\
&\leq Cn^{\frac{p}{2}}\int_0^1 q\left\{\sum_{i=1}^n EE(|\tilde{X}_i^+(r)|^p|\mathcal{F}_{i-1}) + E|\tilde{X}_i^+(r)|^p\right\} \\
&+ (1-q)\sum_{i=1}^n \{EE(|\tilde{X}_i^-(r)|^p|\mathcal{F}_{i-1}) + E|\tilde{X}_i^-(r)|^p\} dr \text{ (CJI)} \\
&= Cn^{\frac{p}{2}}\sum_{i=1}^n E\int_0^1 q|\tilde{X}_i^+(r)|^p + (1-q)|\tilde{X}_i^-(r)|^p dr \text{ (FT)} \\
&= Cn^{\frac{p}{2}}\sum_{i=1}^n E||\tilde{X}_i||_{p,q}^p.
\end{aligned}$$

□

It should be noted that the above theorem is a generalization of Theorem 2.11 of [23] to fuzzy martingales.

Now we state and prove two theorems which are extensions of Theorem 2.13 in [23] and Theorem 9.5 of [24], respectively.

Theorem 3. *If* $\{\tilde{S}_n, \mathcal{F}_n\}$ *is a fuzzy martingale, then there exists a constant C depending only on p such that*

$$ED^2_{2,q}(\oplus_{i=1}^n \tilde{X}_i, \oplus_{i=1}^n E(\tilde{X}_i|\mathcal{F}_{i-1})) \leq C\sum_{i=1}^n ED^2_{2,q}[\tilde{X}_i, E(\tilde{X}_i|\mathcal{F}_{i-1})], \ \forall\, q \in [0,1].$$

Proof. It is obvious that $\left\{\sum_{i=1}^n [\tilde{X}_i^+(r) - E(\tilde{X}_i^+(r)|\mathcal{F}_{i-1})], \mathcal{F}_n\right\}$ and $\left\{\sum_{i=1}^n [\tilde{X}_i^-(r) - E(\tilde{X}_i^-(r)|\mathcal{F}_{i-1})], \mathcal{F}_n\right\}$ constitute ordinary martingales. Also, it is easy to see that

$$E[\tilde{X}_i^+(r) - E(\tilde{X}_i^+(r)|\mathcal{F}_{i-1})]^2 = E[\tilde{X}_i^+(r)]^2 - E[E(\tilde{X}_i^+(r)|\mathcal{F}_{i-1})]^2 \ \ \forall\, r \in [0,1], \tag{2}$$

and

$$E\left[\tilde{X}_i^-(r) - E(\tilde{X}_i^-(r)|\mathcal{F}_{i-1})\right]^2 = E\left[\tilde{X}_i^-(r)\right]^2 - E\left[E(\tilde{X}_i^-(r)|\mathcal{F}_{i-1})\right]^2 \ \ \forall\, r \in [0,1]. \tag{3}$$

By invoking Fubini's theorem, Burkholder's inequality, and relations (2) and (3), we obtain

$$\begin{aligned}
&E\ D^2_{2,q}(\tilde{S}_n, \oplus_{i=1}^n E(\tilde{X}_i^+(r)|\mathcal{F}_{i-1}))\\
&= E\int_0^1 (1-q)\left(\sum_{i=1}^n [\tilde{X}_i^+(r) - E(\tilde{X}_i^+|\mathcal{F}_{i-1})]\right)^2 + q\left(\sum_{i=1}^n [\tilde{X}_i^-(r) - E(\tilde{X}_i^-(r)|\mathcal{F}_{i-1})]\right)^2 dr\\
&= \int_0^1 (1-q)E\left(\sum_{i=1}^n [\tilde{X}_i^+(r) - E(\tilde{X}_i^+(r)|\mathcal{F}_{i-1})]\right)^2 + qE\left(\sum_{i=1}^n [\tilde{X}_i^-(r) - E(\tilde{X}_i^-|\mathcal{F}_{i-1})]\right)^2 dr\\
&\leq C\int_0^1 (1-q)\sum_{i=1}^n E([\tilde{X}_i^+(r) - E(\tilde{X}_i^+(r)|\mathcal{F}_{i-1})])^2 + q\sum_{i=1}^n E([\tilde{X}_i^-(r) - E(\tilde{X}_i^-(r)|\mathcal{F}_{i-1})])^2 dr\\
&= C\sum_{i=1}^n\int_0^1 (1-q)\left\{E\left[\tilde{X}_i^+(r)\right]^2 - E\left[E(\tilde{X}_i^+(r)|\mathcal{F}_{i-1})\right]^2\right\} + q\left\{E\left[\tilde{X}_i^-(r)\right]^2 - E\left[E(\tilde{X}_i^-(r)|\mathcal{F}_{i-1})\right]^2\right\} dr\\
&= C\sum_{i=1}^n E\int_0^1 (1-q)\left\{[\tilde{X}_i^+(r)]^2 - [E(\tilde{X}_i^+(r)|\mathcal{F}_{i-1})]^2\right\} + q\left\{[\tilde{X}_i^-(r)]^2 - [E(\tilde{X}_i^-(r)|\mathcal{F}_{i-1})]^2\right\} dr\\
&= C\sum_{i=1}^n ED^2_{2,q}\left[\tilde{X}_i, E(\tilde{X}_i|\mathcal{F}_{i-1})\right].
\end{aligned}$$

□

Theorem 4. *If* $\{\tilde{S}_n, \mathcal{F}_n\}$ *is a fuzzy martingale and* $1 \leq p \leq 2$*, then*

$$E\max_{1\leq k\leq n} ||\tilde{S}_k||^p_{p,q} \leq C\sum_{i=1}^n E||\tilde{X}_i||^p_{p,q}.$$

Proof. By using monotonicity property of integral (MPI, for short), Fubini's theorem, Burkholder's inequality, and C_r inequality, we obtain

$$\begin{aligned} E\max_{1\le k\le n}||\tilde{S}_k||^p_{p,q} &= E\max_{1\le k\le n}\int_0^1 q|\tilde{S}_k^+(r)|^p + (1-q)|\tilde{S}_k^-(r)|^p dr \\ &\le E\int_0^1 \left\{q\max_{1\le k\le n}|\tilde{S}_k^+(r)|^p + (1-q)\max_{1\le k\le n}|\tilde{S}_k|^p\right\} dr \text{ (MPI)} \\ &= \int_0^1 \left\{qE\max_{1\le k\le n}|\tilde{S}_k^+(r)|^p + (1-q)E\max_{1\le k\le n}|\tilde{S}_k^-(r)|^p\right\} dr \text{ (FT)} \\ &\le C\int_0^1 qE\left(\sum_{i=1}^n(\tilde{X}_i^+(r))^2\right)^{\frac{p}{2}} + (1-q)E\left(\sum_{i=1}^n(\tilde{X}_i^-(r))^2\right)^{\frac{p}{2}} dr \text{ (BI)} \\ &= C\int_0^1 \left\{q\sum_{i=1}^n E|\tilde{X}_i^+(r)|^p + (1-q)\sum_{i=1}^n E|\tilde{X}_i^-(r)|^p\right\} dr \text{ } (C_r I) \\ &= C\sum_{i=1}^n E||\tilde{X}_i||^p_{p,q}. \end{aligned}$$

□

Convergence Theorems

In this section, by invoking theorems of previous section, we establish some weak and strong convergence theorems for fuzzy martingales.

Theorem 5. *If* $\{\tilde{X}_n, \mathcal{F}_n\}$ *is a fuzzy martingale, then*

$$Ed^2_*(\tilde{X}_n, \tilde{X}_m) = E||\tilde{X}_n||^2_* - E||\tilde{X}_m||^2_*, \text{ for } n > m.$$

Proof. By invoking Fubini's theorem and orthogonality lemma of ordinary martingale (OLOM, for short), we obtain

$$\begin{aligned} Ed^2_*(\tilde{X}_n, \tilde{X}_m) &= E\int_0^1 \{|\tilde{X}_n^-(r) - \tilde{X}_m^-(r)|^2 + |\tilde{X}_n^+(r) - \tilde{X}_m^+(r)|^2\} dr \\ &= \int_0^1 \{E|\tilde{X}_n^-(r) - \tilde{X}_m^-(r)|^2 + E|\tilde{X}_n^+(r) - \tilde{X}_m^+(r)|^2\} dr \text{ (FT)} \\ &= \int_0^1 \{E|\tilde{X}_n^-(r)|^2 - E|\tilde{X}_m^-(r)|^2 + E|\tilde{X}_n^+(r)|^2 - E|\tilde{X}_m^+(r)|^2\} dr \text{ (OLOM)} \\ &= E||\tilde{X}_n||^2_* - E||\tilde{X}_m||^2_*. \end{aligned}$$

□

Note that Theorem 5 is a generalization of Lemma 10.4.1 of [24] to fuzzy martingales.

Corollary 2. *If* $\{\tilde{X}_n, n \ge 1\}$ *is a fuzzy martingale such that* $\sup_{n\ge 1} E||\tilde{X}_n||^2_* < \infty$, *then* $\{\tilde{X}_n, n \ge 1\}$ *is Cauchy convergent in* L^2 *with respect to the metric* d_*.

Lemma 2. *If* $\{\tilde{X}_n, \mathcal{F}_n\}$ *is a sequence of FRVs, then*

$$||E(\tilde{X}_{n+1}|\mathcal{F}_n)||^2_* \le E(||\tilde{X}_{n+1}||^2_*|\mathcal{F}_n).$$

Proof. By using conditional Jensen's inequality and Fubini's theorem, we obtain

$$\begin{aligned}||E(\tilde{X}_{n+1}|\mathcal{F}_n)||_*^2 &= \int_0^1 \left\{E^2\left(\tilde{X}_{n+1}^+(r)|\mathcal{F}_n\right) + E^2\left(\tilde{X}_{n+1}^-(r)|\mathcal{F}_n\right)\right\} dr \\ &\leq \int_0^1 \left\{E\left(\{\tilde{X}_{n+1}^+(r)\}^2|\mathcal{F}_n\right) + E\left(\{\tilde{X}_{n+1}^-(r)\}^2|\mathcal{F}_n\right)\right\} dr \text{ (CJI)} \\ &= E\left(||\tilde{X}_{n+1}||_*^2|\mathcal{F}_n\right) \text{ (FT).}\end{aligned}$$

□

Corollary 3. *If $\{\tilde{X}_n, n \geq 1\}$ is a fuzzy martingale, then $\{||\tilde{X}_n||_*^2, n \geq 1\}$ is a real-valued sub-martingale.*

Theorem 6. *Let $\{\tilde{S}_n, \mathcal{F}_n\}$ be a fuzzy martingale and $p > 2$, then $\frac{\tilde{S}_n}{n} \to \tilde{0}$ in probability with respect to the metric $D_{p,q}$ if*

$$\sum_{i=1}^n E||\tilde{X}_i||_{p,q}^p = o(n^{\frac{p}{2}}) \text{ as } n \to \infty.$$

Proof. By using Markov's inequality and Theorem 2, the proof is straightforward. □

Example 2. *Let $\{\tilde{X}_n, n \geq 1\}$ be a sequence of independent fuzzy random variables with the following membership functions:*

$$\mu_{\tilde{X}_n(\omega)}(x) = \begin{cases} \frac{x - X_n(\omega)}{X_n(\omega)}, & X_n(\omega) < x \leq 2X_n(\omega), \\ \frac{3X_n(\omega) - x}{X_n(\omega)}, & 2X_n(\omega) < x < 3X_n(\omega), \\ 0, & \text{otherwise,} \end{cases}$$

where $\{X_n, n \geq 1\}$ is a sequence of independent random variables with the following probability $P\left(X_n = n^\beta\right) = \frac{1}{2} = P\left(X_n = -n^\beta\right)$.

It is easy to see that

$$||\tilde{X}_n||_{p,q}^p = W_{p,q}|X_n|^p,$$

where

$$W_{p,q} = \frac{q}{p+1}\left[2^{p+1} - 1\right] - \frac{1-q}{p+1}\left[2^{p+1} - 3^{p+1}\right].$$

We want to prove that $\sum_{i=1}^n E||\tilde{X}_i||_{p,q}^p = o(n^{\frac{p}{2}})$, i.e., $\frac{1}{n^{\frac{p}{2}}}\sum_{i=1}^n i^{\beta p} \to 0$ as $n \to \infty$. By using Kronecker's lemma, it suffices to show that

$$\sum_{n=1}^\infty \frac{1}{n^{p(\frac{1}{2}-\beta)}} < \infty.$$

If $p > \frac{2}{1-2\beta}$, then

$$\sum_{n=1}^\infty \frac{1}{n^{p(\frac{1}{2}-\beta)}} < \infty.$$

By using Theorem 6, we obtain $\frac{\tilde{S}_n}{n} \to \tilde{0}$ in probability with respect to the metric $D_{p,q}$ as $n \to \infty$.

The following theorem is an extension of Theorem 2.13 in [23] to fuzzy martingales.

Theorem 7. *Let $\{\tilde{S}_n, n \geq 1\}$ be a martingale and $\{b_n\}$ be a sequence of positive constants with $b_n \uparrow \infty$. Then $b_n^{-1}\tilde{S}_n \to \tilde{0}$ in probability with respect to the metric $D_{2,q}$, if*

$$\sum_{i=1}^{n} P(||\tilde{X}_i||_{2,q} > b_n) \to 0, \tag{4}$$

$$b_n^{-1}\sum_{i=1}^{n} ||E(\tilde{X}_{ni}|\mathcal{F}_{i-1})||_{2,q} \to 0 \ \ in \ probability, \tag{5}$$

$$b_n^{-2}\sum_{i=1}^{n}\{ED^2_{2,q}[\tilde{X}_{n,i}, E(\tilde{X}_{n,i}|\mathcal{F}_{i-1})]\} \to 0, \tag{6}$$

where, $\tilde{X}_{ni} = \tilde{X}_i I_{\{||\tilde{X}_i||_{2,q} \leq n\}}$ and $\tilde{S}_n = \oplus_{i=1}^{n}\tilde{X}_i$.

Proof. Suppose that $\tilde{S}_{nn} = \oplus_{i=1}^{n}\tilde{X}_{n,i}$. It is easy to see that

$$b_n^{-1}||\tilde{S}_n||_{2,q} \leq b_n^{-1}D_{2,q}(\tilde{S}_n, \tilde{S}_{nn}) + b_n^{-1}||\tilde{S}_{nn}||_{2,q}. \tag{7}$$

We want to prove $b_n^{-1}||\tilde{S}_n||_{2,q} \to$ 0 in probability. By using relation (7), we should prove that (a) $b_n^{-1}D_{2,q}(\tilde{S}_n, \tilde{S}_{nn}) \to 0$ and (b) $b_n^{-1}||\tilde{S}_{nn}||_{2,q} \to 0$ in probability:

(a) For each $\epsilon > 0$, we have

$$P(b_n^{-1}D_{2,q}(\tilde{S}_n, \tilde{S}_{nn}) > \epsilon) = P(\tilde{S}_n \neq \tilde{S}_{nn}) \leq \sum_{i=1}^{n} P(||\tilde{X}_i||_{2,q} > b_n) \to 0.$$

Thus, $b_n^{-1}D_{2,q}(\tilde{S}_n, \tilde{S}_{nn}) \to 0$ in probability.

(b) By using sub-additivity of the metric $D_{2,q}$, we obtain

$$b_n^{-1}||\tilde{S}_{nn}||_{2,q} \leq b_n^{-1}D_{2,q}(\tilde{S}_{nn}, \oplus_{i=1}^{n}E(\tilde{X}_{ni}|\mathcal{F}_{i-1})) + ||\oplus_{i=1}^{n} E(\tilde{X}_{n,i}|\mathcal{F}_{i-1})||_{2,q}.$$

The sub-additivity of the norm $||.||_{2,q}$ and condition (3) imply that $b_n^{-1}||\oplus_{i=1}^{n} E(\tilde{X}_{ni}|\mathcal{F}_{i-1})||_{2,q} \leq b_n^{-1}\sum_{i=1}^{n}||E(\tilde{X}_{ni}|\mathcal{F}_{i-1})||_{2,q} \to 0$ in probability.

It remains to show that $b_n^{-1}D_{2,q}(\tilde{S}_{nn}, \oplus_{i=1}^{n}E(\tilde{X}_{ni}|\mathcal{F}_{i-1})) \to 0$ in probability.

By using Markov's inequality and Theorem 3, we have

$$P(b_n^{-1}D_{2,q}(\tilde{S}_{nn}, \oplus_{i=1}^{n}E(\tilde{X}_{ni}|\mathcal{F}_{i-1})) > \epsilon) \leq \frac{ED^2_{2,q}(\tilde{S}_{nn}, \oplus_{i=1}^{n}E(\tilde{X}_{ni}|\mathcal{F}_{i-1}))}{b_n^2\epsilon^2} \leq \frac{C\sum_{i=1}^{n}ED^2_{2,q}(\tilde{X}_{ni}, E(\tilde{X}_{ni}|\mathcal{F}_{i-1}))}{b_n^2\epsilon^2} \to 0.$$

This completes the proof. □

The following example is an evidence of Theorem 7.

Example 3. *Suppose that $\{\tilde{X}_n, n \geq 1\}$ is a sequence of independent FRVs with the following probability function:*

$$P(\tilde{X}_n = n^{-\alpha}\tilde{u}) = P(\tilde{X}_n = -n^{-\alpha}\tilde{u}) = \frac{1}{2},$$

where $\tilde{u}$ is a fuzzy number such that $||\tilde{u}||_{p,q} = 1$. It is obvious that

$$(i) \ \sum_{i=1}^{n} P(||\tilde{X}_i||_{2,q} > n) = 0.$$

We want to show that

$$(ii) \ n^{-1} \sum_{i=1}^{n} ||E(\tilde{X}_{ni}|\mathcal{F}_{i-1})||_{2,q} \to 0 \ \text{in probability}.$$

By invoking Markov's inequality and Lemma 2, we obtain

$$\begin{aligned} P(n^{-1} \sum_{i=1}^{n} ||E(\tilde{X}_{ni}|\mathcal{F}_{i-1})||_{2,q} > \epsilon) &\leq \frac{\sum_{i=1}^{n} E||E(\tilde{X}_{ni}|\mathcal{F}_{i-1})||_{2,q}}{n\epsilon} \\ &\leq \frac{\sum_{i=1}^{n} EE^{\frac{1}{2}}(||\tilde{X}_{ni}||_{2,q}^{2}|\mathcal{F}_{i-1})}{n\epsilon} \\ &= \frac{1}{n\epsilon} \sum_{i=1}^{n} \frac{1}{i^{\alpha}}. \end{aligned}$$

By using Kronecker's lemma $\frac{1}{n} \sum_{i=1}^{n} \frac{1}{i^{\alpha}} \to 0$, and consequently

$$n^{-1} \sum_{i=1}^{n} E(\tilde{X}_{ni}|\mathcal{F}_{i-1}) \to 0 \ \text{in probability}.$$

It remains to show that

$$(iii) \ n^{-2} \sum_{i=1}^{n} ED_{2,q}^{2}(\tilde{X}_{ni}, E(\tilde{X}_{ni}|\mathcal{F}_{i-1})) \to 0 \ \text{as} \ n \to \infty.$$

It is easy to see that

$$n^{-2} \sum_{i=1}^{n} ED_{2,q}^{2}(\tilde{X}_{ni}, E(\tilde{X}_{ni}|\mathcal{F}_{i-1})) = n^{-2} \sum_{i=1}^{n} \frac{1}{i^{2\alpha}}.$$

By invoking Kronecker's lemma, we obtain $n^{-2} \sum_{i=1}^{n} \frac{1}{i^{2\alpha}} \to 0$ and thus,

$$n^{-2} \sum_{i=1}^{n} ED_{2,q}^{2}(\tilde{X}_{ni}, E(\tilde{X}_{ni}|\mathcal{F}_{i-1})) \to 0 \ \text{as} \ n \to \infty.$$

Therefore, all of the conditions of the above theorem hold, thus $\frac{\tilde{S}_n}{n} \to \tilde{0}$ in probability with respect to the metric $D_{2,q}$.

Theorem 8. *Let $\{\tilde{S}_n, \mathcal{F}_n\}$ be a fuzzy martingale. If for all $1 \leq p < 2$, $\{||\tilde{X}_n||_{p,q}^{p}, n \geq 1\}$ is uniformly integrable, then $n^{-2}E||\tilde{S}_n||_{p,q}^{p} \to 0$ as $n \to \infty$.*

Proof. By invoking Theorem 2, we obtain

$$E||\tilde{S}_n||_{p,q}^{p} \leq C \sum_{i=1}^{n} E||\tilde{X}_i||_{p,q}^{p}.$$

Since $\{||\tilde{X}_n||_{p,q}^p, n \geq 1\}$ is uniformly integrable, by invoking Theorem 5.4.1 in [24] $\sup_{n\geq 1}E||\tilde{X}_n||_{p,q}^p < \infty$ i.e., $\exists M$ s.t. $\sup_{n\geq 1}E||\tilde{X}_n||_{p,q}^p < M$. Hence,

$$\begin{aligned} E||\tilde{S}_n||_{p,q}^p &\leq C\sum_{i=1}^{n} E||\tilde{X}_i||_{p,q}^p \\ &\leq Cn\sup_{n\geq 1}E||\tilde{X}_n||_{p,q}^p \\ &\leq CnM. \end{aligned}$$

So,

$$n^{-2}E||\tilde{S}_n||_{p,q}^p \leq Cn^{-1}M,$$

which implies that $n^{-2}E||\tilde{S}_n||_{p,q}^p \to 0$. □

The above theorem is a generalization of Theorem 2.22 in [23] to fuzzy martingales.

Example 4. *Suppose that $\{\tilde{X}_n, n \geq 1\}$ is a sequence of independent Fs with the following probability function:*

$$P(\tilde{X}_n = n^{-1}\tilde{u}) = \frac{1}{n}, \; P\left(\tilde{X}_n = \frac{-\tilde{u}}{n(n-1)}\right) = 1 - \frac{1}{n},$$

where $\tilde{u}$ is a fuzzy number with the following membership function:

$$\tilde{u}(x) = 1 - \frac{\sqrt{6}}{3}|x|, \quad -\frac{\sqrt{6}}{2} \leq x \leq \frac{\sqrt{6}}{2}.$$

It is easy to see that $||\tilde{X}_n||_{p,q}^p$ is uniformly bounded and, consequently, is uniformly integrable. Thus, by invoking Theorem 8, $n^{-2}E||\tilde{S}_n||_{p,q}^p \to 0$ as $n \to \infty$.

Theorem 9. *If $\{\tilde{S}_n, \mathcal{F}_n\}$ is a fuzzy martingale such that $\sum_{n=1}^{\infty} E||\tilde{X}_n||_{p,q}^p < \infty$ for all $p \in [1, 2]$, then $\oplus_{n=1}^{\infty}\tilde{X}_n$ converges almost surely with respect to the metric $D_{p,q}$. In particular case, $\oplus_{n=1}^{\infty}\tilde{X}_n$ converges almost surely with respect to the metric d_* if $\sum_{n=1}^{\infty} Var(\tilde{X}_n) < \infty$.*

Proof. The completeness of the metric space $(E, D_{p,q})$ implies that $\oplus_{n=1}^{\infty}\tilde{X}_n$ converges almost surely with respect to the metric $D_{p,q}$ iff $\sup_{k\geq n} D_{p,q}(\tilde{S}_k, \tilde{S}_n) \to 0$ in probability as $n \to \infty$. Let $n < m$, and consider a Cauchy sequence. Theorem 4 implies that

$$\begin{aligned} P(\max_{n\leq k\leq m} D_{p,q}(\tilde{S}_k, \tilde{S}_n) > \epsilon) &\leq \frac{E\max_{n\leq k\leq m} D_{p,q}^p(\tilde{S}_k, \tilde{S}_n)}{\epsilon^p} \\ &= \frac{E\max_{n\leq k\leq m} || \oplus_{i=n+1}^{k} \tilde{X}_i||_{p,q}^p}{\epsilon^p} \\ &\leq C\sum_{i=n+1}^{m} E||\tilde{X}_i||_{p,q}^p. \end{aligned}$$

Now, letting $m \to \infty$, we have

$$P(\sup_{k\geq n} D_{p,q}(\tilde{S}_k, \tilde{S}_n) > \epsilon) \leq \frac{C\sum_{k=n+1}^{\infty} E||\tilde{X}_k||_{p,q}^p}{\epsilon^p} \to 0 \;\; as \;\; n \to \infty.$$

□

Example 5. *Let $\{\tilde{X}_n, n \geq 1\}$ be a sequence of independent fuzzy random variables with the following membership functions:*

$$\mu_{\tilde{X}_n(\omega)}(x) = \begin{cases} \frac{x - X_n(\omega)}{X_n(\omega)}, & X_n(\omega) < x \leq 2X_n(\omega), \\ \frac{3X_n(\omega) - x}{X_n(\omega)}, & 2X_n(\omega) < x < 3X_n(\omega), \\ 0, & otherwise, \end{cases}$$

where $\{X_n, n \geq 1\}$ is a sequence of independent random variables such that $Var(X_n) = n^\beta \sigma^2$. By considering $p = 2$ and $q = \frac{1}{2}$, we have $ED^p_{p,q}(\tilde{X}_n, E\tilde{X}_n) = Var(\tilde{X}_n)$. For establishing a strong convergence of $\{\tilde{S}_n, n \geq 1\}$ with respect to the metric $D_{p,q}$, it remains to show that $\sum_{n=1}^{\infty} Var(\tilde{X}_n) < \infty$. It is easy to see that $Var(\tilde{X}_n) = \frac{19}{6} n^\beta \sigma^2$, and $\sum_{n=1}^{\infty} Var(\tilde{X}_n) = \frac{19\sigma^2}{6} \sum_{n=1}^{\infty} n^\beta < \infty$, where $\beta < -1$. Therefore, Theorem 9 implies that $\oplus_{n=1}^{\infty} \tilde{X}_n$ converges almost surely with respect to the metric $D_{p,q}$.

Conclusions

In this article, we extended some well-known inequalities to fuzzy martingales based on the metric $D_{p,q}$. Also, by invoking these inequalities, we established some limit theorems for fuzzy martingales. It is mentioned that if fuzzy random variables reduce to ordinary (real-valued) random variables, all results hold. The study of weak convergence and strong convergence of fuzzy sub-(super-) martingales is a potential topic for future research.

Acknowledgements

The authors are grateful to referees for their useful and constructive suggestions which lead to an improved version of the paper. The third author would like to thank University of Tehran for partial support of this research.

Author details

[1]Department of Statistics, Faculty of Mathematical Sciences, Ferdowsi University of Mashhad, Mashhad 91775-1159, Iran.
[2]Department of Engineering Science, College of Engineering, University of Tehran, Tehran 11365-4563, Iran.

References

1. Kwakernaak, H: Fuzzy random variables I. Inform. Sci. **15**, 1–29 (1978)
2. Miyakoshi, M, Shimbo, M: A strong law of large numbers for fuzzy random variables. Fuzzy Set. Syst. **12**, 133–142 (1984)
3. Puri, ML, Ralescu, DA: Fuzzy random variables. J. Math. Anal. Appl. **114**, 402–422 (1986)
4. Puri, ML, Ralescu, DA: Convergence theorems for fuzzy martingales. J. Math. Anal. Appl. **160**, 107–122 (1991)
5. Miranda, E, Couso, I, Gil, P: Random sets as imprecise random variables. J. Math. Anal. Appl. **307**, 32–47 (2005)
6. Klement, EP, Puri, ML, Ralescu, DA: Limit theorems for fuzzy random variables. Proc. Roy. Soc. London Ser. A. **407**, 171–182 (1986)
7. Taylor, RL, Seymour, L, Chen, Y: Weak laws of large numbers for fuzzy random sets. Nonlinear Anal. **47**, 114–125 (2001)
8. Joo, SY, Kim, YK, Kwon, JS: On Chung's type law of large numbers for fuzzy random variables. Stat. Probab. Lett. **74**, 67–75 (2005)
9. Fu, K, Zhang, L-X: Strong limit theorems for random sets and fuzzy random sets with slowly varying weights. Inform. Sci. **178**, 2648–2660 (2008)
10. Korner, R: On the variance of fuzzy random variables. Fuzzy Set. Syst. **92**, 83–93 (1997)
11. Fei, W, Wu, R, Shao, S: Doob's stopping theorem for fuzzy (super, sub) martingales with discrete time. Fuzzy Set. Syst. **135**, 377–390 (2003)
12. Ralescu, D: Fuzzy random variables revisited. In: *Mathematics of Fuzzy Sets, Handbook Fuzzy Sets Series*, vol. 3, pp. 701–710, Kluwer, Boston, (1999)
13. Stojakovic, M: Fuzzy martingales-a simple form of fuzzy processes. Stochastic Anal. Appl. **14**, 355–367 (1996)
14. Feng, Y, Hu, L, Shu, H: The variance and covariance of fuzzy random variables and their applications. Fuzzy Set. Syst. **120**, 487–497 (2001)
15. Feng, Y: Convergence theorems for fuzzy random variables and fuzzy martingales. Fuzzy Set. Syst. **103**, 435–441 (1999)
16. Feng, Y: An approach to generalize laws of large numbers for fuzzy random variables. Fuzzy Set. Syst. **128**, 237–245 (2002)
17. Moore, RE, Kearfott, RB, Cloud, MJ: Introduction to Interval Analysis, Siam, Philadelphia (2009)
18. Viertl, R: Statistical Methods for Fuzzy Data. Wiley, Chichester, West Sussex (2011)

19. Sadeghpour Gildeh, B, Rahimpour, S: A fuzzy bootstrap test for the mean with $D_{p,q}$-distance. Fuzzy Inf. Eng. **4**, 351–358 (2011)
20. Wu, HC: The laws of large numbers for fuzzy random variables. Fuzzy Set. Syst. **116**, 245–262 (2000)
21. Aumann, RJ: Integrals of set-valued functions. J. Math. Anal. Appl. **12**, 1–12 (1965)
22. Fei, W: Regularity and stopping theorem for fuzzy martingales with continuous parameters. Inform. Sci. **169**, 175–187 (2005)
23. Hall, P, Heyde, CC: Martingale Limit Theory and Its Application. Academic, New York (1980)
24. Gut, A: Probability: a Graduate Course. Springer, New York (2005)

3

Trapezoid fuzzy linguistic prioritized weighted average operators and their application to multiple attribute group decision-making

Rajkumar Verma* and Bhu Dev Sharma

* Correspondence:
rkver83@gmail.com
Department of Mathematics, Jaypee Institute of Information Technology (Deemed University), Noida–201307, Uttar Pradesh, India

Abstract

The prioritized weighted average (PWA) operator was originally introduced by Yager. The prominent characteristic of the PWA operator is that it takes into account prioritization among attributes and decision makers. Motivated by the idea of PWA operator, we develop some prioritized weighted aggregation operators for aggregating trapezoid fuzzy linguistic information. The properties of the new aggregation operators are studied in detail. Furthermore, based on the proposed operators, some approaches to deal with multiple attribute group decision-making problems under trapezoid fuzzy linguistic environments are developed. Finally, a practical example is provided to illustrate the multiple attribute group decision-making process.

Keywords: Multiple attribute group decision-making (MAGDM); Trapezoid fuzzy linguistic variables; Prioritized weighted average operator; Trapezoid fuzzy linguistic prioritized weighted average operator; Trapezoid fuzzy linguistic prioritized weighted geometric operator; Trapezoid fuzzy linguistic prioritized weighted harmonic average operator

Introduction

In the process of multiple attribute decision-making, information aggregation is an essential process of gathering relevant information from various sources. In the literature, a wide range of aggregation operators are found for aggregating the data information [1-4]. The ordered weighted averaging (OWA) operator, introduced by Yager [5], is a well-known aggregation operator that provides a parameterized family of aggregation operators, including the maximum, the minimum, and the average. Since its appearance, the OWA operator has received increasing attention from many authors and it has been applied across many fields [6-28]. Chiclana et al. [8] and Xu and Da [19] introduced the ordered weighted geometric (OWG) operators, which are based on the OWA operator and on the geometric mean. A further interesting extension of the OWA operator is the generalized OWA (GOWA) operator [26] that uses generalized means [29] in the OWA operator.

However, in some situations, the input arguments take the form of linguistic variables, rather than being real numbers because of time pressure, lack of knowledge, and people's limited expertise related with problem domain. Bordonga et al. [6] utilized the OWA operator to solve the group decision-making problem in linguistic context. Herrera and Martínez [30] established a linguistic 2-tuple computational model for dealing with linguistic information. To aggregate uncertain linguistic information, Xu [31] proposed the uncertain linguistic weighted averaging (ULWA) operator, the uncertain linguistic ordered weighted averaging (ULOWA) operator, and the uncertain linguistic hybrid averaging (ULHA) operator. Xu [22] introduced some uncertain linguistic geometric mean operators including the uncertain linguistic geometric mean (ULGM), the uncertain linguistic weighted geometric mean (ULWGM) operator, the uncertain linguistic ordered weighted geometric mean (ULOWGM) operator, and the induced uncertain linguistic ordered weighted geometric mean (IULOWGM) operator and developed an approach to group decision-making with uncertain multiplicative linguistic relation. Wei [32] defined the uncertain linguistic hybrid geometric mean (ULHGM) and applied it to the group decision-making. Further, Xu [33] proposed some aggregation operators for aggregating triangular fuzzy linguistic information such as the fuzzy linguistic averaging (FLA) operator, the fuzzy linguistic weighted averaging (FLWA) operator, the fuzzy linguistic ordered weighted averaging (FLOWA) operator, and the induced fuzzy linguistic ordered weighted averaging (IFLOWA) operator. The trapezoid fuzzy linguistic variable (TFLV), introduced by Xu [34], generalizes the linguistic variable, the uncertain linguistic variable, and the triangular fuzzy linguistic variable, and research on aggregation operators under trapezoid fuzzy linguistic environment is very significant. Xu [34] and Liang and Chen [35] proposed the trapezoid fuzzy linguistic weighted averaging (TFLWA) operator and applied it to multiple attribute decision-making problems. Wei and Yi [36] introduced the trapezoid fuzzy linguistic weighted geometric mean (TFLWGM) operator and developed an approach to group decision-making with trapezoid fuzzy linguistic information. Liu and Su [37] introduced the trapezoid fuzzy linguistic ordered weighted averaging (TFLOWA) operator and the trapezoid fuzzy linguistic hybrid ordered weighted averaging (TFLHOWA) operator. Further, Liu and Su [38] developed some trapezoid fuzzy linguistic harmonic averaging operators such as the trapezoid fuzzy linguistic weighted harmonic averaging (TFLWHA) operator, the trapezoid fuzzy linguistic ordered weighted harmonic averaging (TFLOWHA) operator, and the trapezoid fuzzy linguistic hybrid harmonic averaging (TFLHHA) operator, and then studied some desirable properties of these operators. Based on the idea of Bonferroni mean [39], Liu and Jin [40] proposed some Bonferroni mean operators such as the trapezoid fuzzy linguistic Bonferroni mean (TFLBM), the trapezoid fuzzy linguistic weighted Bonferroni mean (TFLWBM), the trapezoid fuzzy linguistic Bonferroni OWA (TFLBOWA), and the trapezoid fuzzy linguistic weighted Bonferroni OWA (TFLWBOWA) for aggregating trapezoid fuzzy linguistic correlative information. Recently, on the basis of the idea of the generalized mean [29], Liu and Wu [41] proposed some generalized trapezoid fuzzy linguistic aggregation operators and found their application in multiple attribute group decision-making.

Trapezoid fuzzy linguistic variables are very useful tools to deal with uncertain or fuzzy information. In the last couple of years, many multiple attribute group decision-making theories and methods have been proposed under trapezoid fuzzy linguistic environments with the assumption that the attributes and the decision makers are at the same priority levels. However, in the real life multiple attribute group decision-making problems, attributes and decision makers have different priority levels in general. To overcome this issue, motivated by the idea of prioritized weighted aggregation operators [42,43], in this paper, we propose some trapezoid fuzzy linguistic prioritized weighted aggregation operators such as the trapezoid fuzzy linguistic prioritized weighted average (TFLPWA) operator, the trapezoid fuzzy linguistic prioritized weighted geometric (TLLPWG) operator, and the trapezoid fuzzy linguistic prioritized weighted harmonic (TFLPWH) operator. A prominent characteristic of these proposed operators is that they take into account the prioritization among the attributes and decision makers. Further, we have utilized these operators to develop some approaches to solve multiple attribute group decision-making problems under trapezoid fuzzy linguistic environments.

The paper is organized as follows. In the 'Preliminaries' section, some basic concepts related to trapezoid fuzzy linguistic variables and prioritized weighted average operator are briefly given. In the 'Trapezoid fuzzy linguistic prioritized weighted aggregation operators' section, we introduce some trapezoid fuzzy linguistic prioritized weighted aggregation operators: the trapezoid fuzzy linguistic prioritized weighted average (TFLPWA) operator, the trapezoid fuzzy linguistic prioritized weighted geometric (TFLPWG) operator, and the trapezoid fuzzy linguistic prioritized weighted harmonic average (TFLPWHA) operator. Some properties of proposed operators are also studied here. In the 'An approach to multiple attribute group decision-making with trapezoid fuzzy uncertain linguistic information' section, we have applied these operators to develop some decision models for solving trapezoid fuzzy linguistic multiple attribute group decision-making problems in which the attributes and decision makers are in different priority levels. In the 'Numerical example' section, a numerical example is presented to illustrate the proposed approach to multiple attribute group decision-making with trapezoid fuzzy linguistic information. Our conclusions are presented in the 'Conclusions' section.

Preliminaries

In this section, we briefly review some basic concepts related to trapezoid fuzzy linguistic variables and prioritized weighted average operator, which will be needed in the following analysis.

Let $S = \{s_i | i = 1, 2, \ldots, t\}$ be a discrete linguistic term set with odd cardinality. Any label, s_i, represents a possible value for a linguistic variable, and it must have the following characteristics [31]:

(i) The set is ordered: $s_i \geq s_j$ if $i \geq j$.
(ii) There is the negation operator: $\text{neg}(s_i) = s_j$ such that $j = t - i$.
(iii) Max operator: $\max(s_i, s_j) = s_i$ if $s_i \geq s_j$.
(iv) Min operator: $\min(s_i, s_j) = s_i$ if $s_i \leq s_j$.

For example, S can be defined as

$$S = \{ s_1 = \text{extremely poor}, s_2 = \text{very poor}, s_3 = \text{poor}, s_4 = \text{slightly poor}, s_5 = \text{fair}$$
$$s_6 = \text{slightly good}, s_7 = \text{good}, \ s_8 = \text{very good}, s_9 = \text{extremely good}\}.$$

Further, we extend the discrete term set S to a continuous linguistic term set $\bar{S} = \{s_\alpha | s_1 \leq s_\alpha \leq s_t, \alpha \in [1, t]\}$. If $s_\alpha \in S$, then, we call s_α an original linguistic term, otherwise, we call s_α the virtual linguistic term [22]. In general, the decision makers use the original linguistic term to evaluate alternatives, and the virtual linguistic terms can only appear in calculation [31].

Definition 1. Distance between two linguistic variables [33]: Let s_α and s_β be two linguistic variables, then the distance between s_α and s_β is defined as follows:

$$d(s_\alpha, s_\beta) = |\alpha - \beta|. \tag{1}$$

In some situations, however, the decision makers (DMs) may provide fuzzy linguistic information because of time pressure, lack of knowledge, and their limited expertise related with the problem domain. To handle such type of cases, Xu [34] defined the trapezoid fuzzy linguistic variable and introduced some of the operational laws on them.

Definition 2. Trapezoid fuzzy linguistic variable [34]: Let $\tilde{s} = [s_\alpha, s_\beta, s_\gamma, s_\eta]$, where s_α, s_β, s_γ, $s_\eta \in S$, and the subscripts α, β, γ, and η are non-decreasing numbers and s_β and s_γ indicate the interval in which the membership value is 1, with s_α and s_η indicating the lower and upper values of $\tilde{s}$, respectively. Then, $\tilde{s}$ is called the trapezoid fuzzy linguistic variable, which is characterized by the following membership function (see Figure 1):

$$\mu_{\tilde{s}}(\theta) = \begin{cases} 0, & s_0 \le s_\theta \le s_\alpha, \\ \dfrac{d(s_\theta, s_\alpha)}{d(s_\beta, s_\alpha)}, & s_\alpha \le s_\theta \le s_\beta, \\ 1 & s_\beta \le s_\theta \le s_\gamma, \\ \dfrac{d(s_\theta, s_\eta)}{d(s_\gamma, s_\eta)}, & s_\gamma \le s_\theta \le s_\eta, \\ 0, & s_\eta \le s_\theta \le s_q. \end{cases} \tag{2}$$

Especially, if any two of α, β, γ, and η are equal, then $\tilde{s}$ is reduced to a triangular fuzzy linguistic variable [33], and if any three of α, β, γ, and η are equal, then $\tilde{s}$ is reduced to an uncertain linguistic variable [31].

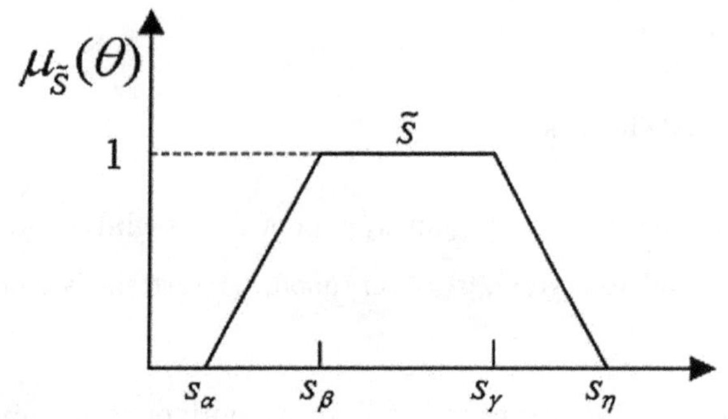

Figure 1 A trapezoid fuzzy linguistic variable.

Definition 3. Arithmetical operations on trapezoid fuzzy linguistic variables [34,37,41]: Let $\tilde{s} = [s_\alpha, s_\beta, s_\gamma, s_\eta]$, $\tilde{s}_1 = [s_{\alpha_1}, s_{\beta_1}, s_{\gamma_1}, s_{\eta_1}]$, and $\tilde{s}_2 = [s_{\alpha_2}, s_{\beta_2}, s_{\gamma_2}, s_{\eta_2}]$ be three trapezoid fuzzy linguistic variables, then some arithmetical operations are defined as follows:

(i) $\tilde{s}_1 \oplus \tilde{s}_2 = [s_{\alpha_1}, s_{\beta_1}, s_{\gamma_1}, s_{\eta_1}] \oplus [s_{\alpha_2}, s_{\beta_2}, s_{\gamma_2}, s_{\eta_2}] = [s_{\alpha_1+\alpha_2}, s_{\beta_1+\beta_2}, s_{\gamma_1+\gamma_2}, s_{\eta_1+\eta_2}]$,

(ii) $\tilde{s}_1 \otimes \tilde{s}_2 = [s_{\alpha_1}, s_{\beta_1}, s_{\gamma_1}, s_{\eta_1}] \otimes [s_{\alpha_2}, s_{\beta_2}, s_{\gamma_2}, s_{\eta_2}] = [s_{\alpha_1\alpha_2}, s_{\beta_1\beta_2}, s_{\gamma_1\gamma_2}, s_{\eta_1\eta_2}]$,

(iii) $\lambda\tilde{s} = \lambda[s_\alpha, s_\beta, s_\gamma, s_\eta] = [s_{\lambda\alpha}, s_{\lambda\beta}, s_{\lambda\gamma}, s_{\lambda\eta}], \quad \lambda \geq 0$

(iv) $\tilde{s}^\lambda = [s_\alpha, s_\beta, s_\gamma, s_\eta]^\lambda = [s_{\alpha^\lambda}, s_{\beta^\lambda}, s_{\gamma^\lambda}, s_{\eta^\lambda}], \quad \lambda \geq 0$

(v) $\tilde{s}^{-1} = [s_\alpha, s_\beta, s_\gamma, s_\eta]^{-1} = \left[\frac{1}{s_\alpha}, \frac{1}{s_\beta}, \frac{1}{s_\gamma}, \frac{1}{s_\eta}\right] = \left[\frac{1}{s_{1/\eta}}, \frac{1}{s_{1/\gamma}}, \frac{1}{s_{1/\beta}}, \frac{1}{s_{1/\alpha}}\right].$

Definition 4 [35]: Let $\tilde{s}_1 = [s_{\alpha_1}, s_{\beta_1}, s_{\gamma_1}, s_{\eta_1}]$ and $\tilde{s}_2 = [s_{\alpha_2}, s_{\beta_2}, s_{\gamma_2}, s_{\eta_2}]$ be two trapezoid fuzzy linguistic variables, then the degree of possibility, $p(\tilde{s}_1 \geq \tilde{s}_2)$, of $(\tilde{s}_1 \geq \tilde{s}_2)$ is defined as follows:

$$p(\tilde{s}_1 \geq \tilde{s}_2) = \min\left\{\max\left\{\frac{(\gamma_1+\eta_1)-(\alpha_2+\beta_2)}{(\gamma_1+\eta_1)-(\alpha_1+\beta_1)+(\gamma_2+\eta_2)-(\alpha_2+\beta_2)}, 0\right\}, 1\right\}. \quad (3)$$

The characteristics of the possibility degree $p(\tilde{s}_1 \geq \tilde{s}_2)$ may be noted as follows [35]:

1. $0 \leq p(\tilde{s}_1 \geq \tilde{s}_2) \leq 1$, $0 \leq p(\tilde{s}_2 \geq \tilde{s}_1) \leq 1$.
2. $p(\tilde{s}_1 \geq \tilde{s}_2) + p(\tilde{s}_2 \geq \tilde{s}_1) = 1$.

Especially, if $p(\tilde{s}_1 \geq \tilde{s}_2) = p(\tilde{s}_2 \geq \tilde{s}_1)$, then $p(\tilde{s}_1 \geq \tilde{s}_2) = p(\tilde{s}_2 \geq \tilde{s}_1) = \frac{1}{2}$.

Definition 5 Expected value of trapezoid fuzzy linguistic variable: Let $\tilde{s} = [s_\alpha, s_\beta, s_\gamma, s_\eta]$ be a trapezoid fuzzy linguistic variable, then the expected value of $\tilde{s}$ is defined as follows:

$$E(\tilde{s}) = \left(\frac{s_\alpha \oplus s_\beta \oplus s_\gamma \oplus s_\eta}{4}\right). \quad (4)$$

The prioritized weighted average (PWA) operator was originally introduced by Yager [42,43] as follows:

Definition 6. PWA operator [42,43]: Let $G = \{G_1, G_2, \ldots, G_n\}$ be a collection of attributes and let there be a prioritization between the attributes expressed by the linear ordering $G_1 \succ G_2 \succ G_3 \ldots \succ G_n$, indicating that attribute G_i has a higher priority than G_j, if $i < j$. Also, let $G_i(x)$ be the performance value of any alternative x under attribute G_i and satisfies $G_i(x) \in [0, 1]$. If

$$\mathrm{PWA}(G_1(x), G_2(x), \ldots, G_n(x)) = \sum_{i=1}^{n} \frac{T_i}{\sum_{i=1}^{n} T_i} G_i(x), \quad (5)$$

where $T_i = \prod_{j=1}^{i-1} G_j(x), \quad i = 2, 3, ..., n,\ T_1 = 1$, then PWA($G_1(x)$, $G_2(x)$,...,$G_n(x)$) is called the PWA operator.

In the next section, to aggregate the trapezoid fuzzy linguistic information, we propose some prioritized weighted aggregation operators such as the trapezoid fuzzy linguistic prioritized weighted average (TFLPWA) operator, the trapezoid fuzzy linguistic prioritized weighted geometric (TFLPWG) operator, and the trapezoid fuzzy linguistic prioritized weighted harmonic average (TFLPWHA) operator with properties.

Trapezoid fuzzy linguistic prioritized weighted aggregation operators

Trapezoid fuzzy linguistic prioritized weighted average operator

Based on Definition 6, we give definition of the TFLPWA operator as follows:

Definition 7. Trapezoid fuzzy linguistic prioritized weighted average operator: Given a set of trapezoid fuzzy linguistic variables, $\tilde{s}_i = \left[s_{\alpha_i}, s_{\beta_i}, s_{\gamma_i}, s_{\eta_i}\right], i = 1, 2, ..., n$, the TFLPWA operator is defined as follows:

$$\text{TFLPWA}(\tilde{s}_1, \tilde{s}_2, ..., \tilde{s}_n) = \mathop{\oplus}_{i=1}^{n} \left(\frac{T_i}{\sum_{i=1}^{n} T_i}\tilde{s}\right)_i = \frac{T_1}{\sum_{i=1}^{n} T_i}\tilde{s}_1 \oplus \frac{T_2}{\sum_{i=1}^{n} T_i}\tilde{s}_2 \oplus \cdots \oplus \frac{T_n}{\sum_{i=1}^{n} T_i}\tilde{s}_n, \tag{6}$$

where $T_i = \prod_{j=1}^{i-1} \frac{I(E(\tilde{s}_j))}{t}, \quad i = 2, 3, ..., n,\ T_1 = 1$ and $E(\tilde{s}_j)$ is the expected value of $\tilde{s}_j = \left[s_{\alpha_j}, s_{\beta_j}, s_{\gamma_j}, s_{\eta_j}\right]$, $I(E(\tilde{s}_j))$ is the subscript of $E(\tilde{s}_j)$.

Note 1: If the priority levels of the aggregated arguments reduce to the same level, then the TFLPWA operator reduces to the trapezoid fuzzy linguistic weighted average (TFLWA) operator [34,35]:

$$\text{TFLPWA}(\tilde{s}_1, \tilde{s}_2, ..., \tilde{s}_n) = (w_1\tilde{s}_1 \oplus w_2\tilde{s}_2 \oplus \cdots \oplus w_n\tilde{s}_n). \tag{7}$$

Next, based on the operational laws of TFLVs, we can easily prove the following theorem:

Theorem 1. Let $\tilde{s}_i = \left[s_{\alpha_i}, s_{\beta_i}, s_{\gamma_i}, s_{\eta_i}\right],\ i = 1, 2, ..., n,$ be a set of trapezoid fuzzy linguistic variables, then the aggregated value by using the TFLPWA operator is also a trapezoid fuzzy linguistic variable, and

$$\begin{aligned} \text{TFLPWA}(\tilde{s}_1, \tilde{s}_2, ..., \tilde{s}_n) &= \mathop{\oplus}_{i=1}^{n} \frac{T_i}{\sum_{i=1}^{n} T_i}\tilde{s}_i, \\ &= \left[\mathop{\oplus}_{i=1}^{n} \frac{T_i}{\sum_{i=1}^{n} T_i} s_{\alpha_i}, \mathop{\oplus}_{i=1}^{n} \frac{T_i}{\sum_{i=1}^{n} T_i} s_{\beta_i}, \mathop{\oplus}_{i=1}^{n} \frac{T_i}{\sum_{i=1}^{n} T_i} s_{\gamma_i}, \mathop{\oplus}_{i=1}^{n} \frac{T_i}{\sum_{i=1}^{n} T_i} s_{\eta_i}\right], \end{aligned} \tag{8}$$

where $T_i = \prod_{j=1}^{i-1} \frac{I(E(\tilde{s}_j))}{t}, i = 2, 3, ..., n$, $T_1 = 1$, $E(\tilde{s}_j)$ is the expected value of $\tilde{s}_j = \left[s_{\alpha_j}, s_{\beta_j}, s_{\gamma_j}, s_{\eta_j}\right]$ and $I(E(\tilde{s}_j))$ is the subscript of $E(\tilde{s}_j)$.

Properties of TFLPWA operator

P1. (*Idempotency*): Let $\tilde{s}_i = \left[s_{\alpha_i}, s_{\beta_i}, s_{\gamma_i}, s_{\eta_i}\right],\ i = 1, 2, ..., n,$ be a set of trapezoid fuzzy linguistic variables, $T_i = \prod_{j=1}^{i-1} \frac{I(E(\tilde{s}_j))}{t}, i = 2, 3, ..., n,\ T_1 = 1,\ E(\tilde{s}_j)$ be the expected value of

$\tilde{s}_j = \left[s_{\alpha_j}, s_{\beta_j}, s_{\gamma_j}, s_{\eta_j}\right]$ and $I(E(\tilde{s}_j))$ be the subscript of $E(\tilde{s}_j)$. If all the trapezoid fuzzy linguistic variables $\tilde{s}_i$, $i = 1, 2,\ldots, n$, are equal, i.e., $\tilde{s}_i = \tilde{s} = [s_\alpha, s_\beta, s_\gamma, s_\eta]\,\forall i$, then

$$\text{TFLPWA}(\tilde{s}_1, \tilde{s}_2, \ldots, \tilde{s}_n) = \tilde{s}. \tag{9}$$

P2. (*Boundedness*): Let $\tilde{s}_i = [s_{\alpha_i}, s_{\beta_i}, s_{\gamma_i}, s_{\eta_i}]$, $i = 1, 2, \ldots, n$, be a set of trapezoid fuzzy linguistic variables, $T_i = \prod_{j=1}^{i-1} \frac{I(E(\tilde{s}_j))}{t}, i = 2, 3, \ldots, n$, $T_1 = 1$, $E(\tilde{s}_j)$ be the expected value of $\tilde{s}_j = \left[s_{\alpha_j}, s_{\beta_j}, s_{\gamma_j}, s_{\eta_j}\right]$ and $I(E(\tilde{s}_j))$ be the subscript of $E(\tilde{s}_j)$. Also, let

$$\tilde{s}^- = \min_i \tilde{s}_i = \left[\min_i s_{\alpha_i}, \min_i s_{\beta_i}, \min_i s_{\gamma_i}, \min_i s_{\eta_i}\right], \tag{10}$$

and

$$\tilde{s}^+ = \max_i \tilde{s}_i = \left[\max_i s_{\alpha_i}, \max_i s_{\beta_i}, \max_i s_{\gamma_i}, \max_i s_{\eta_i}\right]. \tag{11}$$

Then,

$$\tilde{s}^- \leq \text{TFLPWA}(\tilde{s}_1, \tilde{s}_2, \ldots, \tilde{s}_n) \leq \tilde{s}^+. \tag{12}$$

P3. (*Monotonicity*): Let $\tilde{s}_i = [s_{\alpha_i}, s_{\beta_i}, s_{\gamma_i}, s_{\eta_i}]$ and $\tilde{s}_i^{'} = \left[s_{\alpha_i}^{'}, s_{\beta_i}^{'}, s_{\gamma_i}^{'}, s_{\eta_i}^{'}\right], i = 2, 3, \ldots, n,$ be two sets of trapezoid fuzzy linguistic variables, $T_i = \prod_{j=1}^{i-1} \frac{I(E(\tilde{s}_j))}{t}$, $T_i^{'} = \prod_{j=1}^{i-1} \frac{I\left(E\left(\tilde{s}_j^{'}\right)\right)}{t}$, $i = 2, 3, \ldots, n$, $T_1 = T_1^{'} = 1$, $E(\tilde{s}_j)$ be the expected value of $\tilde{s}_j = \left[s_{\alpha_j}, s_{\beta_j}, s_{\gamma_j}, s_{\eta_j}\right]$, $E\left(\tilde{s}_j^{'}\right)$ be the expected value of $\tilde{s}_j^{'} = \left[s_{\alpha_j}^{'}, s_{\beta_j}^{'}, s_{\gamma_j}^{'}, s_{\eta_j}^{'}\right]$, $I(E(\tilde{s}_j))$ be the subscript of $E(\tilde{s}_j)$, $I\left(E\left(\tilde{s}_j^{'}\right)\right)$ be the subscript of $E\left(\tilde{s}_j^{'}\right)$. If $\tilde{s}_i \leq \tilde{s}_i^{'}$ for all i, then

$$\text{TFLPWA}(\tilde{s}_1, \tilde{s}_2, \ldots, \tilde{s}_n) \leq \text{TFLPWA}\left(\tilde{s}_1^{'}, \tilde{s}_2^{'}, \ldots, \tilde{s}_n^{'}\right). \tag{13}$$

Trapezoid fuzzy linguistic prioritized weighted geometric operator

Based on TFLPWA operator and the geometric mean, here, we give the definition of the TFLPWG operator as follows:

Definition 8. Trapezoid fuzzy linguistic prioritized weighted geometric operator: Given a set of trapezoid fuzzy linguistic variables, $\tilde{s}_i = [s_{\alpha_i}, s_{\beta_i}, s_{\gamma_i}, s_{\eta_i}]$, $i = 1, 2, \ldots, n$, the TFLPWG operator is defined as follows:

$$\begin{aligned}\text{TFLPWG}(\tilde{s}_1, \tilde{s}_2, \ldots, \tilde{s}_n) &= \bigotimes_{i=1}^{n} (\tilde{s}_i)^{\frac{T_i}{\sum_{i=1}^{n} T_i}} \\ &= \left((\tilde{s}_1)^{\frac{T_1}{\sum_{i=1}^{n} T_i}} \otimes (\tilde{s}_2)^{\frac{T_2}{\sum_{i=1}^{n} T_i}} \otimes \cdots \otimes (\tilde{s}_n)^{\frac{T_n}{\sum_{i=1}^{n} T_i}}\right),\end{aligned} \tag{14}$$

where $T_i = \prod_{j=1}^{i-1} \frac{I(E(\tilde{s}_j))}{t}$, $i = 2, 3, ..., n$, $T_1 = 1$, $E(\tilde{s}_j)$ is the expected value of $\tilde{s}_j = [s_{\alpha_j}, s_{\beta_j}, s_{\gamma_j}, s_{\eta_j}]$ and $I(E(\tilde{s}_j))$ is the subscript of $E(\tilde{s}_j)$.

Note 2: If the priority levels of the aggregated arguments reduce to the same level, then the TFLPWG operator reduces to the trapezoid fuzzy linguistic weighted geometric (TFLWG) operator [36]:

$$\text{TFLPWG}(\tilde{s}_1, \tilde{s}_2, ..., \tilde{s}_n) = (\tilde{s}_1)^{w_1} \otimes (\tilde{s}_2)^{w_2} \otimes \cdots \otimes (\tilde{s}_n)^{w_n}. \quad (15)$$

Next, based on the operational laws of TFLVs, we can prove a result in the following theorem:

Theorem 2. Let $\tilde{s}_i = [s_{\alpha_i}, s_{\beta_i}, s_{\gamma_i}, s_{\eta_i}]$, $i = 1, 2, ..., n$, be a set of trapezoid fuzzy linguistic variables, then the aggregated value by using the TFLPWG operator is also a trapezoid fuzzy linguistic variable, and

$$\text{TFLPWG}(\tilde{s}_1, \tilde{s}_2, ..., \tilde{s}_n) = \bigotimes_{i=1}^{n} (\tilde{s}_i)^{\frac{T_i}{\sum_{i=1}^{n} T_i}},$$
$$= \left[\bigotimes_{i=1}^{n} (s_{\alpha_i})^{\frac{T_i}{\sum_{i=1}^{n} T_i}}, \bigotimes_{i=1}^{n} (s_{\beta_i})^{\frac{T_i}{\sum_{i=1}^{n} T_i}}, \bigotimes_{i=1}^{n} s_{\gamma_i}^{\frac{T_i}{\sum_{i=1}^{n} T_i}}, \bigotimes_{i=1}^{n} (s_{\eta_i})^{\frac{T_i}{\sum_{i=1}^{n} T_i}} \right], \quad (16)$$

where $T_i = \prod_{j=1}^{i-1} \frac{I(E(\tilde{s}_j))}{t}$, $i = 2, 3, ..., n$, $T_1 = 1$, $E(\tilde{s}_j)$ is the expected value of $\tilde{s}_j = [s_{\alpha_j}, s_{\beta_j}, s_{\gamma_j}, s_{\eta_j}]$ and $I(E(\tilde{s}_j))$ is the subscript of $E(\tilde{s}_j)$.

Properties of TFLPWG operator

P1. (*Idempotency*): Let $\tilde{s}_i = [s_{\alpha_i}, s_{\beta_i}, s_{\gamma_i}, s_{\eta_i}]$, $i = 1, 2, ..., n$, be a set of trapezoid fuzzy linguistic variables, $T_i = \prod_{j=1}^{i-1} \frac{I(E(\tilde{s}_j))}{t}$, $i = 2, 3, ..., n$, $T_1 = 1$, $E(\tilde{s}_j)$ be the expected value of $\tilde{s}_j = [s_{\alpha_j}, s_{\beta_j}, s_{\gamma_j}, s_{\eta_j}]$ and $I(E(\tilde{s}_j))$ be the subscript of $E(\tilde{s}_j)$. If all the trapezoid fuzzy linguistic variables $\tilde{s}_i$, $i = 1, 2, ..., n$, are equal, i.e., $\tilde{s}_i = \tilde{s} = [s_\alpha, s_\beta, s_\gamma, s_\eta]\, \forall i$, then

$$\text{TFLPWG}(\tilde{s}_1, \tilde{s}_2, ..., \tilde{s}_n) = \tilde{s}. \quad (17)$$

P2. (*Boundedness*): Let $\tilde{s}_i = [s_{\alpha_i}, s_{\beta_i}, s_{\gamma_i}, s_{\eta_i}]$, $i = 1, 2, ..., n$, be a set of trapezoid fuzzy linguistic variables, $T_i = \prod_{j=1}^{i-1} \frac{I(E(\tilde{s}_j))}{t}$, $i = 2, 3, ..., n$, $T_1 = 1$, $E(\tilde{s}_j)$ be the expected value of $\tilde{s}_j = [s_{\alpha_j}, s_{\beta_j}, s_{\gamma_j}, s_{\eta_j}]$ and $I(E(\tilde{s}_j))$ be the subscript of $E(\tilde{s}_j)$. Also, let

$$\tilde{s}^- = \min_i \tilde{s}_i = \left[\min_i s_{\alpha_i}, \min_i s_{\beta_i}, \min_i s_{\gamma_i}, \min_i s_{\eta_i} \right], \quad (18)$$

and

$$\tilde{s}^{+} = \max_i \tilde{s}_i = \left[\max_i s_{\alpha_i},\ \max_i s_{\beta_i},\ \max_i s_{\gamma_i},\ \max_i s_{\eta_i} \right]. \tag{19}$$

Then

$$\tilde{s}^{-} \le \mathrm{TFLPWG}(\tilde{s}_1, \tilde{s}_2, \ldots, \tilde{s}_n) \le \tilde{s}^{+}. \tag{20}$$

P3. (*Monotonicity*): Let $\tilde{s}_i = \left[s_{\alpha_i}, s_{\beta_i}, s_{\gamma_i}, s_{\eta_i}\right]$ and $\tilde{s}_i^{'} = \left[s_{\alpha_i}^{'}, s_{\beta_i}^{'}, s_{\gamma_i}^{'}, s_{\eta_i}^{'}\right], i = 2, 3, \ldots, n$, be two sets of trapezoid fuzzy linguistic variables, $T_i = \prod_{j=1}^{i-1} \frac{I(E(\tilde{s}_j))}{t}$, $T_i^{'} = \prod_{j=1}^{i-1} \frac{I\left(E\left(\tilde{s}_j^{'}\right)\right)}{t}$, $i = 2, 3, \ldots, n$, $T_1 = T_1^{'} = 1$, $E(\tilde{s}_j)$ be the expected value of $\tilde{s}_j = \left[s_{\alpha_j}, s_{\beta_j}, s_{\gamma_j}, s_{\eta_j}\right]$, $E\left(\tilde{s}_j^{'}\right)$ be the expected value of $\tilde{s}_j^{'} = \left[s_{\alpha_j}^{'}, s_{\beta_j}^{'}, s_{\gamma_j}^{'}, s_{\eta_j}^{'}\right]$, $I(E(\tilde{s}_j))$ the subscript of $E(\tilde{s}_j)$, $I\left(E\left(\tilde{s}_j^{'}\right)\right)$ be the subscript of $E\left(\tilde{s}_j^{'}\right)$. If $\tilde{s}_i \le \tilde{s}_i^{'}$ for all i, then

$$\mathrm{TFLPWG}(\tilde{s}_1, \tilde{s}_2, \ldots, \tilde{s}_n) \le \mathrm{TFLPWG}\left(\tilde{s}_1^{'}, \tilde{s}_2^{'}, \ldots, \tilde{s}_n^{'}\right). \tag{21}$$

Trapezoid fuzzy linguistic prioritized weighted harmonic average operator

Based on TFLPWA operator and the harmonic average, here, we give the definition of the TFLPWHA operator as follows:

Definition 9. Trapezoid fuzzy linguistic prioritized weighted harmonic average operator: Given a set of trapezoid fuzzy linguistic variables, $\tilde{s}_i = \left[s_{\alpha_i}, s_{\beta_i}, s_{\gamma_i}, s_{\eta_i}\right]$, $i = 1, 2, \ldots, n$, the TFLPWHA operator is defined as follows:

$$\mathrm{TFLPWA}(\tilde{s}_1, \tilde{s}_2, \ldots, \tilde{s}_n) = \left(\mathop{\oplus}_{i=1}^{n} \frac{\frac{T_i}{\sum_{i=1}^{n} T_i}}{\tilde{s}_i} \right)^{-1} = \frac{1}{\frac{\frac{T_1}{\sum_{i=1}^{n} T_i}}{\tilde{s}_1} \oplus \frac{\frac{T_2}{\sum_{i=1}^{n} T_i}}{\tilde{s}_2} \oplus \cdots \oplus \frac{\frac{T_n}{\sum_{i=1}^{n} T_i}}{\tilde{s}_n}}, \tag{22}$$

where $T_i = \prod_{j=1}^{i-1} \frac{I(E(\tilde{s}_j))}{t}, i = 2, 3, \ldots, n$, $T_1 = 1$, $E(\tilde{s}_j)$ is the expected value of $\tilde{s}_j = \left[s_{\alpha_j}, s_{\beta_j}, s_{\gamma_j}, s_{\eta_j}\right]$ and $I(E(\tilde{s}_j))$ is the subscript of $E(\tilde{s}_j)$.

Note 3: If the priority levels of the aggregated arguments reduce to the same level, then the TFLPWHA operator reduces to the TFLWHA operator [38]:

$$\mathrm{TFLPWHA}(\tilde{s}_1, \tilde{s}_2, \ldots, \tilde{s}_n) = \frac{1}{\frac{w_1}{\tilde{s}_1} \oplus \frac{w_2}{\tilde{s}_2} \oplus \cdots \frac{w_n}{\tilde{s}_n}}. \tag{23}$$

Next, based on the operational laws of TFLVs, we can prove a result in the following theorem:

Theorem 3. Let $\tilde{s}_i = \left[s_{\alpha_i}, s_{\beta_i}, s_{\gamma_i}, s_{\eta_i}\right]$, $i = 1, 2, \ldots, n$, be a set of trapezoid fuzzy linguistic variables, then the aggregated value by using the TFLPWHA operator is also a trapezoid fuzzy linguistic variable, and

$$\text{TFLPWA}(\tilde{s}_1, \tilde{s}_2, \ldots, \tilde{s}_n) = \left(\overset{n}{\underset{i=1}{\oplus}} \frac{\frac{T_i}{\sum_{i=1}^{n} T_i}}{\tilde{s}_i} \right)^{-1},$$

$$= \left[\left(\overset{n}{\underset{i=1}{\oplus}} \frac{\frac{T_i}{\sum_{i=1}^{n} T_i}}{s_{\alpha_i}} \right)^{-1}, \left(\overset{n}{\underset{i=1}{\oplus}} \frac{\frac{T_i}{\sum_{i=1}^{n} T_i}}{s_{\beta_i}} \right)^{-1}, \left(\overset{n}{\underset{i=1}{\oplus}} \frac{\frac{T_i}{\sum_{i=1}^{n} T_i}}{s_{\gamma_i}} \right)^{-1}, \left(\overset{n}{\underset{i=1}{\oplus}} \frac{\frac{T_i}{\sum_{i=1}^{n} T_i}}{s_{\eta_i}} \right)^{-1} \right], \tag{24}$$

where $T_i = \prod_{j=1}^{i-1} \frac{I(E(\tilde{s}_j))}{t}, i = 2, 3, \ldots, n$, $T_1 = 1$ and $E(\tilde{s}_j)$ is the expected value of $\tilde{s}_j = \left[s_{\alpha_j}, s_{\beta_j}, s_{\gamma_j}, s_{\eta_j}\right]$, $I(E(\tilde{s}_j))$ is the subscript of $E(\tilde{s}_j)$.

Properties of TFLPWHA operator

P1. (*Idempotency*): Let $\tilde{s}_i = \left[s_{\alpha_i}, s_{\beta_i}, s_{\gamma_i}, s_{\eta_i}\right]$, $i = 1, 2, \ldots, n$, be a set of trapezoid fuzzy linguistic variables, $T_i = \prod_{j=1}^{i-1} \frac{I(E(\tilde{s}_j))}{t}, i = 2, 3, \ldots, n$, $T_1 = 1$, $E(\tilde{s}_j)$ be the expected value of $\tilde{s}_j = \left[s_{\alpha_j}, s_{\beta_j}, s_{\gamma_j}, s_{\eta_j}\right]$ and $I(E(\tilde{s}_j))$ be the subscript of $E(\tilde{s}_j)$. If all the trapezoid fuzzy linguistic variables $\tilde{s}_i$, $i = 1, 2, \ldots, n$, are equal, i.e., $\tilde{s}_i = \tilde{s} = \left[s_\alpha, s_\beta, s_\gamma, s_\eta\right]$ $\forall i$, then

$$\text{TFLPWHA}(\tilde{s}_1, \tilde{s}_2, \ldots, \tilde{s}_n) = \tilde{s}. \tag{25}$$

P2. (*Boundedness*): Let $\tilde{s}_i = \left[s_{\alpha_i}, s_{\beta_i}, s_{\gamma_i}, s_{\eta_i}\right]$, $i = 1, 2, \ldots, n$, be a set of trapezoid fuzzy linguistic variables, $T_i = \prod_{j=1}^{i-1} \frac{I(E(\tilde{s}_j))}{t}, i = 2, 3, \ldots, n$, $T_1 = 1$, $E(\tilde{s}_j)$ be the expected value of $\tilde{s}_j = \left[s_{\alpha_j}, s_{\beta_j}, s_{\gamma_j}, s_{\eta_j}\right]$ and $I(E(\tilde{s}_j))$ be the subscript of $E(\tilde{s}_j)$. Also, let

$$\tilde{s}^- = \min_i \tilde{s}_i = \left[\min_i s_{\alpha_i}, \min_i s_{\beta_i}, \min_i s_{\gamma_i}, \min_i s_{\eta_i}\right], \tag{26}$$

and

$$\tilde{s}^+ = \max_i \tilde{s}_i = \left[\max_i s_{\alpha_i}, \max_i s_{\beta_i}, \max_i s_{\gamma_i}, \max_i s_{\eta_i}\right]. \tag{27}$$

Then

$$\tilde{s}^- \leq \text{TFLPWHA}(\tilde{s}_1, \tilde{s}_2, \ldots, \tilde{s}_n) \leq \tilde{s}^+. \tag{28}$$

P3. (*Monotonicity*): Let $\tilde{s}_i = \left[s_{\alpha_i}, s_{\beta_i}, s_{\gamma_i}, s_{\eta_i}\right]$ and $\tilde{s}_i^{'} = \left[s_{\alpha_i}^{'}, s_{\beta_i}^{'}, s_{\gamma_i}^{'}, s_{\eta_i}^{'}\right]$, $i = 1, 2, \ldots, n$, be two sets of trapezoid fuzzy linguistic variables, $T_i = \prod_{j=1}^{i-1} \frac{I(E(\tilde{s}_j))}{t}$, $T_i^{'} = \prod_{j=1}^{i-1} \frac{I\left(E\left(\tilde{s}_j^{'}\right)\right)}{t}$, $i = 2, 3, \ldots, n$, $T_1 = T_1^{'} = 1$, $E(\tilde{s}_j)$ be the expected value of $\tilde{s}_j = \left[s_{\alpha_j}, s_{\beta_j}, s_{\gamma_j}, s_{\eta_j}\right]$, $E\left(\tilde{s}_j^{'}\right)$ be the expected value of $\tilde{s}_j^{'} = \left[s_{\alpha_j}^{'}, s_{\beta_j}^{'}, s_{\gamma_j}^{'}, s_{\eta_j}^{'}\right]$, $I(E(\tilde{s}_j))$ the subscript of $E(\tilde{s}_j)$, $I\left(E\left(\tilde{s}_j^{'}\right)\right)$ be the subscript of $E\left(\tilde{s}_j^{'}\right)$. If $\tilde{s}_i \leq \tilde{s}_i^{'}$ for all i, then

$$\text{TFLPWHA}(\tilde{s}_1, \tilde{s}_2, \ldots, \tilde{s}_n) \leq \text{TFLPWHA}\left(\tilde{s}_1^{'}, \tilde{s}_2^{'}, \ldots, \tilde{s}_n^{'}\right). \tag{29}$$

In the following section, we suggest the application of the proposed operators to solve multiple-attribute decision-making problems with trapezoid fuzzy linguistic information.

An approach to multiple attribute group decision-making with trapezoid fuzzy uncertain linguistic information

Let us consider a multiple attribute group decision-making problem involving a set of alternatives $X = \{X_1, X_2,\ldots, X_m\}$ to be considered under a set of attributes $G = \{G_1, G_2,\ldots, G_n\}$ and let there be a prioritization between the attributes expressed by the linear ordering $G_1 \succ G_2 \succ \cdots \succ G_n$ (indicating that attribute G_j has a higher priority than G_l, if $j < l$), and let $D = \{D_1, D_2,\ldots, D_q\}$ be the set of decision makers and let there be a prioritization between the decision makers expressed by the linear ordering $D_1 \succ D_2 \succ \cdots \succ D_q$, indicating decision maker D_η has a higher priority than D_ς, if $\eta < \varsigma$. Let $\tilde{R}^{(k)} = \left(\tilde{r}_{ij}^{(k)}\right)_{m\times n} = \left(\left[s_{\alpha_{ij}}^{(k)}, s_{\beta_{ij}}^{(k)}, s_{\gamma_{ij}}^{(k)}, s_{\eta_{ij}}^{(k)}\right]\right)_{m\times n}$ be a trapezoid fuzzy linguistic decision matrix, where $\tilde{r}_{ij}^{(k)} \in \tilde{S}$ is an attribute value, which takes the form of trapezoid fuzzy linguistic variables, provided by the decision maker $D_k \in D$, for the alternative $X_i \in X$ with respect to the attribute $G_j \in G$.

Using the TFLPWA (or TFLPWG or TFLPWHA) operator, we now formulate an algorithm to solve multiple attribute group decision-making problems with trapezoid fuzzy linguistic information:

Step 1. Calculate the values of $T_{ij}^{(k)}$ $(k = 1, 2, \ldots, q)$ as follows:

$$T_{ij}^{(k)} = \prod_{\gamma=1}^{k-1} \frac{I\left(E\left(\tilde{r}_{ij}^{\gamma}\right)\right)}{t} \quad (k = 2, 3, \ldots, q), \tag{30}$$

$$T_{ij}^{(1)} = 1. \tag{31}$$

Step 2. Utilize appropriately the TFLPWA operator:

$$\begin{aligned} \tilde{r}_{ij} &= \left(\tilde{s}_{\alpha_{ij}}, \tilde{s}_{\beta_{ij}}, \tilde{s}_{\gamma_{ij}}, \tilde{s}_{\eta_{ij}}\right) \\ &= \text{TFLPWA}\left(\tilde{r}_{ij}^{(1)}, \tilde{r}_{ij}^{(2)}, \ldots, \tilde{r}_{ij}^{(q)}\right) \\ &= \frac{T_{ij}^{(1)}}{\sum_{k=1}^{q} T_{ij}^{(k)}} \tilde{r}_{ij}^{(1)} \oplus \frac{T_{ij}^{(2)}}{\sum_{k=1}^{q} T_{ij}^{(k)}} \tilde{r}_{ij}^{(2)} \oplus \cdots \oplus \frac{T_{ij}^{(q)}}{\sum_{k=1}^{q} T_{ij}^{(k)}} \tilde{r}_{ij}^{(q)} \\ &= \left[\mathop{\oplus}_{k=1}^{q} \frac{T_{ij}^{(k)} s_{\alpha_{ij}}^{(k)}}{\sum_{k=1}^{q} T_{ij}^{(k)}}, \mathop{\oplus}_{k=1}^{q} \frac{T_{ij}^{(k)} s_{\beta_{ij}}^{(k)}}{\sum_{k=1}^{q} T_{ij}^{(k)}}, \mathop{\oplus}_{k=1}^{q} \frac{T_{ij}^{(k)} s_{\gamma_{ij}}^{(k)}}{\sum_{k=1}^{q} T_{ij}^{(k)}}, \mathop{\oplus}_{k=1}^{q} \frac{T_{ij}^{(k)} s_{\eta_{ij}}^{(k)}}{\sum_{k=1}^{q} T_{ij}^{(k)}}\right], \end{aligned} \tag{32}$$

or the TFLPWG operator

$$\tilde{r}_{ij} = \left(\tilde{s}_{\alpha_{ij}}, \tilde{s}_{\beta_{ij}}, \tilde{s}_{\gamma_{ij}}, \tilde{s}_{\eta_{ij}}\right) = \text{TFLPWG}\left(\tilde{r}_{ij}^{(1)}, \tilde{r}_{ij}^{(2)}, \ldots, \tilde{r}_{ij}^{(q)}\right)$$
$$= \left(\tilde{r}_{ij}^{(1)}\right)^{\frac{T_{ij}^{(1)}}{\sum_{k=1}^{q} T_{ij}^{(k)}}} \otimes \left(\tilde{r}_{ij}^{(2)}\right)^{\frac{T_{ij}^{(2)}}{\sum_{k=1}^{q} T_{ij}^{(k)}}} \otimes \cdots \otimes \left(\tilde{r}_{ij}^{(q)}\right)^{\frac{T_{ij}^{(q)}}{\sum_{k=1}^{q} T_{ij}^{(k)}}}$$
$$= \left[\bigotimes_{k=1}^{q} \left(s_{\alpha_{ij}}^{(k)}\right)^{\frac{T_{ij}^{(k)}}{\sum_{k=1}^{q} T_{ij}^{(k)}}}, \bigotimes_{k=1}^{q} \left(s_{\beta_{ij}}^{(k)}\right)^{\frac{T_{ij}^{(k)}}{\sum_{k=1}^{q} T_{ij}^{(k)}}}, \bigotimes_{k=1}^{q} \left(s_{\gamma_{ij}}^{(k)}\right)^{\frac{T_{ij}^{(k)}}{\sum_{k=1}^{q} T_{ij}^{(k)}}}, \bigotimes_{k=1}^{q} \left(s_{\eta_{ij}}^{(k)}\right)^{\frac{T_{ij}^{(k)}}{\sum_{k=1}^{q} T_{ij}^{(k)}}}\right], \tag{33}$$

or the TFLPWHA operator

$$\tilde{r}_{ij} = \left(\tilde{s}_{\alpha_{ij}}, \tilde{s}_{\beta_{ij}}, \tilde{s}_{\gamma_{ij}}, \tilde{s}_{\eta_{ij}}\right) = \text{TFLPWHA}\left(\tilde{r}_{ij}^{(1)}, \tilde{r}_{ij}^{(2)}, \ldots, \tilde{r}_{ij}^{(q)}\right)$$
$$= \left(\frac{\frac{T_{ij}^{(1)}}{\sum_{k=1}^{q} T_{ij}^{(k)}}}{\tilde{r}_{ij}^{(1)}}\right)^{-1} \oplus \left(\frac{\frac{T_{ij}^{(2)}}{\sum_{k=1}^{q} T_{ij}^{(k)}}}{\tilde{r}_{ij}^{(2)}}\right)^{-1} \oplus \cdots \oplus \left(\frac{\frac{T_{ij}^{(q)}}{\sum_{k=1}^{q} T_{ij}^{(k)}}}{\tilde{r}_{ij}^{(q)}}\right)^{-1}$$
$$= \left[\left(\bigoplus_{k=1}^{q} \frac{\frac{T_{ij}^{(k)}}{\sum_{k=1}^{q} T_{ij}^{(k)}}}{s_{\alpha_{ij}}^{(k)}}\right)^{-1}, \left(\bigoplus_{k=1}^{q} \frac{\frac{T_{ij}^{(k)}}{\sum_{k=1}^{q} T_{ij}^{(k)}}}{s_{\beta_{ij}}^{(k)}}\right)^{-1}, \left(\bigoplus_{k=1}^{q} \frac{\frac{T_{ij}^{(k)}}{\sum_{k=1}^{q} T_{ij}^{(k)}}}{s_{\gamma_{ij}}^{(k)}}\right)^{-1}, \left(\bigoplus_{k=1}^{q} \frac{\frac{T_{ij}^{(k)}}{\sum_{k=1}^{q} T_{ij}^{(k)}}}{s_{\eta_{ij}}^{(k)}}\right)^{-1}\right], \tag{34}$$

to aggregate all the individual trapezoid fuzzy linguistic decision matrices $\tilde{R}^{(k)} = \left(\tilde{r}_{ij}^{(k)}\right)_{m\times n} = \left(\left[s_{\alpha_{ij}}^{(k)}, s_{\beta_{ij}}^{(k)}, s_{\gamma_{ij}}^{(k)}, s_{\eta_{ij}}^{(k)}\right]\right)_{m\times n}$ $(k = 1, 2,\ldots, q)$ into the collective trapezoid fuzzy linguistic decision matrix $\tilde{R}^* = \left(\tilde{r}_{ij}\right)_{m\times n} = \left(\left[s_{\alpha_{ij}}, s_{\beta_{ij}}, s_{\gamma_{ij}}, s_{\eta_{ij}}\right]\right)_{m\times n}$, $i = 1, 2,\ldots, m$; $j = 1, 2,\ldots, n$.

Step 3. Calculate the values T_{ij}, $i = 1, 2,\ldots, m$, $j = 1, 2,\ldots, n$, as follows:

$$T_{ij} = \prod_{v=1}^{j-1} \frac{I(E(\tilde{r}_{1v}))}{t}, \quad i = 1, 2, \ldots, m; \quad j = 2, 3, \ldots, n, \tag{35}$$

$$T_{i1} = 1, \quad i = 1, 2, \ldots, m. \tag{36}$$

Step 4. Aggregate all trapezoid fuzzy linguistic variables $\tilde{r}_{ij}, j = 1, 2, \ldots, n$, for each option X_i, $i = 1, 2,\ldots, m$, by the TFLPWA operator:

$$\tilde{r}_i = \left(\tilde{s}_{\alpha_i}, \tilde{s}_{\beta_i}, \tilde{s}_{\gamma_i}, \tilde{s}_{\eta_i}\right) = \text{TFLPWA}(\tilde{r}_{i1}, \tilde{r}_{i2}, \ldots, \tilde{r}_{in})$$
$$= \frac{T_{i1}}{\sum_{j=1}^{n} T_{ij}} \tilde{r}_{i1} \oplus \frac{T_{i2}}{\sum_{j=1}^{n} T_{ij}} \tilde{r}_{i2} \oplus \cdots \oplus \frac{T_{in}}{\sum_{j=1}^{n} T_{ij}} \tilde{r}_{in} \tag{37}$$
$$= \left[\bigoplus_{j=1}^{n} \frac{T_{ij}\left(s_{\alpha_{ij}}\right)}{\sum_{j=1}^{n} T_{ij}}, \bigoplus_{j=1}^{n} \frac{T_{ij}\left(s_{\beta_{ij}}\right)}{\sum_{j=1}^{n} T_{ij}}, \bigoplus_{j=1}^{n} \frac{T_{ij}\left(s_{\gamma_{ij}}\right)}{\sum_{j=1}^{n} T_{ij}}, \bigoplus_{j=1}^{n} \frac{T_{ij}\left(s_{\eta_{ij}}\right)}{\sum_{j=1}^{n} T_{ij}}\right], \quad i = 1, 2, \ldots, m$$

or the TFLPWG operator:

$$\tilde{r}_i = \left(\tilde{s}_{\alpha_i}, \tilde{s}_{\beta_i}, \tilde{s}_{\gamma_i}, \tilde{s}_{\eta_i}\right) = \text{TFLPWG}(\tilde{r}_{i1}, \tilde{r}_{i2}, ..., \tilde{r}_{in})$$
$$= (\tilde{r}_{i1})^{\frac{T_{i1}}{\sum_{j=1}^{n} T_{ij}}} \otimes (\tilde{r}_{i2})^{\frac{T_{i2}}{\sum_{j=1}^{n} T_{ij}}} \otimes \cdots \otimes (\tilde{r}_{in})^{\frac{T_{in}}{\sum_{j=1}^{n} T_{ij}}}$$
$$= \left[\underset{j=1}{\overset{n}{\otimes}} \left(s_{\alpha_{ij}}\right)^{\frac{T_{ij}}{\sum_{j=1}^{n} T_{ij}}}, \underset{j=1}{\overset{n}{\otimes}} \left(s_{\beta_{ij}}\right)^{\frac{T_{ij}}{\sum_{j=1}^{n} T_{ij}}}, \underset{j=1}{\overset{n}{\otimes}} \left(s_{\gamma_{ij}}\right)^{\frac{T_{ij}}{\sum_{j=1}^{n} T_{ij}}}, \underset{j=1}{\overset{n}{\otimes}} \left(s_{\eta_{ij}}\right)^{\frac{T_{ij}}{\sum_{j=1}^{n} T_{ij}}} \right], \quad i = 1, 2, ..., m \tag{38}$$

or the TFLPWHA operator:

$$\tilde{r}_i = \left(\tilde{s}_{\alpha_i}, \tilde{s}_{\beta_i}, \tilde{s}_{\gamma_i}, \tilde{s}_{\eta_i}\right) = \text{TFLPWHA}(\tilde{r}_{i1}, \tilde{r}_{i2}, ..., \tilde{r}_{in})$$
$$= \left(\frac{\frac{T_{i1}}{\sum_{j=1}^{n} T_{ij}}}{\tilde{r}_{i1}} \right)^{-1} \oplus \left(\frac{\frac{T_{i2}}{\sum_{j=1}^{n} T_{ij}}}{\tilde{r}_{i2}} \right)^{-1} \oplus \cdots \oplus \left(\frac{\frac{T_{in}}{\sum_{j=1}^{n} T_{ij}}}{\tilde{r}_{in}} \right)^{-1}$$
$$= \left[\left(\underset{j=1}{\overset{n}{\oplus}} \frac{\frac{T_{ij}}{\sum_{j=1}^{n} T_{ij}}}{\left(s_{\alpha_{ij}}\right)} \right)^{-1}, \left(\underset{j=1}{\overset{n}{\oplus}} \frac{\frac{T_{ij}}{\sum_{j=1}^{n} T_{ij}}}{\left(s_{\beta_{ij}}\right)} \right)^{-1}, \left(\underset{j=1}{\overset{n}{\oplus}} \frac{\frac{T_{ij}}{\sum_{j=1}^{n} T_{ij}}}{\left(s_{\gamma_{ij}}\right)} \right)^{-1}, \left(\underset{j=1}{\overset{n}{\oplus}} \frac{\frac{T_{ij}}{\sum_{j=1}^{n} T_{ij}}}{\left(s_{\eta_{ij}}\right)} \right)^{-1} \right], \quad i = 1, 2, ..., m \tag{39}$$

to derive the overall trapezoid fuzzy linguistic variables $\tilde{r}_i$, $i = 1, 2,..., m$, of the options X_i, $i = 1, 2,..., m$.

Step 5. Compare each $\tilde{r}_i$ with all $\tilde{r}_k$, $i, k = 1, 2,..., m$, by (3). For simplicity, we let $p_{ik} = p(r_i \geq r_k)$, and then construct the possibility matrix $P = (p_{ik})_{m \times m}$, where $p_{ik} \geq 0$, $p_{ik} + p_{ki} = 1$, $p_{ii} = 0.5$ $\forall i, k = 1, 2,..., m$. Summing all the elements in each row of matrix P, get

$$p_i = \sum_{k=1}^{m} p_{ik}, \quad i = 1, 2, ..., m \tag{40}$$

Then, arrange the collective overall preference values $\tilde{r}_i$, $i = 1, 2,..., m$, in descending order in accordance with the values of p_i, $i = 1, 2,..., m$.

Step 6. Rank all the options X_i, $i = 1, 2,..., m$, by the ranking of r_i, $i = 1, 2,..., m$, and select the best one(s).

Step 7. End.

Numerical example

In order to demonstrate the applicability of the proposed method to multiple attribute group decision-making, we consider below a university faculty recruitment group decision-making problem.

Example: The department of mathematics in a university wants to appoint outstanding mathematics teachers. The appointment is done by a committee of three decision makers, President D_1, Dean of Academics D_2, and Human Resource Officer D_3. After preliminary screening, five teachers X_i, $i = 1, 2, 3, 4, 5$, remain for further evaluation. Panel of decision makers made strict evaluation for five teachers X_i, $i = 1, 2, 3, 4, 5$, according to the following four attributes: (1) G_1, the past experience, (2) G_2, the research capability, (3) G_3, subject knowledge, (4) G_4, the teaching skill. During this process, the university President has the absolute priority for decision-making, Dean of Academics comes next. The prioritization relationship for the attributes is as follows: $G_1 \succ G_2 \succ G_3 \succ G_4$. The three decision makers evaluated the teachers X_i, $i = 1, 2, 3, 4, 5$, with respect to the attributes G_j, $j = 1, 2, 3, 4$ by using the following linguistic scale:

$$S = \{ s_1 = \text{extremely poor}, s_2 = \text{very poor}, s_3 = \text{poor}, s_4 = \text{slightly poor}, s_5 = \text{fair} \\ s_6 = \text{slightly good}, s_7 = \text{good}, s_8 = \text{very good}, s_9 = \text{extremely good}\},$$

and provided their evaluation values in terms of trapezoid fuzzy linguistic variables and constructed the following three trapezoid fuzzy linguistic decision matrices $\tilde{R}^{(q)} = \left(\tilde{r}_{ij}{}^{(q)}\right)_{5\times 4}$, $q = 1, 2, 3$ (see Tables 1, 2, and 3).

Step 1: Utilize the expressions in (30) and (31) to calculate the $T_{ij}^{(1)}$, $T_{ij}^{(2)}$ and $T_{ij}^{(3)}$,

$$\left[T_{ij}^{(1)}\right] = \begin{bmatrix} 1 & 1 & 1 & 1 \\ 1 & 1 & 1 & 1 \\ 1 & 1 & 1 & 1 \\ 1 & 1 & 1 & 1 \\ 1 & 1 & 1 & 1 \end{bmatrix}, \quad \left[T_{ij}^{(2)}\right] = \begin{bmatrix} 0.444 & 0.722 & 0.750 & 0.528 \\ 0.583 & 0.722 & 0.722 & 0.667 \\ 0.750 & 0.611 & 0.833 & 0.500 \\ 0.750 & 0.778 & 0.583 & 0.833 \\ 0.500 & 0.361 & 0.472 & 0.694 \end{bmatrix}$$

$$\left[T_{ij}^{(3)}\right] = \begin{bmatrix} 0.259 & 0.281 & 0.375 & 0.323 \\ 0.389 & 0.201 & 0.441 & 0.333 \\ 0.313 & 0.306 & 0.301 & 0.306 \\ 0.625 & 0.389 & 0.227 & 0.370 \\ 0.222 & 0.151 & 0.341 & 0.270 \end{bmatrix}.$$

Step 2: Utilize the TFLPWA operator (Equation 32) to aggregate all the individual decision matrices $R^{(q)}$, $q = 1, 2, 3$, into the collective decision matrix $R^* = (\tilde{r}_{ij})_{5\times 4} = \left(\left[s_{\alpha_{ij}}, s_{\beta_{ij}}, s_{\gamma_{ij}}, s_{\eta_{ij}}\right]\right)_{5\times 4}$ and we get the following table (Table 4):

Step 3: Using expressions in (35) and (36) to calculate the T_{ij}, $i = 1, 2, \ldots, 5$; $j = 1, 2, \ldots, 4$, we get

$$\left[T_{ij}\right] = \begin{bmatrix} 1 & 0.5058 & 0.2732 & 0.1633 \\ 1 & 0.5697 & 0.3048 & 0.2071 \\ 1 & 0.6417 & 0.3693 & 0.2192 \\ 1 & 0.6886 & 0.4254 & 0.2231 \\ 1 & 0.4803 & 0.1958 & 0.1062 \end{bmatrix}.$$

Step 4: Aggregating all trapezoid fuzzy linguistic variables $\tilde{r}_{ij}, j = 1, 2, 3, 4$, by using the TFLPWA operator (Equation 37) to derive the overall trapezoid fuzzy linguistic variables $\tilde{r}_i, i = 1, 2, \ldots, 5$ of the teachers X_i, $i = 1, 2, \ldots, 5$:

Table 1 Trapezoid fuzzy linguistic decision matrix $\tilde{R}^{(1)}$

	G_1	G_2	G_3	G_4
X_1	$[s_2, s_3, s_5, s_6]$	$[s_4, s_5, s_8, s_9]$	$[s_5, s_6, s_7, s_9]$	$[s_3, s_4, s_5, s_7]$
X_2	$[s_3, s_5, s_6, s_7]$	$[s_5, s_6, s_7, s_8]$	$[s_4, s_5, s_8, s_9]$	$[s_4, s_5, s_7, s_8]$
X_3	$[s_4, s_6, s_8, s_9]$	$[s_4, s_5, s_6, s_7]$	$[s_6, s_7, s_8, s_9]$	$[s_3, s_4, s_5, s_6]$
X_4	$[s_5, s_6, s_7, s_9]$	$[s_4, s_7, s_8, s_9]$	$[s_3, s_5, s_6, s_7]$	$[s_6, s_7, s_8, s_9]$
X_5	$[s_3, s_4, s_5, s_6]$	$[s_1, s_2, s_4, s_6]$	$[s_2, s_4, s_5, s_6]$	$[s_4, s_5, s_7, s_9]$

Table 2 Trapezoid fuzzy linguistic decision matrix $\tilde{R}^{(2)}$

	G_1	G_2	G_3	G_4
X_1	$[s_3, s_5, s_6, s_7]$	$[s_2, s_3, s_4, s_5]$	$[s_3, s_4, s_5, s_6]$	$[s_4, s_5, s_6, s_7]$
X_2	$[s_4, s_5, s_7, s_8]$	$[s_1, s_2, s_3, s_4]$	$[s_4, s_5, s_6, s_7]$	$[s_3, s_4, s_5, s_6]$
X_3	$[s_2, s_3, s_4, s_6]$	$[s_2, s_4, s_5, s_7]$	$[s_1, s_3, s_4, s_5]$	$[s_4, s_5, s_6, s_7]$
X_4	$[s_6, s_7, s_8, s_9]$	$[s_3, s_4, s_5, s_6]$	$[s_2, s_3, s_4, s_5]$	$[s_2, s_3, s_5, s_6]$
X_5	$[s_1, s_3, s_5, s_7]$	$[s_2, s_3, s_4, s_6]$	$[s_5, s_6, s_7, s_8]$	$[s_2, s_3, s_4, s_5]$

$$\begin{aligned}\tilde{r}_1 &= [s_{2.87}, s_{4.01}, s_{5.57}, s_{6.69}],\\ \tilde{r}_2 &= [s_{3.29}, s_{4.58}, s_{5.92}, s_{6.95}],\\ \tilde{r}_3 &= [s_{3.45}, s_{4.83}, s_{6.07}, s_{7.32}],\\ \tilde{r}_4 &= [s_{3.66}, s_{5.15}, s_{6.35}, s_{7.53}],\\ \tilde{r}_5 &= [s_{2.25}, s_{3.55}, s_{4.92}, s_{6.35}].\end{aligned}$$

Step 5: Comparing each r_i with all r_k, $i, k = 1, 2,\ldots, 5$, by Equation (3) and let $p_{ik} = p(r_i \geq r_k)$, and then constructing the possibility matrix, we get

$$P = \begin{bmatrix} 0.5000 & 0.4229 & 0.3794 & 0.3301 & 0.5954 \\ 0.5771 & 0.5000 & 0.4540 & 0.4032 & 0.6753 \\ 0.6206 & 0.5460 & 0.5000 & 0.4499 & 0.7174 \\ 0.6699 & 0.5968 & 0.5501 & 0.5000 & 0.7666 \\ 0.4046 & 0.3247 & 0.2826 & 0.2334 & 0.5000 \end{bmatrix}.$$

Now, summing all the elements in each row of matrix P, we have

$$p_1 = 2.2278, p_2 = 2.6096, p_3 = 2.8339, p_4 = 3.0834, p_5 = 1.7453,$$

then

$$r_4 \succ r_3 \succ r_2 \succ r_1 \succ r_5.$$

Ranking all the teachers X_i, $i = 1, 2, 3, 4, 5$, in accordance with the values of r_i, $i = 1, 2, 3, 4, 5$, we have

$$X_4 \succ X_3 \succ X_2 \succ X_1 \succ X_5.$$

Thus X_4 is most desirable alternative.

Based on the TFLPWG operator, the decision steps are as follows:

Step 1': See step 1.

Step 2': Utilize the TFLPWG operator (Equation 33) to aggregate all the individual decision matrices $R^{(q)}$, $q = 1, 2, 3$, into the collective decision matrix $R^* = \left(\tilde{r}_{ij}\right)_{5\times4} = \left(\left[s_{\alpha_{ij}}, s_{\beta_{ij}}, s_{\gamma_{ij}}, s_{\eta_{ij}}\right]\right)_{5\times4}$ and we get the following table (Table 5):

Table 3 Trapezoid fuzzy linguistic decision matrix $\tilde{R}^{(3)}$

	G_1	G_2	G_3	G_4
X_1	$[s_4, s_5, s_6, s_7]$	$[s_1, s_2, s_3, s_4]$	$[s_2, s_3, s_4, s_5]$	$[s_3, s_4, s_5, s_7]$
X_2	$[s_2, s_3, s_4, s_5]$	$[s_3, s_4, s_5, s_7]$	$[s_4, s_6, s_7, s_8]$	$[s_1, s_3, s_5, s_6]$
X_3	$[s_6, s_7, s_8, s_9]$	$[s_4, s_5, s_6, s_7]$	$[s_2, s_3, s_5, s_6]$	$[s_2, s_3, s_4, s_5]$
X_4	$[s_1, s_3, s_5, s_6]$	$[s_2, s_3, s_5, s_6]$	$[s_4, s_5, s_6, s_7]$	$[s_2, s_3, s_4, s_5]$
X_5	$[s_2, s_4, s_5, s_6]$	$[s_4, s_6, s_7, s_8]$	$[s_3, s_4, s_5, s_6]$	$[s_4, s_5, s_6, s_7]$

Table 4 Collective trapezoid fuzzy linguistic decision matrix R^* (for TFLPWA operator)

	G_1	G_2	G_3	G_4
X_1	$[s_{2.56}, s_{3.83}, s_{5.41}, s_{6.41}]$	$[s_{2.86}, s_{3.86}, s_{5.86}, s_{6.86}]$	$[s_{3.76}, s_{4.76}, s_{5.76}, s_{7.24}]$	$[s_{3.29}, s_{4.29}, s_{5.29}, s_{7.00}]$
X_2	$[s_{3.10}, s_{4.61}, s_{5.90}, s_{6.90}]$	$[s_{3.29}, s_{4.29}, s_{5.29}, s_{6.39}]$	$[s_{4.00}, s_{5.20}, s_{7.13}, s_{8.13}]$	$[s_{3.17}, s_{4.33}, s_{6.00}, s_{7.00}]$
X_3	$[s_{3.58}, s_{5.06}, s_{6.55}, s_{7.91}]$	$[s_{3.36}, s_{4.68}, s_{5.68}, s_{7.00}]$	$[s_{3.48}, s_{4.87}, s_{6.01}, s_{7.01}]$	$[s_{3.11}, s_{4.11}, s_{5.11}, s_{6.11}]$
X_4	$[s_{4.26}, s_{5.53}, s_{6.79}, s_{8.21}]$	$[s_{3.28}, s_{5.20}, s_{6.38}, s_{7.38}]$	$[s_{2.80}, s_{4.36}, s_{5.36}, s_{6.36}]$	$[s_{3.82}, s_{4.82}, s_{6.19}, s_{7.19}]$
X_5	$[s_{2.29}, s_{3.71}, s_{5.00}, s_{6.29}]$	$[s_{1.54}, s_{2.64}, s_{4.30}, s_{6.20}]$	$[s_{2.97}, s_{4.52}, s_{5.52}, s_{6.52}]$	$[s_{3.73}, s_{4.29}, s_{5.80}, s_{7.31}]$

Step 3': Utilize expressions in (35) and (36) to calculate the T_{ij}, $i = 1, 2,..., 5; j = 1, 2,..., 4$, we get

$$[T_{ij}] = \begin{bmatrix} 1 & 0.4986 & 0.2515 & 0.1454 \\ 1 & 0.5597 & 0.2705 & 0.1828 \\ 1 & 0.6131 & 0.3496 & 0.1892 \\ 1 & 0.6525 & 0.3906 & 0.2009 \\ 1 & 0.4728 & 0.1867 & 0.0992 \end{bmatrix}.$$

Step 4': Aggregating all trapezoid fuzzy linguistic variables $\tilde{r}_{ij}, j = 1, 2, 3, 4$, by using the TFLPWG operator (Equation 38) to derive the overall trapezoid fuzzy linguistic variables $\tilde{r}_i, i = 1, 2, ..., 5$ of the teachers X_i, $i = 1, 2,..., 5$:

$$\tilde{r}_1 = [s_{2.68}, s_{3.84}, s_{5.42}, s_{6.55}], \tilde{r}_2 = [s_{2.99}, s_{4.37}, s_{5.70}, s_{6.76}], \tilde{r}_3 = [s_{3.12}, s_{4.62}, s_{5.86}, s_{7.22}],$$
$$\tilde{r}_4 = [s_{3.23}, s_{4.89}, s_{6.21}, s_{7.40}]\tilde{r}_5 = [s_{1.97}, s_{3.39}, s_{4.86}, s_{6.31}].$$

Step 5': Comparing each r_i with all r_k, $i, k = 1, 2,...,5$, by Equation (3) and let $p_{ik} = p(r_i \geq r_k)$, and then constructing the possibility matrix, we get

$$P = \begin{bmatrix} 0.5000 & 0.4370 & 0.3920 & 0.3519 & 0.5870 \\ 0.5630 & 0.5000 & 0.4521 & 0.4098 & 0.6508 \\ 0.6080 & 0.5479 & 0.5000 & 0.4580 & 0.6924 \\ 0.6481 & 0.5902 & 0.5420 & 0.5000 & 0.7301 \\ 0.4130 & 0.3492 & 0.3076 & 0.2699 & 0.5000 \end{bmatrix}.$$

Now, summing all the elements in each row of matrix P, we have

$$p_1 = 2.2679, p_2 = 2.5757, p_3 = 2.8063, p_4 = 3.0122, p_5 = 1.8397,$$

then

$$r_4 \succ r_3 \succ r_2 \succ r_1 \succ r_5.$$

Ranking all the teachers X_i, $i = 1, 2, 3, 4, 5$, in accordance with the values of r_i, $i = 1, 2, 3, 4, 5$, we have

$$X_4 \succ X_3 \succ X_2 \succ X_1 \succ X_5.$$

Hence, X_4 is most desirable alternative.

Table 5 Collective trapezoid fuzzy linguistic decision matrix R^* (for TFLPWG operator)

	G_1	G_2	G_3	G_4
X_1	$[s_{2.47}, s_{3.70}, s_{5.39}, s_{6.39}]$	$[s_{2.57}, s_{3.66}, s_{5.43}, s_{6.50}]$	$[s_{3.55}, s_{4.60}, s_{5.63}, s_{7.03}]$	$[s_{3.26}, s_{4.26}, s_{5.27}, s_{7.00}]$
X_2	$[s_{3.02}, s_{4.52}, s_{5.80}, s_{6.81}]$	$[s_{2.59}, s_{3.81}, s_{4.92}, s_{6.08}]$	$[s_{4.00}, s_{5.18}, s_{7.07}, s_{8.08}]$	$[s_{2.89}, s_{4.26}, s_{5.92}, s_{6.93}]$
X_3	$[s_{3.31}, s_{4.77}, s_{6.22}, s_{7.77}]$	$[s_{3.21}, s_{4.66}, s_{5.66}, s_{7.00}]$	$[s_{2.55}, s_{4.46}, s_{5.71}, s_{6.76}]$	$[s_{3.03}, s_{4.05}, s_{5.06}, s_{6.07}]$
X_4	$[s_{3.47}, s_{5.25}, s_{6.68}, s_{8.09}]$	$[s_{3.19}, s_{4.92}, s_{6.21}, s_{7.23}]$	$[s_{2.73}, s_{4.24}, s_{5.27}, s_{6.28}]$	$[s_{3.29}, s_{4.41}, s_{5.96}, s_{7.00}]$
X_5	$[s_{2.07}, s_{3.68}, s_{5.00}, s_{6.27}]$	$[s_{1.36}, s_{2.46}, s_{4.23}, s_{6.17}]$	$[s_{2.74}, s_{4.45}, s_{5.46}, s_{6.47}]$	$[s_{3.64}, s_{4.17}, s_{5.62}, s_{7.06}]$

Based on the TFLPWHA operator, the decision steps are as follows:

Step 1": See step 1.

Step 2": Utilize the TFLPWHA operator (Equation 34) to aggregate all the individual decision matrices $R^{(q)}$, $q = 1,2,3$, into the collective decision matrix $R^* = (\tilde{r}_{ij})_{5\times4} = \left(\left[s_{\alpha_{ij}}, s_{\beta_{ij}}, s_{\gamma_{ij}}, s_{\eta_{ij}}\right]\right)_{5\times4}$ and we get the following table (Table 6):

Step 3": Utilize expressions in (35) and (36) to calculate the T_{ij}, $i = 1, 2,\ldots,5; j = 1, 2,\ldots,4$, it gives

$$[T_{ij}] = \begin{bmatrix} 1 & 0.4925 & 0.2308 & 0.1289 \\ 1 & 0.5486 & 0.2371 & 0.1603 \\ 1 & 0.5836 & 0.3291 & 0.1635 \\ 1 & 0.6097 & 0.3533 & 0.1781 \\ 1 & 0.4647 & 0.1796 & 0.0936 \end{bmatrix}.$$

Step 4": Aggregating all trapezoid fuzzy linguistic variables $\tilde{r}_{ij}, j = 1, 2, \ldots, n$ by using the TFLPWHA operator (Equation 39) to derive the overall trapezoid fuzzy linguistic variables $\tilde{r}_i, i = 1, 2, \ldots, m$ of the teachers X_i:

$$\tilde{r}_1 = [s_{2.48}, s_{3.68}, s_{5.28}, s_{6.41}], \tilde{r}_2 = [s_{2.68}, s_{4.11}, s_{5.45}, s_{6.56}], \tilde{r}_3 = [s_{2.77}, s_{4.42}, s_{5.66}, s_{7.11}]$$
$$\tilde{r}_4 = [s_{2.75}, s_{4.69}, s_{6.12}, s_{7.29}], \tilde{r}_5 = [s_{1.70}, s_{3.25}, s_{4.82}, s_{6.29}].$$

Step 5": Comparing each r_i with all r_k, $i, k = 1, 2,\ldots,5$, by Equation (3) and let $p_{ik} = p(r_i \geq r_k)$, and then constructing the possibility matrix

$$P = \begin{bmatrix} 0.5000 & 0.4558 & 0.4050 & 0.3696 & 0.5766 \\ 0.5442 & 0.5000 & 0.4463 & 0.4084 & 0.6204 \\ 0.5950 & 0.5537 & 0.5000 & 0.4615 & 0.6661 \\ 0.6304 & 0.5916 & 0.5385 & 0.5000 & 0.6974 \\ 0.4234 & 0.3796 & 0.3339 & 0.3026 & 0.5000 \end{bmatrix}.$$

Now, summing all the elements in each row of matrix P, we have

$$p_1 = 2.3070, p_2 = 2.5193, p_3 = 2.77623, p_4 = 2.9579, p_5 = 1.9395.$$

Ranking all the teachers X_i, $i = 1, 2, 3, 4, 5$, in accordance with the values of p_i, $i = 1, 2, 3, 4, 5$, we have

$$X_4 \succ X_3 \succ X_2 \succ X_1 \succ X_5,$$

then

$$r_4 \succ r_3 \succ r_2 \succ r_1 \succ r_5.$$

Thus, X_4 is still most desirable alternative.

Table 6 Collective trapezoid fuzzy linguistic decision matrix R^* (for TFLPWHA operator)

	G_1	G_2	G_3	G_4
X_1	$[s_{2.39}, s_{3.59}, s_{5.37}, s_{6.38}]$	$[s_{2.25}, s_{3.45}, s_{5.02}, s_{6.15}]$	$[s_{3.33}, s_{4.44}, s_{5.50}, s_{6.83}]$	$[s_{3.29}, s_{4.29}, s_{5.29}, s_{7.00}]$
X_2	$[s_{2.93}, s_{4.42}, s_{5.68}, s_{6.72}]$	$[s_{1.94}, s_{3.33}, s_{4.54}, s_{5.75}]$	$[s_{4.00}, s_{5.18}, s_{7.12}, s_{8.03}]$	$[s_{3.17}, s_{4.33}, s_{6.00}, s_{7.00}]$
X_3	$[s_{3.05}, s_{4.47}, s_{5.87}, s_{7.62}]$	$[s_{3.03}, s_{4.63}, s_{5.64}, s_{7.00}]$	$[s_{1.86}, s_{4.10}, s_{5.42}, s_{6.51}]$	$[s_{3.11}, s_{4.11}, s_{5.11}, s_{6.11}]$
X_4	$[s_{2.50}, s_{4.93}, s_{6.57}, s_{7.95}]$	$[s_{3.08}, s_{4.64}, s_{6.05}, s_{7.09}]$	$[s_{2.66}, s_{4.12}, s_{5.17}, s_{6.20}]$	$[s_{3.82}, s_{4.82}, s_{6.19}, s_{7.19}]$
X_5	$[s_{1.82}, s_{3.65}, s_{5.00}, s_{6.26}]$	$[s_{1.24}, s_{2.34}, s_{4.18}, s_{6.15}]$	$[s_{2.56}, s_{4.38}, s_{5.40}, s_{6.42}]$	$[s_{3.73}, s_{4.29}, s_{5.80}, s_{7.31}]$

Conclusions

In this paper, we explored multiple attribute group decision-making problems in which the attribute and decision makers are at different priority levels, and the decision information provided by decision makers takes the form of trapezoid linguistic variables. Motivated by the idea of prioritized weighted aggregation operators [42,43], we have developed some prioritized weighted aggregation operators for aggregating trapezoid fuzzy linguistic information: the TFLPWA operator, the TFLPWG operator and the TFLPWHA operator. A number of properties of the proposed operators have been proved. Then, we have developed an algorithm to solve the trapezoid fuzzy linguistic multiple attribute decision-making problems in which the attributes and decision makers are in different priority levels. Finally, a numerical example is given to verify the developed approaches and to demonstrate their practicality and effectiveness.

References

1. Beliakov, G, Pradera, A, Calvo, T: Aggregation functions: a guide for practitioners. Springer, Berlin (2007)
2. Calvo, T, Mayor, G, Mesiar, R: Aggregation operators: new trends and applications. Physica-Verlag, Berlin (2002)
3. Xu, ZS: Da, QL: An overview of operators for aggregating information. Int. J. Intell. Syst. **18**(9), 953–969 (2003)
4. Yager, RR, Kacprzyk, J: The ordered weighted averaging operator: theory and applications. Kluwer, Boston (1997)
5. Yager, RR: On ordered weighted averaging aggregation operators in multicriteria decision making. IEEE Trans. Syst. Man Cybern. B Cybern. **18**(1), 183–190 (1988)
6. Bordonga, G, Fedrizzi, M, Passi, G: A linguistic modelling of consensus in group decision making based on OWA operators. IEEE Trans. Syst. Man Cybern. Syst. Hum **27**(1), 126–132 (1997)
7. Chen, H, Liu, C, Sheng, Z: Induced ordered weighted harmonic averaging (IOWHA) operator and its application to combination forecasting method. Chin. J. Manage. Sci. **12**, 35–40 (2004)
8. Chiclana, F, Herrera, F, Viedma, EH: The ordered weighted geometric operator: properties and applications. In: Proceedings of the 8th International Conference on Information Processing and Management of Uncertainty in Knowledge Based Systems, pp. 985–991. Madrid, Spain (2000)
9. Fodor, J, Marichal, JL, Roubens, M: Characterization of the ordered weighted averaging operators. IEEE Trans. Fuzzy Syst. **3**(2), 236–240 (1995)
10. Merigó, JM: A unified model between the weighted average and the induced OWA operator. Expert Syst. Appl. **8**(9), 11560–11572 (2011)
11. Merigó, JM, Gil-Lafuente, AM: The induced generalized OWA operator. Inf. Sci. **179**(6), 729–741 (2009)
12. Merigó, JM, Casanovas, M: The fuzzy generalized OWA operator and its application in strategic decision making. Cybern. Syst. **41**(5), 359–370 (2010)
13. Merigó, JM, Casanovas, M: Induced and uncertain heavy OWA operators. Comp. Ind. Eng. **60**(1), 106–116 (2011)
14. Merigó, JM, Casanovas, M: The uncertain induced quasi-arithmetic OWA operator. Int. J. Intell. Syst. **26**(1), 1–24 (2011)
15. Merigó, JM, Casanovas, M, Martínez, L: Linguistic aggregation operators for linguistic decision making based on the Dempster-Shafer theory of evidence. International Journal Uncertainty, Fuzziness and Knowledge Based Systems **18**(3), 287–304 (2010)
16. Wang, XR, Fan, ZP: Fuzzy ordered weighted averaging operator and its applications. Fuzzy Syst. Math. **17**(4), 67–72 (2003)
17. Wei, GW: FIOWHM operator and its application to multiple attribute group decision making. Expert Syst. Appl. **38**(4), 2984–2989 (2011)
18. Xu, ZS: Fuzzy ordered weighted geometric operator and its applications in fuzzy AHP. Syst. Eng. Electron. **31**(4), 855–858 (2002)
19. Xu, ZS, QL, D: The ordered weighted geometric averaging operators. Int. J. Intell. Syst. **17**(7), 709–716 (2002)
20. Xu, ZS: A priority method for triangular fuzzy number complementary judgment matrix. Syst. Eng. Theor. Pract. **23**(10), 86–89 (2003)
21. Xu, ZS: EOWA and EOWG operators for aggregating linguistic labels based on linguistic preference relations. Int. J. Uncertainty Fuzziness Knowl. Based Syst. **12**(6), 791–810 (2004)
22. Xu, ZS: An approach based on the uncertain LOWG and induced uncertain LOWG operators to group decision making with uncertain multiplicative linguistic preference relations. Decis. Support. Syst. **41**(2), 488–499 (2006)
23. Xu, ZS, Xia, G, Chen, M, Wang, L: Induced generalized intuitionistic fuzzy OWA operator for multi-attribute group decision making. Expert Syst. Appl. **39**(2), 1902–1910 (2012)
24. Yager, RR: Families of OWA operators. Fuzzy Sets Syst. **59**(2), 125–148 (1993)
25. Yager, RR: Quantifier guided aggregation using OWA operators. Int. J. Intell. Syst. **11**(1), 49–73 (1996)
26. Yager, RR: Generalized OWA aggregation operators. Fuzzy Optim. Decis. Making **3**(1), 93–107 (2004)
27. Yager, RR: Centered OWA operators. Soft. Comput. **11**(7), 631–639 (2007)
28. Yager, RR, Kacprzyk, J, Beliakov, G: Recent developments in the ordered weighted averaging operator: theory and practice. Springer, Heidelberg (2011)
29. Dyckhoff, H, Pedrycz, W: Generalized means as model of compensative connectives. Fuzzy Sets Syst. **14**(2), 143–154 (1984)

30. Herrera, F, Martínez, L: An approach for combining numerical and linguistic information based on the 2-tuple fuzzy linguistic representation model in decision making. Int.J. Uncertainty Fuzziness Knowl. Based Syst. **8**(5), 539–562 (2000)
31. Xu, ZS: Uncertain linguistic aggregation operators based approach to multiple attribute group decision making under uncertain linguistic environment. Inf. Sci. **168**(1–4), 171–184 (2004)
32. Wei, GW: Uncertain linguistic hybrid geometric mean operator and its application to group decision making under uncertain linguistic environment. Int. J. Uncertainty Fuzziness Knowl. Based Syst. **17**(2), 251–267 (2009)
33. Xu, ZS: Group decision making with triangular fuzzy linguistic variables, pp. 17–26. Springer, Berlin, Heidelberg (2007)
34. Xu, ZS: An approach based on similarity measure to multiple attribute decision making with trapezoid fuzzy linguistic variables, pp. 110–117. Springer, Berlin, Heidelberg (2005)
35. Liang, XC, Chen, SF: Multiple attribute decision making method based on trapezoid fuzzy linguistic variables. J. Southeast Univ. (English Edition) **24**(4), 478–481 (2008)
36. Wei, GW, Yi, WD: Trapezoid fuzzy linguistic weighted geometric averaging operator. In: Proceedings of the 4th International Conference on Fuzzy System and Knowledge Discovery, pp. 300–304. Haikou (2007)
37. Liu, P, Su, Y: The multiple attribute decision making method based on the TFLHOWA operator. Comput. Math. Appl. **60**(9), 2609–2615 (2010)
38. Liu, P, Su, Y: Multiple attribute decision making method based on the trapezoid fuzzy linguistic hybrid harmonic averaging operator. Informatica **36**(1), 83–90 (2012)
39. Bonferroni, C: Sulle medie multiple di potenze. Bollettino dell'Unione Matematica Italiana **5**(3), 267–270 (1950)
40. Liu, P, Jin, Y: The trapezoid fuzzy linguistic Bonferroni mean operators and their application to multiple attribute decision making. Scientia Iranica **19**(6), 1947–1959 (2012)
41. Liu, P, Wu, X: Multiple attribute decision making method based on the generalized aggregation operators in trapezoid fuzzy linguistic variables. J. Comput. Anal. Appl. **15**(5), 807–816 (2013)
42. Yager, RR: Prioritized aggregation operators. Int. J. Approx. Reason. **48**(1), 263–274 (2008)
43. Yager, RR: Prioritized OWA aggregation. Fuzzy Optim. Decis. Making **8**(3), 245–262 (2009)

4

Stochastic hierarchical multiobjective routing model in MPLS networks with two service classes: an experimental study on imprecision and uncertainty issues

Rita Girão-Silva[1,2]*, José Craveirinha[1,2] and João Clímaco[2]

*Correspondence: rita@deec.uc.pt
[1] Department of Electrical and Computer Engineering, University of Coimbra, Pólo II, Pinhal de Marrocos, Coimbra 3030-290, Portugal
[2] Institute of Computers and Systems Engineering of Coimbra (INESC-Coimbra), R. Antero de Quental, 199, Coimbra 3000-033, Portugal

Abstract

The paper begins by reviewing a two-level hierarchical multicriteria routing model for Multiprotocol Label Switching networks with two service classes (Quality of Service and Best Effort services) and alternative routing, previously proposed by the authors. The features of the considered resolution heuristic are described. Some key issues raised by its complexity are discussed, as well as the major factors that constitute the sources of imprecision, inaccuracy, and uncertainty of the model and the way in which they are dealt with in the adopted resolution approach. Analytic and stochastic discrete-event simulation experiments are performed for different test networks, including experiments with a dynamic version of the routing method. This case study allows for the evaluation of the inaccuracies intrinsic to the analytic/numerical resolution procedures and of the uncertainty associated with the estimates of the mean of the stochastic traffic flows. An analysis focused on key robustness aspects of the model is also carried out.

Keywords: Routing models; Multiobjective optimization; Telecommunication networks; Simulation; Sources of uncertainty

AMS Subject Classification: Primary 90B50; secondary 90B18; 90B15

Introduction and motivation

In modern multiservice networks, multiple and heterogeneous Quality of Service (QoS) routing requirements have to be taken into account. Therefore, the routing models, which are designed to calculate and select sequences of network resources, have to satisfy certain QoS constraints while seeking to optimize route-related objectives. Formulating routing problems in these types of networks as multiple objective optimization problems is potentially advantageous, as the trade-offs among distinct performance metrics and other network cost function(s) can be pursued in a consistent manner when these multiple objective formulations are used.

Reviews on multicriteria routing models for communication networks are presented in [1] (in a broader context of multicriteria analysis applications) and in [2], which also discusses some key methodological issues and includes multiple research papers in this area. More recently, a complete conceptual framework in multiple criteria routing models in QoS/IP networks, considering a reference point-based approach is proposed in [3].

A meta-model for hierarchical multiobjective network-wide routing in Multiprotocol Label Switching (MPLS) networks is presented in [4]. In this approach, two classes of services, QoS and BE (Best Effort) type flows, and different types of traffic flows in each class are considered. A hierarchical optimization with two optimization levels, including fairness objectives, is considered: the first-priority objective functions (o.f.s) refer to the network-level objectives of QoS flows, namely the total expected revenue and the maximal value of the mean blocking of all types of QoS flows; the second-priority o.f.s refer to performance metrics for the different types of QoS services and the total expected revenue associated with the BE traffic flows.

In [5], a heuristic approach (HMOR-S2$_{PAS}$ - *H*ierarchical *M*ultiobjective *R*outing considering *2* classes of *S*ervice with a *P*areto *A*rchive *S*trategy), devised to find 'good' solutions in the sense of multiobjective optimization (see [6]) to this hierarchical multiobjective routing optimization problem, is proposed and applied to two test networks used in a benchmarking case study, for various traffic matrices for each network. Remember that in the context of multiobjective optimization, the concept of optimal solution (usually unfeasible) is replaced by the concept of non-dominated (or Pareto optimal) solutions. A non-dominated solution may be defined as a feasible solution such that it is not possible to improve the value of an o.f. without worsening the value of at least one of the other o.f.s.

In [7], the same heuristic approach is applied to two other networks with more connectivity than the networks in [5]. The results of computational experiments using an analytical model and stochastic discrete-event simulation (with a static routing model where the network routing plan never changes) are presented, in order to evaluate the performance of the proposed heuristic in terms of the effect of the used route calculation and selection procedures.

Throughout the study of the routing model and the implementation of the resolution procedure, some factors of imprecision, inaccuracy (or inaccurate determination), and uncertainty (IIU, for short) of the routing model and their effects on the results of the routing method are taken into account by the authors. In this work, these factors are systematized and analyzed, having in mind their great importance in the context of the model. A broad discussion of modeling issues concerning IIU aspects in relation to multicriteria analysis is in [8].

In this work, a new set of stochastic simulation experiments with the heuristic approach proposed in [5] are presented in order to evaluate the inaccuracies intrinsic to the analytic/numerical resolution procedures. A state-dependent periodic-type dynamic routing model (where the network routing plan is updated as a function of the measured offered traffic in the network) is considered in these experiments. Different simulation parameters have a direct influence on the robustness of the results obtained with the dynamic model simulation. In particular, the effects in the model results of the uncertainty associated with the moving average estimates of the offered traffic in the network and the influence of the routing plan update time interval are analyzed, allowing for a robustness analysis focused on key robustness aspects of the model.

The paper is organized as follows. The two-level hierarchical multiobjective alternative routing model with two service classes and the features of the considered heuristic are briefly reviewed in the next section. Still in the same section, an analysis of the IIU factors associated with key instances of the model and its resolution and their potential effects on the heuristic results, is provided. The results obtained with this procedure, by using analytic and discrete-event simulation experiments for four test networks, are shown and analyzed in relation to the IIU factors, in the following section. These experiments provide some preliminary conclusions on the robustness of the method concerning some of the IIU factors. Conclusions are drawn in the last section. The paper ends with an Appendix with the specification of the notation used in the model and an Appendix with the formalization of the heuristic resolution approach.

Review of the multiobjective routing model and the heuristic resolution approach

The multiobjective routing model

The model addressed in this paper may be considered an application of the multiobjective modeling framework (or 'meta-model') for MPLS networks proposed in [4], as previously noted. It is a network-wide routing optimization approach (that is, the main o.f.s depend explicitly on all traffic flows in the network), in the form of a hierarchical multiobjective optimization model, which takes into account the nature and relations between the adopted o.f.s related to the different types of traffic flows associated with different services.

Two classes of services are considered: QoS, corresponding to services with certain guaranteed QoS levels (represented through the set $\mathcal{S}_Q$) and considered in the model as first-priority flows; BE (represented through the set $\mathcal{S}_B$), regarded as second-priority flows, where the corresponding traffic flows are routed seeking the best possible quality of service to be obtained but not at the cost of the quality of service of the traffic flows belonging to $\mathcal{S}_Q$. The traffic flows of each service type $s \in \mathcal{S}_Q$ or $s \in \mathcal{S}_B$ may differ in important attributes, such as the required bandwidth.

The hierarchical multiobjective routing optimization model tackled here has two levels with several o.f.s in each level. At the first level, the first-priority o.f.s are considered: W_Q (the total expected network revenue associated with QoS traffic flows) and $B_{Mm|Q}$ (the worst average performance among QoS services, represented by the maximal average blocking probability among all QoS service types). These objectives are formulated at the network level for the QoS traffic and allow us to take into account the combined effect of all types of traffic flows in the network. The second level includes the second-priority o.f.s concerned with average performance metrics of the QoS traffic flows associated with the different types of QoS services (represented by $B_{ms|Q}$, the mean blocking probabilities for flows of type $s \in \mathcal{S}_Q$, and $B_{Ms|Q}$, the maximal blocking probability defined over all flows of type $s \in \mathcal{S}_Q$) as well as W_B, which represents the total expected network revenue associated with BE traffic flows. At both levels of optimization, min – max objectives constituting 'fairness' objectives are explicitly considered: $\min_{\overline{R}}\{B_{Mm|Q}\}$ at the first level, and $\min_{\overline{R}}\{B_{Ms|Q}\}$ at the second level, where $\overline{R}$ denotes the routing plan (set of feasible node-to-node loopless paths) for all offered flows.

The considered *two-level hierarchical optimization problem for two service classes* P-M2-S2 ('*P*roblem - *M*ultiobjective with *2* optimization hierarchical levels - with *2* *S*ervice classes') is

Problem P-M2-S2

$$\begin{array}{ll} \text{1st level} & \left\{ \begin{array}{ll} \text{QoS: Network objectives} & \max_{\overline{R}}\{W_Q\} \\ & \min_{\overline{R}}\{B_{Mm|Q}\} \end{array} \right. \\ \text{2nd level} & \left\{ \begin{array}{ll} \text{QoS: Service objectives} & \min_{\overline{R}}\{B_{ms|Q}\} \\ & \min_{\overline{R}}\{B_{Ms|Q}\} \\ & \forall s \in \mathcal{S}_Q \\ \text{BE: Network objective} & \max_{\overline{R}}\{W_B\} \end{array} \right. \end{array}$$

subject to equations of the underlying traffic model.

The notation and basic formulae for the calculation of these o.f.s are in an Appendix at the end of the paper.

Note that in the formulation of P-M2-S2, W_Q (together with $B_{Mm|Q}$) is a first-priority o.f., while W_B is a second-level o.f. This assures that the routing of BE traffic, in a quasi-stationary situation, will not be made at the cost of the decrease in revenue or at the expense of an increase in the maximal blocking probability of QoS traffic flows. Nevertheless, it is important to note that while QoS and BE traffic flows are treated separately in terms of o.f.s so as to take into account their different priority in the routing optimization, the interactions among all traffic flows are fully represented in the model. In fact, the contributions of all the traffic flows which may use every link of the network are used to obtain the blocking probabilities $B(f_s)$.

A full description of the traffic modeling stochastic approach used in the routing model can be seen in [4]. In the considered basic teletraffic model, the blocking probabilities B_{ks}, for micro-flows of service type s in link l_k, are calculated by $B_{ks} = \mathcal{B}_s\left(\overline{d_k}, \overline{\rho_k}, C_k\right)$, with $\mathcal{B}_s$ representing the function (implicit in the teletraffic stochastic model) that expresses the marginal blocking probabilities, B_{ks}, in terms of equivalent effective bandwidths d_{ks} for all service types, reduced traffic loads ρ_{ks} offered by flows of type s to l_k and the link capacity C_k. With this type of approximation (see [9]), the calculation of $\{B_{ks}\}$ can be made by efficient and robust numerical algorithms, which are essential in a network-wide routing optimization model of this type, for tractability reasons. In this situation, the classical Kaufman (or Roberts) algorithm ([10,11]) can be used to calculate the functions $\mathcal{B}_s$ for small values of C_k. For larger values of C_k, approximations based on the uniform asymptotic approximation (UAA) in [12] are used, having in mind its efficiency in these situations.

The decision variables $\overline{R} = \cup_{s=1}^{|\mathcal{S}|} R(s)$ represent the network routing plans, that is, the set of all the feasible routes (i.e., node-to-node loopless paths) for all traffic flows, with $R(s) = \cup_{f_s \in \mathcal{F}_s} R(f_s), s \in \mathcal{S}_Q \cup \mathcal{S}_B$ and $R(f_s) = (r^p(f_s)), p = 1, \cdots, M$ with $M = 2$ in our model. That is, an alternative routing principle is used: for each flow f_s the connection request attempts the first choice route $r^1(f_s)$; if it is blocked the second choice route $r^2(f_s)$ is tried. A request will be blocked only if $r^2(f_s)$ is also blocked.

The very high complexity of the routing problem P-M2-S2 stems from two major factors: all o.f.s are strongly interdependent (via the $\{B(f_s)\}$), and all the o.f. parameters and (discrete) decision variables $\overline{R}$ (network route plans) are also interdependent in terms of their effects.

Considering the form of P-M2-S2, one may conclude on the great intractability of this problem. Note that there are possible conflicts between the o.f.s in P-M2-S2: in many routing situations, the maximization of W_Q leads to a deterioration on some $B(f_s), s \in \mathcal{S}_Q$, for certain traffic flows with low intensity and this tends to increase $B_{Ms|Q}$ and $B_{ms|Q}$, and consequently, $B_{Mm|Q}$. This is a major factor that justifies the interest and potential advantage in using multiobjective approaches when dealing with this type of routing methods.

The heuristic resolution approach

The theoretical foundations of a specialized heuristic strategy for finding 'good' compromise solutions to the very complex bi-level hierarchical multiobjective alternative routing optimization problem, are presented in [13]. In [5], a heuristic procedure (HMOR-S2$_{\text{PAS}}$), devised to seek non-dominated solutions to this problem, is proposed and applied to test networks used in a benchmarking case study, for various traffic matrices. This heuristic resolution approach is briefly reviewed in this section.

This heuristic is based on the recurrent calculation of solutions to an auxiliary constrained bi-objective shortest path problem $\mathcal{P}_{s2}^{(2)}$, formulated for every end-to-end flow f_s,

$$\min_{r(f_s)\in\mathcal{D}(f_s)} \left\{ m^n(r(f_s)) = \sum_{l_k\in r(f_s)} m_{ks}^n \right\}_{n=1;2}$$

The two metrics m^n to be minimized are the marginal implied costs $m_{ks}^1 = c_{ks}^{Q(B)}$ and the marginal blocking probabilities $m_{ks}^2 = -\log(1 - B_{ks})$ for flows of service type s on link l_k. These metrics are chosen because the metric blocking probability tends, at a network level, to minimize the maximal node-to-node blocking probabilities $B(f_s)$, while the metric implied cost tends to maximize the total average revenue W_T in a single class multiservice loss network ([14,15]). The set of all feasible loopless paths for flow f_s, $\mathcal{D}(f_s)$, satisfies specific traffic engineering constraints for flows of type s. By using this approach, the efficiency of different candidate routes in the context of a multicriteria routing framework of this type can be compared by taking into account both the loss probabilities experienced along the candidate routes and the knock-on effects upon the other routes in the network, effects which are associated with the acceptance of a call on that given route. Such effects can be measured exactly through the marginal implied costs for QoS(BE) traffic, $c_{ks}^{Q(B)}$, associated with the acceptance of a connection (or 'call') of traffic f_s of any service type $s \in \mathcal{S}$ on a link l_k, that can be defined as the expected value of the traffic loss induced on all QoS(BE) traffic flows resulting from the capacity decrease in link l_k (see [13]).

In the heuristic, the auxiliary constrained shortest path problem $\mathcal{P}_{s2}^{(2)}$ is solved by an algorithmic approach, MMRA-S2 (*M*odified *M*ultiobjective *R*outing *A*lgorithm for multiservice networks, considering *2* classes of *S*ervice) in [13], which aims at finding a 'best' compromise path from a set of non-dominated solutions, according to a system of preferences. The path computation and selection are fully automated; therefore, the system of preferences is embedded in the working of the algorithm. This is implemented by defining preference regions in the o.f. space obtained from aspiration and reservation levels (preference thresholds) defined for the two o.f.s ([14,15]), as will be explained later.

The candidate solutions $(r^1(f_s), r^2(f_s))$ for each f_s are generated using MMRA-S2. They are selected (or rejected) according to specific criteria, to be 'tuned' throughout the execution of the heuristic. A maximal number of arcs D_s per route for each service type s is previously defined and a feasible route set $D(f_s)$ is obtained for each f_s.

Notice that the successive application of MMRA-S2 to every traffic flow does not lead to an effective resolution approach to the network routing problem P-M2-S2, as anticipated by the theoretical analysis of the model and confirmed by experimentation. This results from an instability phenomenon arising in such path selection procedure, expressed by the fact that the route sets $\overline{R}$ often tend to oscillate between specific solutions some of which may lead to poor global network performance under the prescribed metrics. This instability phenomenon is associated with the complexity and interdependencies in the addressed problem P-M2-S2, namely the interdependencies between $\left\{c_{ks}^{Q(B)}\right\}$ and $\{B_{ks}\}$ and between these two sets and the current network route set $\overline{R}$.

Therefore, it is necessary to search for the subset of the path set, the elements of which should be possibly changed in the next route improvement cycle, which constitutes another core idea of the heuristic approach. A criterion for choosing candidate paths for possible routing improvement by increasing order of a function $\xi(f_s)$ of the current $\left(r^1(f_s), r^2(f_s)\right)$ is proposed in [16]. Preference (concerning the potential value in changing the second choice route when seeking to improve W_Q or W_B) is given to the flows for which the route $r^1(f_s)$ has a low implied cost and the route $r^2(f_s)$ has a high implied cost, or to the flows which currently have worse end-to-end blocking probability.

In a basic version of the heuristic, HMOR-S2, each new solution is obtained by processing the current best solution. A basic searching strategy is to seek for routing solutions $R(s)$ for each service $s \in \mathcal{S}$, in order to achieve a better performance in terms of W_B (if $s \in \mathcal{S}_B$) or $B_{ms|Q}$ and $B_{Ms|Q}$ (if $s \in \mathcal{S}_Q$), while respecting the hierarchy of o.f.s. This also means that network resources are left available for traffic flows of other services so that the solutions selected at each step of the procedure may improve the first-priority o.f.s W_Q and $B_{Mm|Q}$. The heuristic is designed in order to seek, firstly for each QoS service and, secondly, for each BE service, solutions which dominate the current one, in terms of $B_{ms|Q}$ and $B_{Ms|Q}$ for QoS services and in terms of W_B for BE services. These solutions are accepted only if they do not lead to the worsening of any of the network functions W_Q and $B_{Mm|Q}$.

Due to these strict limitations imposed on the acceptance of a new solution, there is the realization that some potentially interesting solutions to the routing problem are not further pursued. Therefore, throughout the execution of the basic heuristic some possibly interesting solutions are stored and later checked, seeking to find a 'best' possible solution to the problem in hand [5].

The steps of this heuristic resolution approach are explicitly written out in an Appendix at the end of the paper.

Dealing with imprecision, inaccuracy, and uncertainty in the model

The development of the routing model and the implementation and application of the resolution procedure put in evidence the importance of IIU factors of the routing model and their impacts on the results of the routing method. In this subsection, an analysis and a systematization of these factors in the context of the model are presented. In Table 1, a summary of the sources of IIU is presented. Also, the way these aspects are dealt with in our resolution approach are discussed (and summarized in Table 1) or a brief explanation as to why these aspects do not have a significant impact on the final results is provided.

Table 1 Sources of IIU and how they are dealt with in the heuristic

Sources of IIU	Effect	Dealing with IIU
High complexity of the routing problem P-M2-S2	Inaccuracy (the solutions are inherently approximate)	Different mechanisms of the heuristic resolution in its present version (HMOR-$S2_{PAS}$)
Simplifications and approximations assumed in the stochastic model for the traffic in the links, leading to an approximate model, unavoidable for computational tractability reasons	Imprecision	The focus is on the relative value of the results of the traffic model; small differences between o.f. values can be disregarded when comparing solutions (this is a achieved by using adequate numerical traffic calculation procedures); scenario evaluation with different traffic matrices
Instability phenomenon potentially arising in the path selection procedure if the MMRA-S2 is applied successively to all the end-to-end flows of each service type	Imprecision and uncertainty	Criterion for choosing candidate paths for possible routing improvement, embedded in the main heuristic 'optimization' cycle
Numerical errors in the calculations of marginal implied costs and blocking probabilities of all the flows, propagating throughout the resolution procedure	Imprecision	'Robust' and well-tested numerical algorithms (namely the Kaufman/Roberts algorithm and fixed-point iterators)
Stochastic nature of the traffic offered to the network	External uncertainty	Periodical update of traffic flows means via a statistical estimate (first-order moving average) based on real-time measurements; sensitivity/robustness analysis; scenarios evaluation.
Uncertainties in the identification of the 'virtual decision maker' in a fully automated decision application environment	Specific form of internal uncertainty	Definition of dynamic preference thresholds in the o.f. space, combined with the use of reference points

Imprecision factors

The imprecision factors stem mainly from the approximations inherent to the analytic traffic model (underlying the optimization model), namely a superposition of independent Poisson flows and independent occupations of the links. A general description of the stochastic traffic models associated with this issue can be seen in ([17], Chapter 6). An exact model which might be in principle applied to these networks is in [18], which is based on the consideration of MMPPs (Markov Modulated Poisson Processes) for representing the superposition of the overflows (resulting from the alternative routing) from independent Poisson processes. The type of considered simplification leads inevitably to intrinsic imprecisions in the values of the traffic parameters which are reflected in the calculation of the o.f. values. Still, if a more accurate and realistic representation of the traffic flows is used, better estimates of the blocking probabilities will be achieved. Nonetheless, the approximations in our model can be considered appropriate and, above all, absolutely necessary in this context, for practical reasons. In fact, if more exact stochastic models are used to represent the traffic flows and to calculate the blocking probabilities in overflow conditions such as the one in [18], the computational burden will be too heavy since the analytical model has to be numerically solved many times during the execution of the heuristic and the routing method will become intractable in terms of memory and processing time requirements.

Some imprecision also arises from the numerical errors, associated with the resolution of the system of equations of the traffic model which propagate throughout the resolution procedure, as the resolution of the traffic model has to be performed many times (in the calculations of marginal implied costs and blocking probabilities of all the flows). To minimize the latter imprecision effects, some 'robust' and well-tested classical numerical algorithms (namely the Kaufman/Roberts algorithm ([10,11]) and fixed-point iterators ([19])) are used to estimate the blocking probabilities in the system. Moreover, these two types of errors do not compromise the inequality relations between the o.f. values, as the aim of the routing optimization procedure is just the comparison of routing solutions in terms of the values of the o.f.s. That is, the focus of the tackled multiobjective optimization model is on the relative value of the results of the traffic model rather than on the absolute accuracy of such values. Also, small differences between o.f. values can be disregarded when comparing solutions.

Further imprecision effects stem from an instability phenomenon which may potentially arise in the path selection procedure. In fact, the route sets $\overline{R}$, if obtained by successive application of MMRA-S2 to every traffic flow, often tend to oscillate between certain solutions some of which may lead to poor global network performance under the prescribed metrics, thus leading to uncertainty in the results. The experimentations confirm that the successive application of MMRA-S2 to every traffic flow does not lead to an effective resolution approach to the network routing problem P-M2-S2. For dealing with this issue in a successful manner, detailed analysis and extensive experimentation with the heuristic have led to the proposal of a criterion for choosing candidate paths for possible routing improvement by increasing order of a function $\xi(f_s)$ of the current routes of a flow f_s, giving preference (concerning the potential value in changing the routes when seeking to improve the QoS(BE) traffic revenue) to the flows for which the first route has a low implied cost and the second route has a high implied cost, or to the flows which currently have worse end-to-end blocking probability.

Inaccuracy factors

The inaccuracies are a consequence of the high complexity of the routing problem P-M2-S2 stemming from the combinatorial nature of the global routing multiobjective optimization problem and are related to two major factors specific to the described model: all o.f.s are strongly interdependent (via the $B(f_s)$) and all the o.f. parameters and (discrete) decision variables $\overline{R}$ (network route plans) are also interdependent in terms of their effects in the problem solutions. These interdependencies result from the fact that all the specified o.f.s depend on all traffic flow patterns in the global network, which may change significantly with any alteration in any route choice for any given node-to-node flow. This mechanism generates potential instability in the global routing solutions, as analyzed in a similar yet simpler model in [15]. Note that all these interdependencies are defined explicitly or implicitly through the underlying traffic model.

Also, note that even in the simplest degenerated case (single service with single-criterion optimization and no alternative routing), the problem is NP-complete in the strong sense, as proved in [20]. This high complexity leads to inaccurate results, because the solutions of the routing problem are inherently approximate. In fact, when heuristic solution techniques are employed, the resolution method does not calculate all the non-dominated solutions and even for those selected solutions which are computed there is no certainty that they are not potentially dominated by other solutions. Concerning the possibility of not detecting the condition of certain weakly dominated solutions[a] this may be explained by small variations in the values of some o.f.s for certain solution(s) close to the current one (for example, by changing a single route for a given flow), solution(s) which, in some cases, may not be detected by the heuristic. It must be remarked that this occurrence is rare but in principle may arise. The mechanisms of the heuristic resolution in its present version (HMOR-$S2_{PAS}$), as described in the previous sub-section, are devised to deal with this issue in order to minimize the impact of these inaccuracies in the quality of the obtained compromise solutions.

Uncertainty factors

As for the uncertainty issues, they are raised mainly by the stochastic nature of the offered traffic and the related estimation procedures performed throughout the stochastic simulation experiments with the proposed heuristic. In the classification framework discussed in [21], this type of uncertainty may be considered as an 'external uncertainty', as it relates to a form of uncertainty that results from environmental conditions (the stochastic traffic load conditions in the network) of the model that cannot be controlled by the decision maker.

A particular form of 'internal uncertainty' (in the sense discussed in [21]) may be identified in the context of this model. This type of 'uncertainty' has to do with the identification and representation of the decision maker preferences. Noting that the model is supposed to be applied in an automated manner and following our previous experience in this area [5], we have developed a solution selection procedure based on the definition of dynamic preference thresholds in the o.f. space, combined with the use of reference points, inspired by the methodology described in [22]. This corresponds to imbedding the preferences of a 'virtual decision maker' in the form of a specific, problem-oriented, procedure of selection of non-dominated solutions. The effectiveness of such solutions is tested *a posteriori* using a simulation test-bed, applied in various typical application scenarios. For further details on the solution selection procedure, see [5].

In order to evaluate the effect of these uncertainty factors in the model results, an experimental study consisting of stochastic simulation experiments is performed. In the discrete event stochastic simulation experiments performed with a static routing model, a measurement of the degree of uncertainty of the o.f. values can be obtained by applying a classical statistical procedure (batch means with independent replications). As for the state-dependent periodic-type dynamic routing model, the traffic flow means are periodically updated via a statistical estimate, namely a first-order moving average, based on real-time measurements, dependent on a parameter b (fixed *a priori*), as explained in the 'Simulation experiments' section, which is another source of statistical uncertainty. This requires a sensitivity/robustness analysis, for the evaluation of the influence of this parameter and of the routing plan update time interval on the final global routing solution.

Note that the problem that is being considered here is not a 'classical' problem of decision under uncertainty as discussed in depth in [21]. A classical approach to that type of problem is the use of expected utility theory to planning under uncertainty, while our model may be considered as a specific type of multicriteria optimization model with stochastic variables. In the model, the o.f.s are means or probabilities defined in a classical probability theory framework.

Experimental study

In this section, the analytical and simulation results obtained with the HMOR-S2$_{PAS}$ heuristic in four different networks are presented. The analytical results are obtained by a single run of the heuristic and the results for the QoS flows revenue W_Q may be compared with theoretical bounds for the same function. As for the simulation results, they allow for a more realistic assessment of the results of the heuristic having in mind the combined effects of the analyzed IIU factors.

This set of experiments represents a set of scenarios [21], in the sense that each experiment has its specific features and allows for the analysis of relevant working conditions that can be encountered in actual networks. Notice however that the use of scenarios in this context is not for the evaluation and development of alternative solutions (a technique often used in multiattribute models, as extensively analyzed in [21]), but rather to explore the different possibilities that may arise in the routing context, having in mind to test the robustness and effectiveness (in terms of performance) of the solution selection procedure. This is achieved by considering scenarios of implementation of static and dynamic versions of the model and the consideration of three situations of load/overload conditions.

Two types of simulation are considered, one corresponding to a *static routing model* where the routing plan calculated by the heuristic is never changed regardless of the random variations in offered traffic throughout the simulation, for a given matrix of mean traffic offered in statistical equilibrium. The other corresponds to a *periodic type state-dependent dynamic routing model,* where the routing plans are updated periodically as a function of real-time traffic measurements, by using the heuristic HMOR-S2 repeatedly. Dynamic routing in a telecommunications network is a well-known routing principle where the most recent information on the network conditions is taken into account in order to find appropriate paths for the connection requests in the network. This is especially important when there are significant fluctuations of the offered traffic in various parts of the network, in particular as a result of overload or network failures. A comprehensive text on dynamic routing in telecommunication networks can be seen in [23].

For each of these two types of simulation, we consider three relevant network scenarios regarding the random fluctuations of traffic that are typical of stochastic traffic models: a deterministic scenario; a scenario where calls arrive according to a Poisson process, service times follow an exponential distribution and the network is critically loaded; a scenario where traffic flows have a higher 'variability.' The analysis of the results of each of these scenarios in each of the types of simulation gives an insight on the possible effects of inaccuracies intrinsic to the model and to the analytic/numerical resolution method on the results of the heuristic resolution procedure and also enables the verification of the effectiveness of the selected solutions, hence implementing a specific form of robustness analysis.

Application of the model to a network case study

The benchmarking case study considered here is based on the one in [24], where a model for traffic routing and admission control in multiservice, multipriority networks supporting traffic with different QoS requirements, is proposed. Deterministic models are used in the calculation of paths, in particular mathematical programming models based on

Multicommodity Flows (MCFs). These models can be adapted to a stochastic traffic environment by using a simple technique: the requested values of the flow bandwidths in the MCF model are compensated with a factor $\alpha \geq 0.0$, so as to model the effect of the random fluctuations of the traffic that are typical of stochastic traffic models. In the application example in [24], three values of α are proposed: $\alpha = 0.0$ corresponds to the deterministic approach; $\alpha = 0.5$ is the compensation factor when calls arrive according to a Poisson process, service times follow an exponential distribution and the network is critically loaded; and $\alpha = 1.0$ for traffic flows with higher 'variability.'

The o.f.s of this problem, to be maximized, are the revenues W_Q and W_B, associated with QoS and BE flows. A bi-criteria lexicographic optimization formulation including admission control for BE traffic is considered, concerning the revenues W_Q and W_B, so that the improvements in W_B are to be found under the constraint that the optimal value of W_Q is maintained.

In the deterministic flow-based model in [24], a base matrix $T = [T_{ij}]$ with offered bandwidth values from node i to node j [Mbps] is given. A multiplier $m_s \in [0.0; 1.0]$ with $\sum_{s \in \mathcal{S}} m_s = 1.0$ is applied to these matrix values to obtain the offered bandwidth of each flow f_s with service type s. The transformation of this type of matrix into a matrix of offered traffic $A(f_s)$, used in our stochastic traffic model, is achieved by $A(f_s) \approx \frac{m_s T_{ij}}{d_s u_0} - \alpha\sqrt{\frac{m_s T_{ij}}{d_s u_0}}$ [Erl] if $\frac{m_s T_{ij}}{d_s u_0} > \alpha^2$ and both $T(f_s) = m_s T_{ij}$ and $A(f_s)$ are high. Otherwise, $A(f_s) \approx \frac{m_s T_{ij}}{d_s u_0}$ [Erl] where $u_0 = 16$ kbps is a basic unit of transmission [bit/s].

Network $\mathcal{M}$

The routing model in [24] and the one considered here can be applied to the test network $\mathcal{M}$ depicted in Figure 1. It has $|\mathcal{N}| = 8$ nodes, with ten pairs of nodes linked by a direct arc and a total of $|\mathcal{L}| = 20$ unidirectional arcs, giving an average node degree for this network of $\delta_{\mathcal{M}} = 2.5$. The bandwidth of each arc C'_k [Mbps] is shown in Figure 1. The number of channels C_k is $C_k = \left\lceil \frac{C'_k}{u_0} \right\rceil$, with basic unit capacity u_0. A total of $|\mathcal{S}| = 4$ service types with the features displayed in Table 2 are considered. The values of the required effective bandwidths $d_s = \frac{d'_s}{u_0}$ [channels] $\forall s \in \mathcal{S}$ are also in the table (where d'_s is the required bandwidth in kbps). The expected revenue for a call of type s is assumed to be $w_s = d_s, \forall s \in \mathcal{S}$. The average duration of a type s call is h_s and D_s represents the maximum

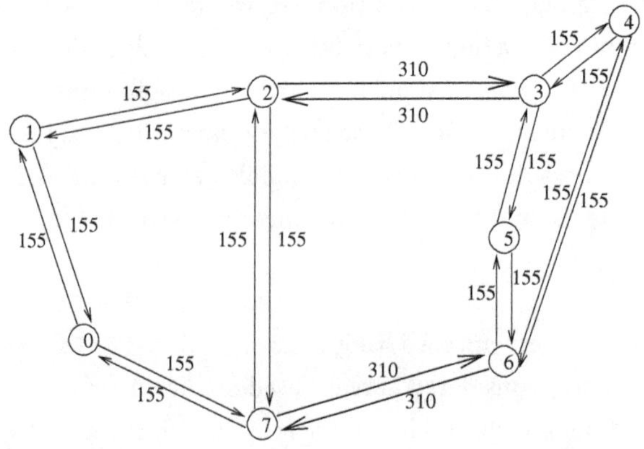

Figure 1 Test network $\mathcal{M}$. This test network (with the indication of the bandwidth of each arc C'_k, in Mbps) is in [24].

number of arcs for a route of a type s call. The application example in [24] provides upper bounds $W_Q^{\max}$ for the optimal value of the QoS traffic revenue W_Q in our model.

Network $\mathcal{E}$

In another set of tests, the considered routing model is applied to another test network $\mathcal{E}$ (depicted in Figure 2), so as to further grasp the potential of the proposed heuristic strategy. Network $\mathcal{E}$ has $|\mathcal{N}| = 10$ nodes, with 12 pairs of nodes linked by a direct arc and a total of $|\mathcal{L}| = 24$ unidirectional arcs, giving an average node degree for this network of $\delta_{\mathcal{E}} = 2.4$. The bandwidth of each arc C'_k (Mbps) is shown in Figure 2. This test network is based on the network $\mathcal{O}$, in [25], dimensioned for extremely low blocking probabilities. This network $\mathcal{O}$ has a topology similar to the one in Figure 3, with a capacity of $C'_k = 50$ Mbps for each arc, which is equivalent to a capacity of $C_k = \frac{C'_k}{u_0} = 3125$ channels. The changes in the network by eliminating a few links allow for the achievement of slightly higher blocking probabilities. The traffic matrix remains the same as in the original reference ([25]). The information on the network and on the traffic matrix is used as an input to the routing model considered here. A comparison of the traffic that is offered to each of the networks $\mathcal{M}$ and $\mathcal{E}$ for each α shows that the offered load is lower in $\mathcal{E}$.

The service features are the same as for the tests with network $\mathcal{M}$ and they are displayed in Table 2. Notice that the network diameter of network $\mathcal{E}$ (equal to 4) is higher than the network diameter of network $\mathcal{M}$ (equal to 3), which will influence the value of D_s. These networks have similar density since their average node degree δ is practically the same.

In [25], no results concerning any of the o.f.s considered here are provided, as their multiobjective routing model is radically different from the one considered here. The only results that can be extracted from the proposed model in [25] are approximate ideal values for the QoS flows revenue, W_Q^{ideal}.

Networks $\mathcal{G}$ and $\mathcal{H}$

For the final set of tests, the considered routing model is applied to test networks $\mathcal{G}$ and $\mathcal{H}$, for which the topology is depicted in Figure 3. They have $|\mathcal{N}| = 10$ nodes, with 16 pairs of nodes linked by a direct arc and a total of $|\mathcal{L}| = 32$ unidirectional arcs, giving an average node degree for these networks of $\delta_{\mathcal{G}} = \delta_{\mathcal{H}} = 3.2$. These networks have greater density than networks $\mathcal{M}$ and $\mathcal{E}$. The bandwidth of each arc C'_k (Mbps) for each of the networks (obtained by employing a very simple network dimensioning algorithm) is shown in Tables 3 and 4.

The test networks $\mathcal{G}$ and $\mathcal{H}$ can be obtained after a redimensioning of the original network $\mathcal{O}$ given in [25]. The offered traffic matrix remains the same as in [25]. A value β_s for the mean blocking probabilities for flows of type s, B_{ms}, is established, with a possible variation of Δ_B.

Table 2 Service features on the test networks

Service	Class	d'_s (kbps)	d_s (ch.)	w_s	h_s (s)	D_s (arcs) $\mathcal{M}$	$\mathcal{E}$	$\mathcal{G}, \mathcal{H}$	m_s
1 - video	QoS	640	40	40	600	3	4	3	0.1
2 - Premium data	QoS	384	24	24	300	4	5	4	0.25
3 - voice	QoS	16	1	1	60	3	4	3	0.4
4 - data	BE	384	24	24	300	7	9	9	0.25

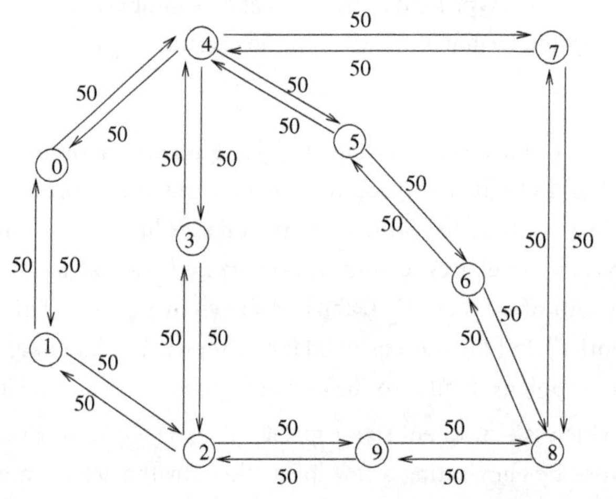

Figure 2 Test network $\mathcal{E}$. This test network (with the indication of the bandwidth of each arc C'_k, in Mbps) is in [25].

Considering a routing method for network $\mathcal{O}$ using only the shortest path direct routing (typical of Internet conventional routing algorithms), the mean blocking probabilities B_{ms} are calculated. Once these values are known, they are compared with the established values at the beginning of the algorithm. If for service s, $B_{ms} > \beta_s$ then the links in paths for flows of service s have their capacity increased. On the contrary, if for service s, $B_{ms} < \Delta_B \beta_s$, then the links in paths for flows of service s have their capacity decreased. The algorithm proceeds iteratively until it converges (which means that $\Delta_B \beta_s < B_{ms} < \beta_s, \forall s \in \mathcal{S}$). Sometimes, the algorithm oscillates between two different solutions, preventing it from converging. Therefore, a maximum number of runs has to be established, so as to avoid this situation.

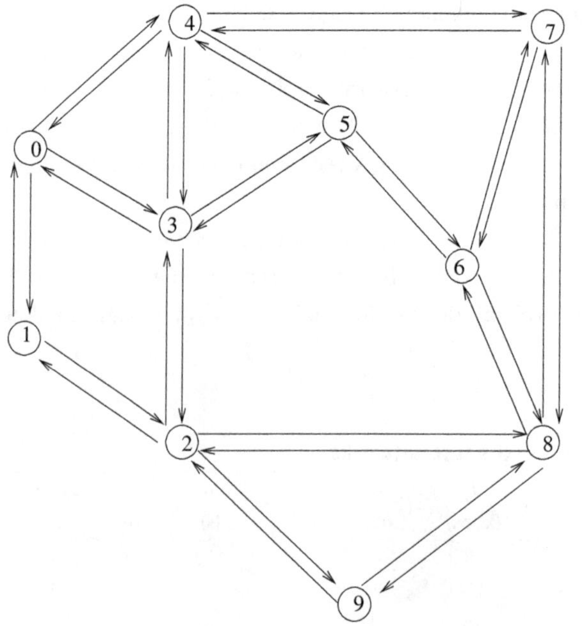

Figure 3 Network topology for test networks $\mathcal{G}$ and $\mathcal{H}$. For information on this network topology, see [25].

Table 3 Bandwidth of each arc C'_k, in Mbps, for the test network $\mathcal{G}$

↱	0	1	2	3	4	5	6	7	8	9
0		40.64		44.384	40.64					
1	40.64		35.024							
2		35.024		35.024					36.896	38.768
3	44.384		35.024		42.512	38.768				
4	40.64			42.512		44.384		40.64		
5				38.768	44.384		38.768			
6						38.768		46.256	40.64	
7					40.64		46.256		38.768	
8			36.896				40.64	38.768		44.384
9			38.768						44.384	

This very simple network dimensioning algorithm can be used for the dimensioning of test networks $\mathcal{G}$ and $\mathcal{H}$, for $\beta_s = 0.1$ and $\beta_s = 0.12$, respectively, with $\Delta_B = 0.9$, that is, a situation of very high blocking is considered, associated with traffic overload for all services, for $\alpha = 0.0$. The purpose is to carry out comparisons of the performance of the considered static and dynamic routing methods in overload conditions ($\alpha = 0.0$) and in low and very low blocking conditions for the QoS traffic for $\alpha = 0.5$ and $\alpha = 1.0$. The original network $\mathcal{O}$ is not used in this study because it is dimensioned for extremely low blocking probabilities.

The traffic matrix $T = [T_{ij}]$ with offered total bandwidth values from node i to node j [Mbps] remains the same as in the original reference ([25]). As with the tests performed for network $\mathcal{E}$, the modified version of the network and the traffic matrix are the only data taken from [25]. This information is used as an input to the routing model considered here.

These two networks $\mathcal{G}$ and $\mathcal{H}$ have more connectivity than networks $\mathcal{M}$ and $\mathcal{E}$. The service features are the same as for the tests with the other networks. The ideal optimal values for the QoS flows revenue, W_Q^{ideal}, are calculated from the data in [25] and can be used for comparison purposes, as done for network $\mathcal{E}$.

Analytical experiments

In the analytical study, the HMOR-S2 heuristic (basic version of the heuristic without storage of non-dominated solutions) is run once, followed by a run of the HMOR-S2$_{\text{PAS}}$ heuristic.

Table 4 Bandwidth of each arc C'_k, in Mbps, for the test network $\mathcal{H}$

↱	0	1	2	3	4	5	6	7	8	9
0		39.6		43.76	39.6					
1	39.6		33.36							
2		33.36		33.36					35.44	37.52
3	43.76		33.36		41.68	37.52				
4	39.6			41.68		43.76		39.6		
5				37.52	43.76		37.52			
6						37.52		45.84	39.6	
7					39.6		45.84		37.52	
8			35.44				39.6	37.52		43.76
9			37.52						43.76	

The initial solution is typical of Internet routing conventional algorithms: only one path for each flow (i.e., without an alternative path) is considered; the initial solution is the same for all the services $s \in \mathcal{S}$ and the paths are symmetrical; the path for every flow f_s is the shortest one (that is, the one with minimum number of arcs); if there is more than one shortest path, the one with maximal bottleneck bandwidth (i.e., the minimal capacity of its arcs) is chosen; if there is more than one shortest path with equal bottleneck bandwidth, the choice is arbitrary.

The routing plan obtained at the end of the HMOR-S2 run for each specific α is the initial solution of HMOR-S2$_{PAS}$. This heuristic is run only once. For the archived routing plans obtained at the end of this single run, values for all the o.f.s are computed and the 'best' possible solution in the best possible preference region is chosen to be the final solution of the algorithm, using a reference point-based procedure as the solution selection mechanism, as in [5].

For the experiments with HMOR-S2$_{PAS}$, an archive of size 5 is considered, chosen empirically after extensive experimentation. Here, the practical conclusion is that an increase in the archive size will not necessarily lead to better final results because at the end of the heuristic run (when the final solution is chosen from those in the archive) the top 5 solutions tend to be the same regardless of the archive size (≥ 5).

Simulation experiments

Simulation experiments, with static and dynamic routing methods using the solutions provided by the heuristic, can also be carried out. The purpose of this simulation study is the validation of the routing model results and the evaluation of the errors intrinsic to the analytical model which provides the estimates for the o.f.s.

In a first set of experiments, the discrete-event stochastic simulation is applied to a static routing model, where the routing plan is the final solution obtained after the HMOR-S2$_{PAS}$ run. This routing plan does not change throughout the simulation regardless of the random variations of traffic offered to the network. After an initialization phase that lasts for a time $t_{warm-up}$ (that should be long enough to guarantee that the system state at the end of the initialization phase is representative of the steady-state behavior of the system), information on the number of offered calls and carried calls in the network for each flow $f_s, s \in \mathcal{S}$, is gathered, until the end of the simulation. With this information, $B(f_s), s \in \mathcal{S}$ and subsequently, the values of the upper- and lower-level o.f.s related to blocking probabilities can be estimated. The calculation of the expected revenues is based on the number of carried calls in the network.

In the periodic and state-dependent dynamic version of the routing method considered here, the network state is assessed periodically and the gathered information on that state is used to periodically choose the most appropriate paths in the network, according to the HMOR-S2 routing algorithm. In the time interval $[n\tau; (n+1)\tau[$, the estimate of the average traffic offered to the network by the flow f_s is given by $\tilde{x}_n(f_s)$, obtained from a first order moving average iteration of the type

$$\tilde{x}_n(f_s) = (1-b)\tilde{x}_{n-1}(f_s) + b\tilde{X}_{n-1}(f_s),$$

where $\tilde{X}_{n-1}(f_s)$ is an estimator of the average value of the traffic offered by f_s to the network in the previous interval $[(n-1)\tau; n\tau[$. The value of $b \in]0.0; 1.0[$ is a compromise between the need to obtain a quick response of the estimator to rapid fluctuations in

$\tilde{X}(f_s)$ and the stability of the long-run variations and should be settled by extensive experimentation with the simulation model. The parameter τ is both the update period of the estimates of the offered traffic and the update period of the network routing plans. Note that the process of path choice should take a short time, when compared to τ.

The routing method seeks to obtain new routing plans adapted to the changing network working conditions resulting from the random fluctuations of traffic intensities. Taking into account the features and great complexity of the routing model, the choice of the 'start-up' routing solution of the dynamic method is of great importance concerning its performance. The start-up solution is the final solution obtained after the HMOR-S2$_{\text{PAS}}$ is run. The availability of a good estimate of the initial nominal traffic matrix is a necessary requirement of this dynamic routing method. Having in mind the periodically updated characteristics of the offered traffic, the 'best' possible set of paths is chosen so as to improve the multidimensional network performance, as specified by the multiobjective routing model.

A first phase of simulation, the initialization phase, lasts for a time $t_{\text{warm-up}} = t_0 + t_1$. In a first stage that lasts t_0, only periodical updates of the estimate of the offered traffic are performed, with period τ. After that time t_0, the offered traffic estimates are assumed to be representative of a steady-state behavior. Afterwards, during a time t_1, the estimate of the offered traffic is still performed with a period τ, along with periodical updates of the routing plan, with the same period τ.

After the warm-up time, both updates are still performed with the indicated period and based on information on the number of offered calls and effectively carried calls in the network for each flow $f_s, s \in \mathcal{S}$ gathered from real-time measurements, until the end of the simulation. Using this information, a calculation of $B(f_s)$ estimates ($s \in \mathcal{S}$) can be made, as well as a calculation of the values of all the upper- and lower-level o.f.s related to blocking probabilities. As for the revenues, the knowledge of the effectively carried calls in the network allows for the calculation of the carried traffic estimates, hence the calculation of revenues follows straightforwardly.

Comments on the experimental results

In Tables 5, 6, 7, and 8, the analytical values of each o.f. are displayed, together with the simulation results (average value $\pm$ half length of the 95% confidence interval) for these functions. The revenue values have two decimal places and the blocking probability values have three significant figures.

In the simulation experiments, a total of six seed files for random number generation are used, so the number of the performed independent runs is $R = 6$, for each α. To illustrate the way in which the 95% confidence interval is calculated, take the example of the QoS revenue, W_Q. An estimate of its average value is $\hat{W}_Q = \frac{1}{R}\sum_{i=1}^{R} W_{Q|i}$ and an estimate of its variance is $\hat{\sigma}^2(\hat{W}_Q) = \frac{\sum_{i=1}^{R}(W_{Q|i}-\hat{W}_Q)^2}{R(R-1)}$ where $W_{Q|i}$ is the QoS revenue value for the i-th run, $i = 1, \cdots, R$. Considering a two-sided Student t-distribution, the confidence interval for W_Q is $\hat{W}_Q \pm t_{0.025;R-1}\hat{\sigma}(\hat{W}_Q)$ where the critical value of t is $t_{0.025;R-1} = t_{0.025;5} = 2.57$ (see [26], Table A.4).

For the static model simulation, different values for the warm-up time can be tried, and the results displayed in the tables correspond to those obtained with $t_{\text{warm-up}} = 8$ h for a total simulation time of 48 h. For the dynamic model simulation (which represents an innovative aspect of the extensive study conducted with this routing model in these

Table 5 Average o.f. values for the simulation of the static and the dynamic routing model

α	O.f.	Initial solution	Analytical results	Static routing model results	Dynamic routing model results		
					$\tau = 10$ m	$\tau = 20$ m	$\tau = 30$ m
0.0	W_Q	54,803.69	64,905.26[a]	64,774.12 ± 68.28	64,776.24 ± 76.03	64,774.68 ± 68.46	64,750.27 ± 61.42
	$B_{Mm\|Q}$	0.413	0.0752	0.0773 ± 0.00356	0.0774 ± 0.00363	0.0771 ± 0.00333	0.0793 ± 0.00302
	$B_{m1\|Q}$	0.413	0.0752	0.0773 ± 0.00356	0.0774 ± 0.00363	0.0771 ± 0.00333	0.0793 ± 0.00302
	$B_{m2\|Q}$	0.314	0.0184	0.0236 ± 0.000576	0.0235 ± 0.000655	0.0237 ± 0.000601	0.0238 ± 0.000696
	$B_{m3\|Q}$	0.0198	0.00184	0.00200 ± 0.0000499	0.00200 ± 0.0000445	0.00200 ± 0.0000567	0.00204 ± 0.0000592
	$B_{M1\|Q}$	0.912	0.708	0.706 ± 0.00912	0.702 ± 0.0145	0.704 ± 0.0115	0.682 ± 0.0198
	$B_{M2\|Q}$	0.766	0.103	0.110 ± 0.00600	0.108 ± 0.0154	0.111 ± 0.00488	0.0916 ± 0.00800
	$B_{M3\|Q}$	0.0585	0.0301	0.0303 ± 0.000146	0.0300 ± 0.000287	0.0303 ± 0.000384	0.0292 ± 0.000978
	W_B	15,106.57	17,039.20	17,017.10 ± 39.32	17,030.28 ± 57.28	17,015.51 ± 39.62	17,059.73 ± 42.35
0.5	W_Q	51,785.21	60,739.76[b]	60,676.12 ± 61.43	60,659.89 ± 53.91	60,675.17 ± 66.08	60,287.27 ± 57.93
	$B_{Mm\|Q}$	0.413	0.0278	0.0306 ± 0.00145	0.0317 ± 0.00146	0.0308 ± 0.00149	0.0533 ± 0.00158
	$B_{m1\|Q}$	0.413	0.0278	0.0306 ± 0.00145	0.0317 ± 0.00146	0.0308 ± 0.00149	0.0533 ± 0.00158
	$B_{m2\|Q}$	0.296	0.00230	0.00463 ± 0.000355	0.00511 ± 0.000556	0.00460 ± 0.000674	0.0163 ± 0.000714
	$B_{m3\|Q}$	0.0174	0.000857	0.000922 ± 0.0000167	0.000904 ± 0.0000179	0.000912 ± 0.0000156	0.00105 ± 0.0000369
	$B_{M1\|Q}$	0.882	0.629	0.626 ± 0.0196	0.622 ± 0.0243	0.628 ± 0.0207	0.481 ± 0.0280
	$B_{M2\|Q}$	0.722	0.00959	0.0158 ± 0.00216	0.0166 ± 0.00142	0.0155 ± 0.00130	0.0552 ± 0.00348
	$B_{M3\|Q}$	0.0517	0.0244	0.0245 ± 0.000261	0.0239 ± 0.000315	0.0244 ± 0.000236	0.0171 ± 0.000325
	W_B	13,787.49	16,685.60	16,696.08 ± 40.87	16,757.72 ± 80.12	16,702.89 ± 72.29	17,562.04 ± 41.26
1.0	W_Q	49,010.41	56,106.51[c]	56,036.04 ± 45.53	56,044.88 ± 66.47	56,036.66 ± 49.07	55,895.41 ± 61.80
	$B_{Mm\|Q}$	0.405	0.0256	0.0274 ± 0.00174	0.0271 ± 0.00319	0.0273 ± 0.00121	0.0375 ± 0.00219
	$B_{m1\|Q}$	0.405	0.0256	0.0274 ± 0.00174	0.0271 ± 0.00319	0.0273 ± 0.00121	0.0375 ± 0.00219
	$B_{m2\|Q}$	0.275	0.00499	0.00805 ± 0.000619	0.00772 ± 0.00123	0.00803 ± 0.000747	0.0131 ± 0.00110
	$B_{m3\|Q}$	0.0150	0.000567	0.000643 ± 0.0000157	0.000590 ± 0.0000498	0.000640 ± 0.0000154	0.000686 ± 0.0000270
	$B_{M1\|Q}$	0.841	0.556	0.552 ± 0.0304	0.495 ± 0.0826	0.555 ± 0.0262	0.354 ± 0.0517
	$B_{M2\|Q}$	0.667	0.0186	0.0310 ± 0.00318	0.0298 ± 0.00504	0.0310 ± 0.00446	0.0492 ± 0.00497
	$B_{M3\|Q}$	0.0446	0.0200	0.0201 ± 0.000295	0.0168 ± 0.00444	0.0202 ± 0.000248	0.0110 ± 0.000972
	W_B	12,445.64	16,465.58	16,436.45 ± 17.45	16,443.61 ± 81.71	16,438.56 ± 16.88	16,690.18 ± 50.64

Average o.f. values, and 95% confidence intervals, for the simulation of the static (with a warm-up time of 8 h) and the dynamic routing model (with $t_0 = t_1 = 4$ h, update period τ and $b = 0.3$), a total simulation time of 48 h on the test network $\mathcal{M}$, for different values of α, when the HMOR-S2 is used to update the routing plan. [a]99.62%; [b]99.85%; [c]99.59% of the upper bounds W_Q^{max} for the optimal value of the QoS traffic revenue W_Q in [24].

Table 6 Average o.f. values for simulation of the static and the dynamic routing model

α	O.f.	Initial solution	Analytical results	Static routing model results	Dynamic routing model results		
					$\tau = 10$ m	$\tau = 20$ m	$\tau = 30$ m
	W_Q	20,636.78	21,596.83[a]	21,587.85 ± 34.78	21,605.44 ± 37.10	21,617.39 ± 38.95	21,617.09 ± 35.60
	$B_{Mm\|Q}$	0.138	0.0220	0.0238 ± 0.00130	0.0245 ± 0.00202	0.0224 ± 0.00148	0.0219 ± 0.00140
	$B_{m1\|Q}$	0.138	0.0220	0.0238 ± 0.00130	0.0245 ± 0.00202	0.0224 ± 0.00148	0.0219 ± 0.00140
	$B_{m2\|Q}$	0.0872	0.00655	0.00779 ± 0.000317	0.00507 ± 0.000480	0.00432 ± 0.000205	0.00457 ± 0.000923
0.0	$B_{m3\|Q}$	0.00394	0.000390	0.000399 ± 0.0000138	$0.000423 \pm 2.79\cdot10^{-5}$	$0.000375 \pm 2.00\cdot10^{-5}$	$0.000365 \pm 2.32\cdot10^{-5}$
	$B_{M1\|Q}$	0.410	0.130	0.137 ± 0.0156	0.165 ± 0.0250	0.156 ± 0.0195	0.138 ± 0.0122
	$B_{M2\|Q}$	0.268	0.0920	0.0952 ± 0.0105	0.0162 ± 0.00137	0.0209 ± 0.00379	0.0303 ± 0.0225
	$B_{M3\|Q}$	0.0127	0.00312	0.00296 ± 0.000193	0.00377 ± 0.000302	0.00336 ± 0.000281	0.00325 ± 0.000220
	W_B	6,606.51	6,940.64	6,954.27 ± 7.20	7,114.88 ± 25.83	7,053.08 ± 14.05	7,046.89 ± 43.31
	W_Q	17,599.15	17,685.84[b]	17,683.50 ± 15.54	17,683.45 ± 15.52	17,683.54 ± 15.54	17,683.51 ± 15.54
	$B_{Mm\|Q}$	0.0194	$3.22\cdot10^{-5}$	$2.23\cdot10^{-5} \pm 5.46\cdot10^{-5}$	$6.35\cdot10^{-5} \pm 7.22\cdot10^{-5}$	$5.02\cdot10^{-7} \pm 3.66\cdot10^{-7}$	$2.16\cdot10^{-5} \pm 5.45\cdot10^{-5}$
	$B_{m1\|Q}$	0.0194	$3.22\cdot10^{-5}$	$2.14\cdot10^{-5} \pm 5.50\cdot10^{-5}$	$6.32\cdot10^{-5} \pm 7.26\cdot10^{-5}$	0	$2.13\cdot10^{-5} \pm 5.47\cdot10^{-5}$
	$B_{m2\|Q}$	0.0107	$1.47\cdot10^{-8}$	0	0	0	0
0.5	$B_{m3\|Q}$	0.000419	$2.50\cdot10^{-6}$	$1.09\cdot10^{-6} \pm 7.54\cdot10^{-7}$	$7.08\cdot10^{-7} \pm 2.97\cdot10^{-7}$	$5.02\cdot10^{-7} \pm 3.66\cdot10^{-7}$	$4.08\cdot10^{-7} \pm 2.13\cdot10^{-7}$
	$B_{M1\|Q}$	0.0911	0.000701	0.00235 ± 0.00603	0.00277 ± 0.00335	0	0.00347 ± 0.00892
	$B_{M2\|Q}$	0.0534	$6.05\cdot10^{-8}$	0	0	0	0
	$B_{M3\|Q}$	0.00215	$1.31\cdot10^{-5}$	$2.17\cdot10^{-5} \pm 1.81\cdot10^{-5}$	$3.35\cdot10^{-5} \pm 2.18\cdot10^{-5}$	$1.60\cdot10^{-5} \pm 1.08\cdot10^{-5}$	$1.74\cdot10^{-5} \pm 6.86\cdot10^{-6}$
	W_B	5,239.94	5,296.26	5,297.08 ± 12.94	5,297.31 ± 12.77	5,297.23 ± 12.72	5,297.24 ± 12.86
	W_Q	16,027.90	16,028.14[c]	16,077.61 ± 15.03	16,077.61 ± 15.03	16,077.61 ± 15.03	16,077.61 ± 15.03
	$B_{Mm\|Q}$	$5.04\cdot10^{-5}$	$9\cdot10^{-10}$	0	0	0	0
	$B_{m1\|Q}$	$5.04\cdot10^{-5}$	$9\cdot10^{-10}$	0	0	0	0
	$B_{m2\|Q}$	$3.75\cdot10^{-5}$	$< 1\cdot10^{-10}$	0	0	0	0
1.0	$B_{m3\|Q}$	$1.21\cdot10^{-6}$	$< 1\cdot10^{-10}$	0	0	0	0
	$B_{M1\|Q}$	0.000400	$8.7\cdot10^{-10}$	0	0	0	0
	$B_{M2\|Q}$	0.000209	$< 1\cdot10^{-10}$	0	0	0	0
	$B_{M3\|Q}$	$7.17\cdot10^{-6}$	$< 1\cdot10^{-10}$	0	0	0	0
	W_B	3,355.75	3,355.88	3,350.97 ± 24.92	3,350.97 ± 24.92	3,350.97 ± 24.92	3,350.97 ± 24.92

Average o.f. values, and 95% confidence intervals, for simulation of the static (with a warm-up time of 8 h) and the dynamic routing model (with $t_0 = t_1 = 4$ h, update period τ and $b = 0.3$), a total simulation time of 48 h on the test network $\boldsymbol{\mathcal{E}}$, for different values of $\boldsymbol{\alpha}$, when the HMOR-S2 is used to update the routing plan. [a]99.47%; [b]100%; [c]99.75% of the approximate ideal values for the QoS flows revenue, W_Q^{ideal}, from the data in [25].

Table 7 Average o.f. values for simulation of the static and the dynamic routing model

α	O.f.	Initial solution	Analytical results	Static routing model results	Dynamic routing model results $\tau = 10$ m	Dynamic routing model results $\tau = 20$ m	Dynamic routing model results $\tau = 30$ m
	W_Q	20,859.85	21,690.16[a]	21,690.52 ± 37.22	21,688.67 ± 37.15	21,693.07 ± 37.19	21,693.46 ± 37.81
	$B_{Mm\|Q}$	0.110	0.00545	0.00619 ± 0.00119	0.00713 ± 0.000937	0.00591 ± 0.000643	0.00574 ± 0.00133
	$B_{m1\|Q}$	0.110	0.00545	0.00619 ± 0.00119	0.00713 ± 0.000937	0.00591 ± 0.000643	0.00574 ± 0.00133
	$B_{m2\|Q}$	0.0689	0.000465	0.000828 ± 0.000124	0.000944 ± 0.000213	0.000849 ± 0.000140	0.000816 ± 0.000193
0.0	$B_{m3\|Q}$	0.00308	0.000275	$0.000288 \pm 2.64 \cdot 10^{-5}$	$0.000141 \pm 2.11 \cdot 10^{-5}$	$0.000125 \pm 2.68 \cdot 10^{-5}$	$0.000155 \pm 2.85 \cdot 10^{-5}$
	$B_{M1\|Q}$	0.555	0.0613	0.0771 ± 0.0117	0.0723 ± 0.0282	0.0520 ± 0.00627	0.0509 ± 0.0214
	$B_{M2\|Q}$	0.378	0.00699	0.00794 ± 0.00174	0.00522 ± 0.00139	0.00475 ± 0.00129	0.00430 ± 0.000795
	$B_{M3\|Q}$	0.0190	0.00287	0.00288 ± 0.000312	0.00126 ± 0.000142	0.00103 ± 0.000274	0.000909 ± 0.000228
	W_B	6,738.68	7,158.14	7,161.10 ± 11.67	7,205.94 ± 19.96	7,172.37 ± 16.34	7,173.99 ± 11.71
	W_Q	17,611.81	17,685.89[b]	17,683.53 ± 15.54	17,683.54 ± 15.54	17,683.51 ± 15.56	17,683.51 ± 15.47
	$B_{Mm\|Q}$	0.0160	$1.04 \cdot 10^{-5}$	$8.59 \cdot 10^{-7} \pm 8.23 \cdot 10^{-7}$	0	$2.15 \cdot 10^{-5} \pm 5.52 \cdot 10^{-5}$	$2.09 \cdot 10^{-5} \pm 5.36 \cdot 10^{-5}$
	$B_{m1\|Q}$	0.0160	$1.04 \cdot 10^{-5}$	0	0	$2.15 \cdot 10^{-5} \pm 5.52 \cdot 10^{-5}$	$2.09 \cdot 10^{-5} \pm 5.36 \cdot 10^{-5}$
	$B_{m2\|Q}$	0.00926	$7.2 \cdot 10^{-9}$	0	0	0	0
0.5	$B_{m3\|Q}$	0.000371	$6.25 \cdot 10^{-7}$	$8.59 \cdot 10^{-7} \pm 8.23 \cdot 10^{-7}$	0	$5.64 \cdot 10^{-8} \pm 1.27 \cdot 10^{-7}$	$1.38 \cdot 10^{-7} \pm 3.00 \cdot 10^{-7}$
	$B_{M1\|Q}$	0.147	0.000128	0	0	0.00292 ± 0.00751	0.000891 ± 0.00229
	$B_{M2\|Q}$	0.0866	$4.45 \cdot 10^{-7}$	0	0	0	0
	$B_{M3\|Q}$	0.00353	$4.46 \cdot 10^{-6}$	$2.24 \cdot 10^{-5} \pm 1.72 \cdot 10^{-5}$	0	$2.26 \cdot 10^{-6} \pm 4.94 \cdot 10^{-6}$	$5.68 \cdot 10^{-6} \pm 9.32 \cdot 10^{-6}$
	W_B	5,247.65	5,296.57	5,297.18 ± 12.84	5,297.31 ± 12.80	5,297.33 ± 12.79	5,297.23 ± 12.86
	W_Q	16,025.69	16,028.14[c]	16,077.61 ± 15.03	16,077.61 ± 15.03	16,077.61 ± 15.03	16,077.61 ± 15.03
	$B_{Mm\|Q}$	0.000577	$5 \cdot 10^{-10}$	0	0	0	0
	$B_{m1\|Q}$	0.000577	$5 \cdot 10^{-10}$	0	0	0	0
	$B_{m2\|Q}$	0.000334	$< 1 \cdot 10^{-10}$	0	0	0	0
1.0	$B_{m3\|Q}$	$1.16 \cdot 10^{-5}$	$< 1 \cdot 10^{-10}$	0	0	0	0
	$B_{M1\|Q}$	0.00650	$1.27 \cdot 10^{-8}$	0	0	0	0
	$B_{M2\|Q}$	0.00347	$< 1 \cdot 10^{-10}$	0	0	0	0
	$B_{M3\|Q}$	0.000123	$2 \cdot 10^{-10}$	0	0	0	0
	W_B	3,354.76	3,355.88	3,350.97 ± 24.92	3,350.97 ± 24.92	3,350.97 ± 24.92	3,350.97 ± 24.92

Average o.f. values, and 95% confidence intervals, for simulation of the static (with a warm-up time of 8 h) and the dynamic routing model (with $t_0 = t_1 = 4$ h, update period τ and $b = 0.3$), a total simulation time of 48 h on the test network $\mathcal{G}$, for different values of α, when the HMOR-S2 is used to update the routing plan. [a]99.90%; [b]100%; [c]99.75% of the approximate ideal values for the QoS flows revenue, W_Q^{ideal}, from the data in [25].

Table 8 Average o.f. values for simulation of the static and the dynamic routing model

α	O.f.	Initial solution	Analytical results	Static routing model results	Dynamic routing model results $\tau = 10$ m	Dynamic routing model results $\tau = 20$ m	Dynamic routing model results $\tau = 30$ m
0.0	W_Q	20,358.90	21,616.01[a]	21,597.91 ± 30.13	21,606.57 ± 44.40	21,616.80 ± 35.49	21,610.35 ± 34.71
	$B_{Mm\|Q}$	0.169	0.0224	0.0245 ± 0.00175	0.0245 ± 0.00173	0.0216 ± 0.00141	0.0225 ± 0.00146
	$B_{m1\|Q}$	0.169	0.0224	0.0245 ± 0.00175	0.0245 ± 0.00173	0.0216 ± 0.00141	0.0225 ± 0.00146
	$B_{m2\|Q}$	0.111	0.00341	0.00580 ± 0.000523	0.00492 ± 0.000528	0.00456 ± 0.000432	0.00506 ± 0.000633
	$B_{m3\|Q}$	0.00536	0.000596	$0.000601 \pm 4.99{\cdot}10^{-5}$	$0.000417 \pm 6.38{\cdot}10^{-5}$	$0.000461 \pm 7.60{\cdot}10^{-5}$	$0.000491 \pm 9.88{\cdot}10^{-5}$
	$B_{M1\|Q}$	0.711	0.146	0.157 ± 0.0231	0.161 ± 0.0326	0.148 ± 0.0176	0.151 ± 0.0277
	$B_{M2\|Q}$	0.518	0.0145	0.0215 ± 0.00229	0.0192 ± 0.00232	0.0203 ± 0.00429	0.0183 ± 0.00445
	$B_{M3\|Q}$	0.0293	0.00362	0.00376 ± 0.000476	0.00339 ± 0.000239	0.00299 ± 0.000433	0.00315 ± 0.000510
	W_B	6,434.17	6,927.67	6,935.83 ± 10.55	7,027.38 ± 34.64	6,991.64 ± 8.57	6,971.23 ± 20.73
0.5	W_Q	17,419.40	17,685.82[b]	17,683.45 ± 15.55	17,683.50 ± 15.56	17,683.50 ± 15.55	17,683.49 ± 15.55
	$B_{Mm\|Q}$	0.0558	$4.86{\cdot}10^{-5}$	$4.55{\cdot}10^{-5} \pm 0.000109$	$2.23{\cdot}10^{-5} \pm 5.38{\cdot}10^{-5}$	$2.23{\cdot}10^{-5} \pm 5.46{\cdot}10^{-5}$	$2.27{\cdot}10^{-5} \pm 5.44{\cdot}10^{-5}$
	$B_{m1\|Q}$	0.0558	$4.86{\cdot}10^{-5}$	$4.28{\cdot}10^{-5} \pm 0.000110$	$2.12{\cdot}10^{-5} \pm 5.44{\cdot}10^{-5}$	$2.14{\cdot}10^{-5} \pm 5.50{\cdot}10^{-5}$	$2.14{\cdot}10^{-5} \pm 5.50{\cdot}10^{-5}$
	$B_{m2\|Q}$	0.0335	$1.78{\cdot}10^{-7}$	0	0	0	0
	$B_{m3\|Q}$	0.00143	$2.27{\cdot}10^{-6}$	$3.28{\cdot}10^{-6} \pm 1.41{\cdot}10^{-6}$	$1.27{\cdot}10^{-6} \pm 7.82{\cdot}10^{-7}$	$1.21{\cdot}10^{-6} \pm 8.71{\cdot}10^{-7}$	$1.66{\cdot}10^{-6} \pm 5.64{\cdot}10^{-7}$
	$B_{M1\|Q}$	0.327	0.000910	0.00273 ± 0.00702	0.000850 ± 0.00219	0.00179 ± 0.00461	0.00179 ± 0.00461
	$B_{M2\|Q}$	0.205	$9.58{\cdot}10^{-7}$	0	0	0	0
	$B_{M3\|Q}$	0.00906	$1.61{\cdot}10^{-5}$	$3.56{\cdot}10^{-5} \pm 9.78{\cdot}10^{-6}$	$4.35{\cdot}10^{-5} \pm 1.69{\cdot}10^{-5}$	$2.93{\cdot}10^{-5} \pm 1.54{\cdot}10^{-5}$	$2.94{\cdot}10^{-5} \pm 1.61{\cdot}10^{-5}$
	W_B	5,119.13	5,295.76	5,296.16 ± 13.30	5,297.13 ± 12.82	5,296.73 ± 12.83	5,296.48 ± 13.08
1.0	W_Q	15,998.35	16,028.14[c]	16,077.61 ± 15.03	16,077.61 ± 15.03	16,077.61 ± 15.03	16,077.61 ± 15.03
	$B_{Mm\|Q}$	0.00678	$1.78{\cdot}10^{-8}$	0	0	0	0
	$B_{m1\|Q}$	0.00678	$1.78{\cdot}10^{-8}$	0	0	0	0
	$B_{m2\|Q}$	0.00416	$< 1 \cdot 10^{-10}$	0	0	0	0
	$B_{m3\|Q}$	0.000153	$1.6{\cdot}10^{-9}$	0	0	0	0
	$B_{M1\|Q}$	0.0530	$4.76{\cdot}10^{-7}$	0	0	0	0
	$B_{M2\|Q}$	0.0298	$< 1 \cdot 10^{-10}$	0	0	0	0
	$B_{M3\|Q}$	0.00113	$1.10{\cdot}10^{-8}$	0	0	0	0
	W_B	3,341.90	3,355.88	3,350.97 ± 24.92	3,350.97 ± 24.92	3,350.97 ± 24.92	3,350.97 ± 24.92

Average o.f. values, and 95% confidence intervals, for simulation of the static (with a warm-up time of 8 h) and the dynamic routing model (with $t_0 = t_1 = 4$ h, update period τ and $b = 0.3$), a total simulation time of 48 h on the test network $\mathcal{H}$, for different values of α, when the HMOR-S2 is used to update the routing plan. [a]99.56%; [b]100%; [c]99.75% of the approximate ideal values for the QoS flows revenue, W_Q^{ideal}, from the data in [25].

networks), different values for the times t_0 and t_1, for τ and for b can be tried, and the results displayed in the tables correspond to those obtained with $t_0 = t_1 = 4$ h; different values of τ; $b = 0.3$; total simulation time of 48 h.

Analytical results

The resolution heuristic manages to start off with an initial solution with poor values for the o.f.s and still finish with a solution with significantly better values. The values for all the o.f.s for all values of α are improved through the heuristic. The QoS revenue of the final solutions is only slightly worse than that of the ideal optimal solutions, as expected. Therefore, we can consider that the resolution heuristic has managed to find an adequate 'good' compromise routing solution to the routing problem P-M2-S2. In fact, these experimental results for three traffic matrices for each network show that the expected QoS revenue obtained with our heuristic is never less than 99.47% of that upper bound while a substantial improvement on the other o.f.s can be obtained with respect to the initial solution, using only shortest path first choice routing, typical of Internet routing conventional algorithms.

Simulation results

The analytical results and the corresponding static routing model simulation results are of similar magnitude, but the analytical results tend to be better, as expected. In particular, considering the results obtained for network $\mathcal{M}$, the analytical results for the QoS flows revenue W_Q are clearly better than the corresponding static routing model simulation results, for all the values of α. Generally speaking, for this network, the analytical results are not inside the 95% confidence interval of the static routing model simulation results and only for $\alpha = 0.0$ do we get a result where a first-level o.f. ($B_{Mm|Q}$) analytical value is in the corresponding confidence interval. Remember that the simulation results for the routing model are average values of performance in a great number of routing update intervals, while the analytical results are obtained in ideal steady-state traffic conditions and use approximate teletraffic sub-models.

For the other networks, the analytical and the simulation results for W_Q are much closer and the analytical result for that o.f. is inside the 95% confidence interval for $\alpha = 0.0$ and $\alpha = 0.5$. For $\alpha = 1.0$, the analytical value of W_Q is actually worse than the corresponding simulation result. In this situation of lower traffic load in networks $\mathcal{E}, \mathcal{G}, \mathcal{H}$, there are many instances throughout the executed simulations where the blocking estimate for certain services is 0, meaning that all the offered calls of those services are actually carried. This is the reason why so high values of the estimate of the QoS traffic revenue are obtained, surpassing the analytical values. Note that these are situations where the occurrence of blocking is a rare event. It is well known in statistics that in these cases, the uncertainty in the estimates is very high, as reflected in the very high relative half length of the calculated 95% confidence intervals of the blocking probabilities.

The differences between the analytical and the simulation results for the static routing model are mainly due to the imprecision effects intrinsic to the analytic/numerical solution, in particular those associated with the simplifications of the traffic model, and the associated error propagation. The analytical model is a simplification which tends to underestimate the blocking probabilities in the network (and therefore to overestimate the revenues), because the overflow traffic is treated as Poisson traffic. This error propagates throughout the complex and lengthy numerical calculations associated with the solution of the traffic model, involving the solution of large systems of implicit non-linear equations. Further simplifications assumed in the stochastic model for the traffic in the links are a superposition of independent Poisson flows and an independent occupation of the links. Although we envisage that a more precise and realistic representation of the traffic flows allows for better estimates of the blocking probabilities, the approximations in our model can be deemed appropriate and actually unavoidable in this context. This results from a compromise between the precision of the representation of the traffic flows and the computational burden of the numerical resolutions throughout the execution of the heuristic algorithm. For this reason, some robust and well-tested classical numerical algorithms (the Kaufman/Roberts algorithm and fixed-point iterators) are used to estimate the blocking probabilities of this system. It is important to note that as previously mentioned the type of error introduced by this approach does not compromise the inequality relations between the o.f. values. In fact, the focus of the optimization model is on the relative value of the results of the traffic model rather than on the absolute accuracy of such values, as explained in the sub-section on Dealing with IIU in the model.

The stochastic nature of the traffic offered to the network leads to some uncertainty in the results. In particular, for the state-dependent periodic-type dynamic routing model, the traffic flow means are periodically updated (with period τ) via a statistical estimate (first-order moving average) based on real-time measurements, dependent on a parameter b. The influence of these parameters on the final global routing solution can be analyzed.

In a first set of experiments, the update period τ is fixed and different values are tested for b. This parameter has to be set in order to reflect a compromise between the stability of the estimate and the quick response to variations in the partial estimate of the average value of the traffic offered by a flow to the network in the previous interval, $\tilde{X}$. The best results for the first-level o.f.s can be obtained with $b = 0.3$ (those are the values displayed in Tables 5, 6, 7, and 8). For $b = 0.4$, the results are only slightly worse. However, for smaller values and for higher values of b, the results for W_Q and $B_{Mm|Q}$ are worse than those displayed in the tables. An increase(decrease) in b means that the estimate of the average traffic offered to the network by a flow, $\tilde{x}$, gives more(less) importance to information on the previous interval and less(more) importance to the previous estimates of traffic obtained throughout the duration of the experiment. These results show that a balance between these two aspects is clearly desirable and no excessive weight should be attributed to either of them. However, as the best results are obtained with $b < 0.5$, it appears that the stability of the estimate is slightly more important than a rapid response to variations in the offered traffic. Note that these results are in accordance with the traffic engineering recommendations in [27] . Notice that the possibility of network changes or sudden strong alterations in traffic patterns are not being considered in this study. This type of events would have required the traffic estimate to be able to respond better to very rapid variations in the offered traffic, so a value of $b > 0.5$ would have been expected to be more appropriate in that situation. Probably, that would have been the case if the network traffic in the simulation model was not modeled as Poisson traffic but rather as self-similar traffic, with its typical bursts of traffic (see for instance [28]), a situation out of the scope of the present study. In a second set of experiments, the parameter b is kept at 0.3 and different values are tested for the update period τ. The best results for the first-level o.f.s are those obtained with $\tau = 10, 20, 30$ m (these values are displayed in

Tables 5, 6, 7, and 8). Only for these parameter values is the dynamic routing model capable of attaining the performance values corresponding to the analytic upper bounds for the static solution. For other parameter values, the experimental results are not so good.

A final remark on the confidence intervals for each o.f.: their length is of the same order for both the static and the dynamic routing model.

The results for network $\mathcal{M}$, which is not dimensioned for low blocking probabilities, have shown that for a smaller value of τ (5 m), the results for W_Q and $B_{Mm|Q}$ are slightly worse than those displayed in Table 5. However, for a higher value of τ (30 m), the results are also worse (see Table 5). Therefore, in situations of high blocking probability, there is no need to update the traffic estimate too often (with update periods of the order of only a few minutes), but a very long update period is also undesirable. Notice that a value of $b = 0.3$ means that the update information focuses more on the medium/long term rather than on the short term, as the estimate of the average traffic offered to the network by a flow incorporates more information on the estimates of traffic that have been obtained throughout the duration of the experiment, rather than on the previous interval. In very long intervals, slight changes in the offered traffic pattern are more likely to occur and these tend to be disregarded, if the update periods are very long (30 min or more) because of the lower importance given to the information on the specific previous interval, in the traffic estimate.

For higher values of α (corresponding to lower load), we realize that the simulation results for each of the networks $\mathcal{E}, \mathcal{G}, \mathcal{H}$ are the same regardless of the value of τ. As mentioned earlier, these networks, for $\alpha = 0.5$ and $\alpha = 1.0$, are dimensioned for low blocking probabilities, and the blocking estimates are close to 0 in many cases, as all the offered calls tend to be carried. In this case, the total expected network revenues associated with QoS and BE traffic flows tend to have maximum values regardless of the value of τ.

In global terms and as expected, the results obtained with the dynamic routing model are better (or approximately the same in the worst case) than those obtained with the static routing model. This is especially noticeable for $\alpha = 0.0$ for all the networks, which is the situation of higher load. This shows that in situations of higher load, the dynamic model is well calibrated for these networks, in terms of the choice of the initial routing solutions to be used by the heuristic and the choice of the routing updating period. In the dynamic routing model, the routing plan is adjusted throughout the simulation run, in accordance with the traffic random fluctuations around the average values corresponding to the nominal traffic matrix defined in steady-state conditions.

The o.f. values are intrinsically imprecise, due to the simplifications and approximations assumed in the stochastic model for the traffic in the links, and to the numerical errors associated with the resolution of the system of equations of the traffic model which propagate throughout the resolution procedure. Still, the representation of the traffic flows as independent Poisson processes and the independence in the occupations of the links may be considered a good compromise between the exactness of the traffic model and the computational burden for solving the analytical model.

Other imprecision effects are due to an instability phenomenon which may potentially arise in the path selection procedure, when all the network routes are liable to change. To avoid oscillations between certain solutions that can possibly lead to a poor global network performance, the core algorithm that seeks new routing solutions (MMRA-S2) is applied only to specific traffic flows that are carefully chosen (according to the value of $\xi(f_s)$, as mentioned earlier), so as to try and improve the o.f. values.

The inaccuracy in the o.f. values may be considered inevitable, due to the very great combinatorial complexity of the optimization model including strong interdependencies among key mathematical entities of the model. The heuristic resolution has different mechanisms, throughout the improvement cycles of the o.f.s, which try to deal with these complex interdependencies, in order to minimize the impact of these inaccuracies in the quality of the obtained compromise solutions.

Conclusions

In this paper, a stochastic two-level hierarchical multiobjective routing model for MPLS networks with two service classes and alternative routing was reviewed and analyzed concerning sources of imprecision, inaccuracy (or inaccurate determination), and uncertainty (IIU). Key issues raised by its high complexity were discussed in a systematic manner, as well as the major factors that constitute the sources of IIU of the model. The mechanisms used by the developed resolution heuristic approach to deal with these issues were also described.

The most important innovative aspect of this paper was the presentation of analytical and stochastic simulation experiments (both for static and dynamic versions of the routing model), enabling the evaluation of inaccuracies intrinsic to the model and to the analytic/numerical resolution method, as well as the evaluation of a particular form of 'internal uncertainty' associated with the necessity of representing a system of preferences (of a 'virtual decision maker') in a fully automated application. The possible effects of these inaccuracies on the results of the heuristic resolution procedure were discussed as well as the forms of minimizing their impacts on the heuristic effectiveness. Furthermore, the experimental study, using discrete-event stochastic simulation, enabled the validation of the routing model results and the evaluation of effects in the model results of the uncertainty associated with the offered traffic estimates, in a dynamic version of the routing method.

The analysis of these types of IIU factors, their effects on the results of the resolution approach of the routing optimization model, and the described general type of procedures for dealing with these issues are in our opinion relevant to other routing models with similar features. That is, we think that dealing with these issues in a proper manner is an important task in the context of multiobjective routing models, based on network-wide optimization approaches (where the combined effect and interactions among the traffic flows have to be explicitly represented) and using a stochastic representation of the traffic flows. Having in mind the high complexity and heavy numerical costs of the addressed model, probably several of the methods and techniques used for dealing with these IIU issues, may be adapted to various network routing models of similar nature. This is naturally an important task that justifies future works on IIU issues (and the ways to deal with these issues) focused on specific types of stochastic network-wide routing optimization models.

Endnote

[a] A weakly dominated solution is a feasible solution such that there is at least another feasible solution with better values for some of the o.f.s and equal values for the other o.f.s.

Appendix

Notation used in the model

The notations used in the model are shown in Table 9.

Table 9 Notation used in the model

Notations	Description
O.f. calculation	
$W_{Q(B)} = \sum_{s\in\boldsymbol{\mathcal{S}}_{Q(B)}} A_s^c w_s$	Total expected network revenue associated with QoS(BE) traffic flows
$B_{Mm\|Q} = \max_{s\in\boldsymbol{\mathcal{S}}_Q}\{B_{ms}\}$	Maximal average blocking probability among all QoS service types
$B_{ms\|Q} = \frac{1}{A_s^o}\sum_{f_s\in\boldsymbol{\mathcal{F}}_s} A(f_s)B(f_s)$	Mean blocking probabilities for flows of type $s \in \boldsymbol{\mathcal{S}}_Q$
$B_{Ms\|Q} = \max_{f_s\in\boldsymbol{\mathcal{F}}_s}\{B(f_s)\}$	Maximal blocking probability defined over all flows of type $s \in \boldsymbol{\mathcal{S}}_Q$
Blocking probabilities calculation	
$B(f_s)$	Node-to-node blocking probability for all flows $f_s \in \boldsymbol{\mathcal{F}}_s$
$B_{ks} = \boldsymbol{\mathcal{B}}_s\left(\overline{d_k}, \overline{\rho_k}, C_k\right)$	Blocking probabilities for micro-flows of service type s in link l_k
$\boldsymbol{\mathcal{B}}_s$	Basic function (implicit in the teletraffic analytical model) to calculate B_{ks}
Decision variables	
$\overline{R} = \cup_{s=1}^{\|\boldsymbol{\mathcal{S}}\|} R(s)$	Network routing plans
$R(s) = \cup_{f_s\in\boldsymbol{\mathcal{F}}_s} R(f_s), s \in \boldsymbol{\mathcal{S}}_Q \cup \boldsymbol{\mathcal{S}}_B$	Set of all the feasible routes for the traffic flows of type s
$R(f_s) = (r^p(f_s)), p = 1, \cdots, M$	First, second, $\cdots$, M-th choice route for flow f_s
Path metrics and auxiliary parameters - MMRA-S2	
$m_{ks}^1 = c_{ks}^{Q(B)}$	Marginal implied costs
$m_{ks}^2 = -\log(1 - B_{ks})$	Marginal blocking probabilities
$\boldsymbol{\mathcal{D}}(f_s)$	Set of all feasible loopless paths for flow f_s
Simulation parameters	
$T = [T_{ij}]$	Base matrix with offered bandwidth values from node i to node j (Mbps)
α	Compensation parameter
t_0	Duration of the first stage of the initialization phase, where only periodical updates of the estimate of the offered traffic are performed
t_1	Duration of the second stage of the initialization phase, where periodical updates of the estimate of the offered traffic and of the routing plan are performed
$t_{\text{warm-up}} = t_0 + t_1$	Duration of the initialization phase
τ	Update period of the estimates of the offered traffic and of the network routing plans
$\tilde{x}_n(f_s) = (1-b)\tilde{x}_{n-1}(f_s) + b\tilde{X}_{n-1}(f_s)$	Estimate of the average traffic offered to the network by the flow f_s in the time interval $[n\tau; (n+1)\tau[$
$\tilde{X}_{n-1}(f_s)$	Estimator of the average value of the traffic offered by f_s to the network in the previous interval $[(n-1)\tau; n\tau[$
b	Compromise value between the need to obtain a quick response of the estimator to rapid fluctuations in $\tilde{X}(f_s)$ and the stability of the long-run variations
Miscellany of auxiliary parameters	
f_s	Flow of service type s
$\boldsymbol{\mathcal{S}}_{Q(B)}$	Set of QoS(BE) service types
A_s^o	Total traffic offered by flows of type s
A_s^c	Carried traffic for service type s

Table 9 Notation used in the model ***(continued)***

$A(f_s)$	Mean traffic offered associated with $f_s \in \mathcal{F}_s$
w_s	Expected revenue per call of service type s
ρ_{ks}	Reduced traffic loads offered by flows of type s to l_k
$\overline{\rho_k} = (\rho_{k1}, \cdots, \rho_{k\vert\mathcal{S}\vert})$	Vector of reduced traffic loads
d_{ks}	Equivalent effective bandwidths for flows of type s in l_k
$\overline{d_k} = (d_{k1}, \cdots, d_{k\vert\mathcal{S}\vert})$	Vector of equivalent effective bandwidths
d'_s	Required bandwidth for service s (kbps)
$d_s = \frac{d'_s}{u_0}$	Required effective bandwidth for service s (channels)
$\mathcal{M}, \mathcal{E}, \mathcal{G}, \mathcal{H}$	Test networks
$\vert\mathcal{N}\vert$	Number of nodes in the network
$\vert\mathcal{L}\vert$	Number of unidirectional links in the network
C'_k	Link bandwidth (Mbps)
$C_k = \left\lceil \frac{C'_k}{u_0} \right\rceil$	Link capacity (channels)
u_0	Basic unit capacity
h_s	Average duration of a type s call
D_s	Maximum number of arcs for a type s call
$\xi(f_s)$	Function for choosing candidate paths for flow f_s for possible routing improvement
δ	Average node degree of a network

Formalization of the heuristic resolution approach

The formalization of the heuristic approach (HMOR-S2$_{\mathrm{PAS}}$ – *H*ierarchical *M*ulti*o*bjective *R*outing considering *2* classes of *S*ervice with a *P*areto *A*rchive *S*trategy) follows.

The version without the Pareto archive, i.e., HMOR-S2, is equivalent to this one, without steps VII; VIII.11b•– Add $\overline{R}_a$ to the archive $\cdots$; VIII.11b• Else {and instructions therein}; VIII.11b•– Add $\overline{R}_a$ to the archive $\cdots$; VIII.11b• Else {and instructions therein}; X.

As for the 'core' algorithm MMRA-S2, it is basically an adaptation to the present model of the bi-objective constrained shortest path algorithm in [29] which is an extension of the algorithm in [14] to a multiservice environment.

HMOR-S2$_{PAS}$

I. $\overline{R}_a \leftarrow \overline{R}_o$
II. Compute $\overline{B}$ and $W_Q, B_{Mm|Q}$ for $\overline{R}_a$
III. $W_Q^o \leftarrow W_Q, B_{Mm|Q}^o \leftarrow B_{Mm|Q}$
IV. $\overline{R}_* \leftarrow \overline{R}_a$
V. Compute $\overline{B}$ for $\overline{R}_a$
  Compute $W_Q, B_{Mm|Q}, B_{ms|Q}, B_{Ms|Q} (\forall s \in \mathcal{S}_Q)$, W_B for $\overline{R}_a$
VI. $\max\{W_Q\} \leftarrow W_Q, \min\{B_{Mm|Q}\} \leftarrow B_{Mm|Q}$
  $\min\{B_{ms|Q}\} \leftarrow B_{ms|Q}, \min\{B_{Ms|Q}\} \leftarrow B_{Ms|Q} (\forall s \in \mathcal{S}_Q)$ and $\max\{W_B\} \leftarrow W_B$
VII. Add $\overline{R}_a$ to the archive
VIII. For $nPaths = |\overline{\mathcal{F}}|$ to $nPaths = 1$

1. For $ape = 0$ to $ape = 1$
   (a) If $ape = 0$, $z_{APR} \leftarrow 1.0$
       Else, $z_{APR} \leftarrow 0.01 \cdot nPaths$
   (b) For $s = 1$ to $s = |\mathcal{S}|$

i. For $nCycles = 1$ to $nCycles = 0$

A. Compute $\overline{B}$ and $\overline{c}^Q, s \in \mathcal{S}_Q$ or $\overline{c}^B, s \in \mathcal{S}_B$ for $\overline{R}_a$

B. Compute and order the values of the function $\xi(f_s)$, with $\xi(f_s) = F_L(f_s)$ if $nCycles = 1$ and $\xi(f_s) = F_C^{Q(B)}(f_s)$ if $nCycles = 0$

C. Find the $nPaths$ flows with lower value of $\xi(f_s)$

D. Compute with MMRA-S2 new candidate paths for the corresponding O-D pairs and define a new set of first and second choice paths for the service s, $\overline{R}_a(s)$, according to the rules established for each service

E. Compute $\overline{B}$ for $\overline{R}_a$

Compute $B_{ms|Q}, B_{Ms|Q}$ if $s \in \mathcal{S}_Q$ or W_B if $s \in \mathcal{S}_B$ for $\overline{R}_a$

Compute $W_Q, B_{Mm|Q}$

F. If $s \in \mathcal{S}_Q$ then

- If $[(B_{ms|Q} < \min\{B_{ms|Q}\}$ and $B_{Ms|Q} < \min\{B_{Ms|Q}\})$ and $(W_Q > \max\{W_Q\}$ and $B_{Mm|Q} < \min\{B_{Mm|Q}\})]$ then
  - $\min\{B_{ms|Q}\} \leftarrow B_{ms|Q}, \min\{B_{Ms|Q}\} \leftarrow B_{Ms|Q}$
  - $\max\{W_Q\} \leftarrow W_Q, \min\{B_{Mm|Q}\} \leftarrow B_{Mm|Q}$
  - $\overline{R}_*(s) \leftarrow \overline{R}_a(s)$
  - Add $\overline{R}_a$ to the archive (If it is already full, the priority regions of the solutions in the archive must be evaluated and the first solution found in the worst region of the archive should be removed first.)

- Else,
  - If $[(B_{ms|Q} > \min\{B_{ms|Q}\}$ and $B_{Ms|Q} > \min\{B_{Ms|Q}\})$ and $(W_Q < \max\{W_Q\}$ and $B_{Mm|Q} > \min\{B_{Mm|Q}\})]$ then
    * (Discard $\overline{R}_a$)
  - Else,
    * If there is at least one solution X in the archive for which $[(B_{ms|Q} > B_{ms|Q}(X)$ and $B_{Ms|Q} > B_{Ms|Q}(X))$ and $(W_Q < W_Q(X)$ and $B_{Mm|Q} > B_{Mm|Q}(X))]$, i.e. X dominates $\overline{R}_a$ in terms of the o.f. of interest,
      † (Discard $\overline{R}_a$)
    * Else ($\overline{R}_a$ and the solutions in the archive are non-dominated)
      † If the archive is not full,

‡ Add $\overline{R}_a$ to the archive

† Else

‡ Evaluate the priority regions of $\overline{R}_a$ and the solutions in the archive;

‡ If $\overline{R}_a$ is in the worst priority region

o (Discard $\overline{R}_a$)

‡ Else,

o Remove the first solution found in the worst region of the archive;

o Add $\overline{R}_a$ to the archive

G. Else ($s \in \mathcal{S}_B$)

- If $[(W_B > \max\{W_B\})$ and $(W_Q > \max\{W_Q\}$ and $B_{Mm|Q} < \min\{B_{Mm|Q}\})]$ then

  - $\max\{W_B\} \leftarrow W_B$
  - $\max\{W_Q\} \leftarrow W_Q, \min\{B_{Mm|Q}\} \leftarrow B_{Mm|Q}$
  - $\overline{R}_*(s) \leftarrow \overline{R}_a(s)$
  - Add $\overline{R}_a$ to the archive (If it is already full, the priority regions of the solutions in the archive must be evaluated and the first solution found in the worst region of the archive should be removed first.)

- Else,

  - If $[(W_B < \max\{W_B\})$ and $(W_Q < \max\{W_Q\}$ and $B_{Mm|Q} > \min\{B_{Mm|Q}\})]$

    * (Discard $\overline{R}_a$)
  - Else,

    * If there is at least one solution X in the archive for which $[(W_B < W_B(X))$ and $(W_Q < W_Q(X)$ and $B_{Mm|Q} > B_{Mm|Q}(X))]$, i.e. X dominates $\overline{R}_a$ in terms of the o.f. of interest,

      † (Discard $\overline{R}_a$)

    * Else ($\overline{R}_a$ and the solutions in the archive are non-dominated)

      † If the archive is not full,

      ‡ Add $\overline{R}_a$ to the archive

      † Else

      ‡ Evaluate the priority regions of $\overline{R}_a$ and the solutions in the archive;

‡ If $\overline{R}_a$ is in the worst priority region
o (Discard $\overline{R}_a$)
‡ Else,
o Remove the first solution found in the worst region of the archive;
o Add $\overline{R}_a$ to the archive

H. $\overline{R}_a(s) \leftarrow \overline{R}_*(s)$

End of the cycle *For (nCycles)*

End of the cycle *For (s)*

End of the cycle *For (ape)*

End of the cycle *For (nPaths)*

IX. If $W_Q^o > \max\{W_Q\}$ or $B_{Mm|Q}^o < \min\{B_{Mm|Q}\}$ then

1. The best solution is $\overline{R}_o$.

X. Else,

1. Evaluate the priority regions of the solutions in the archive;
2. The final solution is found in the best region of the archive, using a reference point-based procedure.

XI. Compute the o.f. values for the final solution.

MMRA-S2

I. For each link l_k in the network, compute the path metrics: marginal implied costs $m_{ks}^1 = c_{ks}^{Q(B)}$ and marginal blocking probabilities $m_{ks}^2 = -\log(1 - B_{ks})$, where $s \in \mathcal{S}_{Q(B)}$.
II. Compute average values $c_{av}^{Q(B)}(s) = \frac{1}{|\mathcal{L}|}\sum_{l_k \in \mathcal{L}} c_{ks}^{Q(B)}$ and $B_{av_{\log}}(s) = \frac{1}{|\mathcal{L}|}\sum_{l_k \in \mathcal{L}} (-\log(1 - B_{ks}))$.
III. Compute the weights $\epsilon_1^{Q(B)}(s) = \frac{B_{av_{\log}}(s)}{c_{av}^{Q(B)}(s) + B_{av_{\log}}(s)}$ and $\epsilon_2^{Q(B)}(s) = \frac{c_{av}^{Q(B)}(s)}{c_{av}^{Q(B)}(s) + B_{av_{\log}}(s)}$.
IV. For each link l_k in the network, compute a cost of the link given by a weighted sum $\epsilon_1^{Q(B)}(s) c_{ks}^{Q(B)} + \epsilon_2^{Q(B)}(s)(-\log(1 - B_{ks}))$.
V. Solve the problem $\min_{r(f_s) \in \mathcal{D}(f_s)} \left\{ \sum_{l_k \in r(f_s)} \left(\epsilon_1^{Q(B)}(s) c_{ks}^{Q(B)} + \epsilon_2^{Q(B)}(s)(-\log(1 - B_{ks})) \right) \right\}$, $f_s \in \mathcal{F}_s, s \in \mathcal{S}$ using MPS [30], which allows for the computation of a set of κ paths for each flow $f_s \in \mathcal{F}_s$ for which the paths may change, ordered according to this cost function.
VI. Identify the priority region to which each of the possible paths belongs to, where the first priority region is A, then B_2 and B_1, followed by C and finally D. The limits of the priority regions (Figure 4) are given by

$$-\log(1 - B_{req}(s)) = -D_s \log(1 - B^-(s)) \qquad -\log(1 - B_{ac}(s)) = -D_s \log(1 - B^+(s))$$
$$c_{req}(s) = D_s c^-(s) \qquad c_{ac}(s) = D_s c^+(s)$$

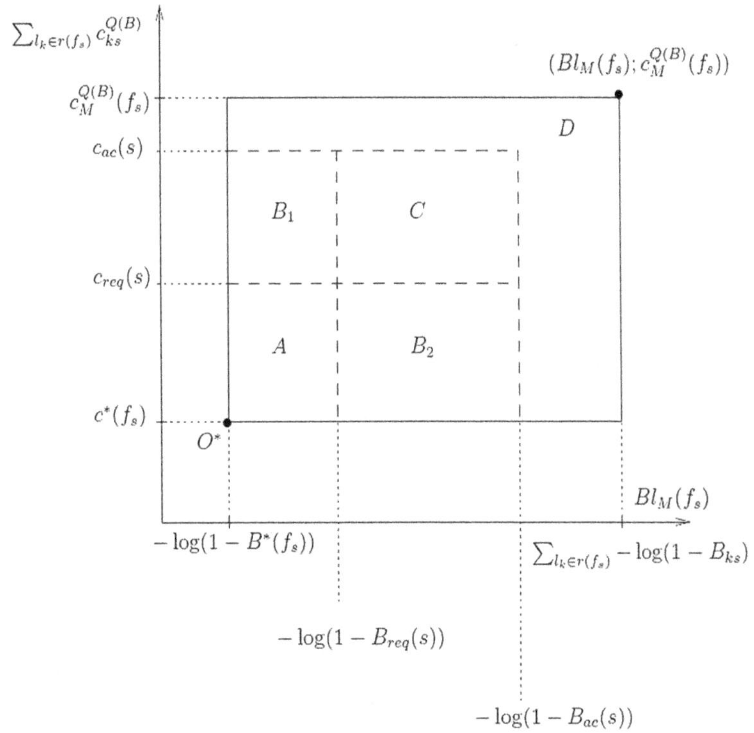

Figure 4 Priority regions. These priority regions are used to rank the possible paths.

where

$$B^-(s) = B_{av}(s) - \Delta B(s) \qquad B^+(s) = B_{av}(s) + \Delta B(s)$$

$$B_{av}(s) = \frac{1}{|\mathcal{L}|}\sum_{l_k \in \mathcal{L}} B_{ks} \qquad \Delta B(s) = \frac{B_{av}(s) - \min_{l_k \in \mathcal{L}}\{B_{ks}\}}{2}$$

$$c^-(s) = c_{av}(s) - \Delta c(s) \qquad c^+(s) = c_{av}(s) + \Delta c(s)$$

$$c_{av}(s) = \begin{cases} c_{av}^Q(s) & \text{(QoS flow)} \\ c_{av}^B(s) & \text{(BE flow)} \end{cases} \qquad \Delta c(s) = \begin{cases} \frac{c_{av}(s) - \min_{l_k \in \mathcal{L}}\{c_{ks}^Q\}}{2} & \text{(QoS flow)} \\ \frac{c_{av}(s) - \min_{l_k \in \mathcal{L}}\{c_{ks}^B\}}{2} & \text{(BE flow)} \end{cases}$$

VII. The first choice route for each flow, $r^1(f_s)$, is a non-dominated solution in the best possible priority region with the lowest value of implied cost.

VIII. Consider all the arc-disjoint solutions from $r^1(f_s)$.

IX. The second choice route for each flow, $r^2(f_s)$, is a non-dominated solution in the best possible priority region with the lowest value of implied cost.

Acknowledgements
The authors would like to thank the anonymous referees for their helpful comments and remarks. This work was financially supported by programme COMPETE of the EC Community Support Framework III and cosponsored by the EC fund FEDER and national funds (FCT - PTDC/EEA-TEL/101884/2008 and PEst-C/EEI/UI0308/2011).

References

1. Clímaco, JCN, Craveirinha, JMF: Multicriteria analysis in telecommunication network planning and design – Problems and issues. In: Figueira, J, Greco, S, Ehrgott, M (eds.) Multiple Criteria Decision Analysis: State of the Art Surveys, International Series in Operations Research & Management Science, *vol. 78*, pp. 899–951. Springer, New York (2005)
2. Clímaco, JCN, Craveirinha, JMF, Pascoal, MMB: Multicriteria routing models in telecommunication networks – Overview and a case study. In: Shi, Y, Olson, DL, Stam, A (eds.) Advances in Multiple Criteria Decision Making and Human Systems Management: Knowledge and Wisdom, pp. 17–46. IOS Press, Amsterdam (2007)

3. Wierzbicki, AP, Burakowski, W: A conceptual framework for multiple-criteria routing in QoS IP networks. Int. Trans. Oper. Res. **18**(3), 377–399 (2011)
4. Craveirinha, J, Girão-Silva, R, Clímaco, J: A meta-model for multiobjective routing in MPLS networks. Central Eur. J. Oper. Res. **16**(1), 79–105 (2008)
5. Girão-Silva, R, Craveirinha, J, Clímaco, J: Hierarchical multiobjective routing model in Multiprotocol Label Switching networks with two service classes – a Pareto archive strategy. Eng. Optimization **44**(5), 613–635 (2012)
6. Steuer, RE: Multiple Criteria Optimization: Theory, Computation and Application. Probability and Mathematical Statistics. John Wiley & Sons, New York (1986)
7. Girão-Silva, R, Craveirinha, J, Clímaco, J: Hierarchical multiobjective routing in MPLS networks with two service classes - Report on a simulation case study. Research Report 11/2011 (ISSN 1645-2631), INESC-Coimbra (2011). Accessed 15 January 2014 at http://www.inescc.pt/documentos/11_2011.pdf
8. Bouyssou, D: Modelling inaccurate determination, uncertainty, imprecision using multiple criteria. In: Lockett, AG, Islei, G (eds.) Improving Decision, Making in Organisations. LNEMS, *vol. 335*, pp. 78–87. Springer, Berlin Heidelberg (1989)
9. Mitra, D, Morrison, JA, Ramakrishnan, KG: Optimization and design of network routing using refined asymptotic approximations. Perform. Eval. **36–37**, 267–288 (1999)
10. Kaufman, JS: Blocking in a shared resource environment. IEEE Trans. Commun. **COM-29**(10), 1474–1481 (1981)
11. Roberts, JW: Teletraffic models for the Telecom 1 integrated services network. In: Proceedings of 10th International Teletraffic Congress. Montreal (1983)
12. Mitra, D, Morrison, JA: Erlang capacity and uniform approximations for shared unbuffered resources. IEEE/ACM Trans. Netw. **2**(6), 558–570 (1994)
13. Craveirinha, J, Girão-Silva, R, Clímaco, J, Martins, L: A hierarchical multiobjective routing model for MPLS networks with two service classes. In: Korytowski, A, Malanowski, K, Mitkowski, W, Szymkat, M (eds.) Revised Selected Papers of the 23rd IFIP TC7 Conference on System Modeling and Optimization, Cracow, July 23–27 2007. IFIP Advances in Information and Communication Technology, *vol. 312*, pp. 196–219. Springer, New York (2009)
14. Craveirinha, J, Martins, L, Gomes, T, Antunes, CH, Clímaco, JN: A new multiple objective dynamic routing method using implied costs. J. Telecommunications Inf. Technol. **3**, 50–59 (2003)
15. Martins, L, Craveirinha, J, Clímaco, JN, Gomes, T: Implementation and performance of a new multiple objective dynamic routing method for multiexchange networks. J. Telecommunications Inf. Technol. **3**, 60–66 (2003)
16. Girão-Silva, R, Craveirinha, J, Clímaco, J: Hierarchical multiobjective routing in Multiprotocol Label Switching networks with two service classes - a heuristic solution. Int. Trans. Oper. Res. **16**(3), 275–305 (2009)
17. Akimaru, H, Kawashima, K: Teletraffic: Theory and Applications, Telecommunication Networks and Computer Systems. Springer, New York (1993)
18. Tralhão, L, Craveirinha, J, Paixão, J: A study on a stochastic system with multiple MMPP inputs subject to access functions. In: New Progress in Probability and Statistics - 4th International Symposium on Statistics of the Base Country, pp. 415–428. VSP - International Publishers, Zeist (1994)
19. Kelly, FP: Blocking probabilities in large circuit-switched networks. Adv. Appl. Probability **18**(2), 473–505 (1986)
20. Sayed, HME, Mahmoud, MS, Bilal, AY, Bernussou, J: Adaptive alternate-routing in telephone networks: Optimal and equilibrium solutions. Inf. Decis. Technol. **14**(1), 65–74 (1988)
21. Durbach, IN, Stewart, TJ: Modeling uncertainty in multi-criteria decision analysis. Eur. J. Oper. Res. **223**, 1–14 (2012)
22. Clímaco, JCN, Craveirinha, JMF, Pascoal, MMB: An automated reference point-like approach for multicriteria shortest path problems. J. Syst. Sci. Syst. Eng. **15**(3), 314–329 (2006)
23. Ash, GR: Traffic Engineering and QoS Optimization of Integrated Voice & Data Networks. Elsevier, Morgan Kaufmann (2007)
24. Mitra, D, Ramakrishnan, KG: Techniques for traffic engineering of multiservice, multipriority networks. Bell Labs Tech. J. **6**(1), 139–151 (2001)
25. Erbas, SC, Erbas, C: A multiobjective off-line routing model for MPLS networks. In: Charzinski, J, Lehnert, R, Tran-Gia, P (eds.) Proceedings of the 18th, International Teletraffic Congress (ITC-18), pp. 471–480. Elsevier, Amsterdam, Berlin (2003)
26. Banks, J, Carson, II, JS: Discrete-Event System Simulation. International Series in Industrial and Systems Engineering. Prentice-Hall, Upper Saddle River (1984)
27. Awduche, D, Chiu, A, Elwalid, A, Widjaja, I, Xiao, X: Overview and principles of Internet traffic engineering. RFC 3272, Network Working Group. The Internet Engineering Task Force, California (2002)
28. Leland, W, Taqqu, M, Willinger, W, Wilson, D: On the self-similar nature of Ethernet traffic (Extended version). IEEE/ACM Trans. Netw. **2**(1), 1–15 (1994)
29. Martins, L, Craveirinha, J, Clímaco, J: A new multiobjective dynamic routing method for multiservice networks: modelling and performance. Comput. Manage. Sci. **3**(3), 225–244 (2006)
30. Gomes, T, Martins, L, Craveirinha, J: An algorithm for calculating k shortest paths with a maximum number of arcs. Investigação Operacional **21**, 235–244 (2001)

5

Similarity in fuzzy systems

Asim Pal*†, Banibrata Mondal†, Namrata Bhattacharyya† and Swapan Raha†

*Correspondence:
itsasimpal@rediffmail.com
†Equal contributors
Department of Mathematics,
Visva-Bharati, Santiniketan, 731235,
India

Abstract

This paper proposes to study the concept of similarity and its use in the design of fuzzy system. The concept of similarity relation is effectively used in fuzzification of crisp values. Similarity index is used in measuring approximate (graded) equality of fuzzy sets over a given universe of discourse. It is proposed to use such an index in modifying a fuzzy relation. Different similarity measures in the literature are elucidated, and a comparative study between different pairs of fuzzy sets is presented. One of these similarity measures has been used successfully in rule selection and modification of a fuzzy relation. In the process, a number of modification schemes based on different logic have been extensively studied for different reasoning mechanisms based on the same data and results are tabulated. A specificity-based approach to defuzzification is also presented, which is found to be suitable for similarity-based fuzzy systems. The results are illustrated with the behaviour of a direct current (DC) shunt motor.

Keywords: Approximate reasoning; Similarity; Similarity-based reasoning

Introduction

A system is defined as an integrated set of interacting elements/objects/components that accomplish a defined objective. Usually, the description of a system can be made at different levels of details. Any subsystem may, itself, be considered as a system consisting of subsystems at a lower level of detail. Human beings reason consistently from incomplete/imprecise knowledge of a system with reasonably good results. For instance, control is often exercise by a human operator who has the requisite skill to control the system successfully but cannot explain sufficiently well how he does it. Often, it becomes difficult to find even a mathematical model for the same. The tentative nature of human thinking involves frequently imprecise concepts (qualitative statements made on the input-output behaviour of such a system) which may not even have an underlying metric. Therefore, reasoning mechanisms should be modified to include provisions for handling such imprecisions in the system.

One may represent human expertise in the form of 'if *statement1*, then *statement2*'. This deductive form is commonly referred to as a rule - with *statement1* as *antecedent* and *statement2* as its *consequence*. A system, whose behaviour is described with a number of such rules, is referred to as a rule-based system. It expresses typically an inference in such a way that, with a known fact and a given rule, one can infer/derive a piece of information on the behaviour of the system which is referred to as the *conclusion*. This form of knowledge representation (processing) is found to be quite appropriate in expressing human understanding in a natural language of communication.

A collection of imprecise information given by human experts often forms the basis of a fuzzy system. The task of a fuzzy system is to exploit experts' knowledge and model the world with it. A fuzzy system reasons with its knowledge. A fuzzy rule-based system is one of the most important areas of research. It is a dominating platform for the development of a precise mathematical model for an imprecisely known system. Many believe that human beings take a similar approach to perceive the world around them in a robust way. In the real world, almost everything is incompletely defined. A fuzzy rule-based system is, therefore, expected to achieve a performance better than any crisp model in dealing with ambiguity, incompleteness and imprecision. A fuzzy rule-based system consists of a set of fuzzy If-THEN rules together with an inference engine, a fuzzifier and a defuzzifier.

Different patterns of reasoning in human beings indicate a need for similarity matching in situations where there are no directly applicable knowledge to come up with a plausible conclusion. In such cases, the confidence in a conclusion may be determined, based on a degree of similarity between the fact(s) and the antecedent of a rule. In order to capture this, our model should have the required flexibility. Specifically, we need means to handle graded information on the one hand and the concept of similarity on the other hand. Conventional approximate reasoning does not consider the concept of similarity measure in deriving a consequence. Existing similarity-based reasoning methods modify the consequence of a rule based on a measure of similarity, thereby making the consequence independent of the conditionals. To satisfy both requirements simultaneously, we need to integrate conventional approximate reasoning and similarity-based reasoning for an adequate theory of similarity-based approximate reasoning.

The objectives of the paper are to study the different modules of a fuzzy system and to observe the role of similarity and subsequently the possibility of a step-by-step introduction of the concept of similarity into approximate reasoning methodology which forms its core.

A formal method of fuzzification is presented in this paper. A thorough exposition on the use of different translating rules is considered. The modification mechanism will be such that every change in the conditional (general) statement and in the fact is incorporated in the fuzzy relation between the variables defining the condition. This procedure ensures the deduction/inference as a function of the concerned change. The more the change, the less specific will be the conclusion. A proposal to formulate different schema for the modification of the conditional fuzzy relation is presented. This paper proposes to demonstrate how one can avoid the use of certainty factor concept for rule misfiring. We thus have to modify the inference mechanism in a way such that a significant change will make the conclusion less specific. This can be done if an expansion type of inference scheme is chosen. Explicitly, when the similarity value becomes low, i.e. when the change is significant, the reasoning process is such that the inference becomes unknown. At the same time, when there is no change, i.e. a perfectly matching case, it is possible to derive the expected consequence (the consequence of the condition in a rule-based system). In other cases, the consequence will be no better than what the condition allows. Different interesting results in this direction are discussed extensively. It seeks to show that the concept of specificity measure of fuzzy sets is inherent in such similarity-based approximate reasoning methodology. Examples are considered to demonstrate the computations under the procedure. A comparison of the result with other existing approaches to similarity-based reasoning and Zadeh's compositional rule of inference is also presented. Defuzzification, a basic operation, used in the development of fuzzy system is discussed in the light of new similarity-based approximate reasoning mechanism. A new scheme for defuzzification, suitable for similarity-based approximate reasoning, is defined. This defuzzification method will then be used in problems of classification, diagnosis and control. Simulation is performed with some real data and results are tabulated.

The paper consists of seven sections. The introductory section is followed by a discussion on the similarity relation which again is followed by a proposal for constructing such an equivalence relation induced by fuzzy sets and its subsequent use in fuzzification in the section 'Similarity relation - fuzzification'. Similarity measure of two fuzzy sets is considered in the section 'Similarity measure - inference'. Approximate reasoning is discussed in the context of similarity. Different schemes are presented and examples considered to illustrate the problem. The section 'Specificity measure - defuzzification' is devoted to defuzzification of fuzzy outputs. Specificity measure of fuzzy sets can be used to determine the anxiety in decision making. A new defuzzification scheme is defined that works on specificity measures of fuzzy sets. Application of the same in different models are presented in the section 'Application in different models' followed by result of a case study on a direct current (DC) shunt motor. The paper is concluded in the section 'Conclusions'. A list of references is provided in the last section.

Similarity relation - fuzzification

Similarity is an important concept for which a crisp model is often found to be inadequate. In [1], the authors showed that the notion of membership is a gradual property of fuzzy sets. They have considered a fuzzy equivalence relation to describe the similarity between elements of a fuzzy set. There, they showed how a crisp set induced a fuzzy set as its extensional hull with respect to a fuzzy equivalence relation. Assigning similarity modelled by a fuzzy equivalence relation as the basis, fuzzy sets were viewed as induced concept. Two elements cannot be distinguished by a fuzzy set if they are both either elements of the same set or its complement [2]. They have shown how membership functions of fuzzy sets can be calculated from the fuzzy equivalence relation as in the following:

Definition 1. A fuzzy equivalence relation (with respect to the conjunction operation $*$, here, a t-norm) on the set U is a mapping $E : U \times U \longrightarrow [0, 1]$ satisfying

$$
\begin{array}{lll}
(E1) & E(u,u) = 1, u \in U & \text{(reflexivity)} \\
(E2) & E(u_1,u_2) = E(u_2,u_1), u_1,u_2 \in U & \text{(symmetry)} \\
(E3) & E(u_1,u_2) * E(u_2,u_3) \le E(u_1,u_3), u_1,u_2,u_3 \in U & \text{(transitivity).}
\end{array}
$$

where $[0, 1]$ is the unit interval with the usual ordering. Sometimes, E is also called a similarity relation [1]. Contextually, some definitions and theorems cited in [1] are hereby recalled.

Definition 2. A fuzzy set $A \in [0, 1]^U$ is called extensional with respect to (w.r.t.) the fuzzy equivalence relation E on U if and only if $\mu_A(u_1) * E(u_1, u_2) \le \mu_A(u_2)$ holds for all $u_1,\ u_2 \in U$.

Definition 3. Let E be a fuzzy equivalence relation on U and let $A \in [0,1]^U$.

The fuzzy set $\hat{A} = \bigcap\{B | A \subseteq B$ and B is extensional w.r.t. $E\}$ is called the extensional hull of A w.r.t. E.

Theorem 1. $\hat{A} = \bigcup \{\mu_A(u_1) * E(u_1, u_2) \mid u_1, u_2 \in U\}$.

Theorem 2. *Let $\mathcal{F} \subseteq [0,1]^U$ be a collection of fuzzy sets and $\leftrightarrow$: $[0,1] \times [0,1] \rightarrow [0,1]$ be a biimplication. Then,*

$$E_{\mathcal{F}}(u_1, u_2) = \bigwedge_{A \in \mathcal{F}} (\mu_A(u_1) \leftrightarrow \mu_A(u_2)) \tag{1}$$

is the coarsest fuzzy equivalence relation on U such that all fuzzy sets in $\mathcal{F}$ are extensional w.r.t. $E_{\mathcal{F}}$.

$E_{\mathcal{F}}$ is reflexive, symmetric and satisfy transitivity relation. The fuzzy equivalence relation (1) can be interpreted in the following way - two elements 'cannot be distinguished by a (fuzzy) set' if they are both elements of the same set or its complement, but not one in the set and the other one in its complement. Thus, $\mu_A(u_1) \leftrightarrow \mu_A(u_2)$ represents the degree to which the elements u_1 and u_2 cannot be distinguished by the fuzzy set A. Therefore, $E_{\mathcal{F}}(u_1, u_2)$ is the degree to which u_1 and u_2 cannot be distinguished by the set $\mathcal{F}$ of fuzzy sets.

We develop a new fuzzy equivalence relation on a universe of discourse U, which is generated by a family $\mathcal{F}$ of fuzzy subsets of U. Accordingly, we define a fuzzy relation E as given below:

$$E(u_1, u_2) = 1 - \sqrt{\frac{\Sigma_A(\mu_A(u_1) - \mu_A(u_2))^2}{n}}, \tag{2}$$

where n is the number of fuzzy sets.

Theorem 3. *The fuzzy relation defined in (2) is a fuzzy equivalence relation.*

From this fuzzy equivalence relation $E(u_1, u_2)$, we can fuzzify any point 'a' on the universe of discourse U by setting

$$\begin{aligned} \mu_a(u_1) &= E(u_1, a), \text{ in an interval } a - \delta \leq u_1 \leq a + \delta, \delta > 0, \\ &= 0, \qquad \text{otherwise.} \end{aligned} \tag{3}$$

Observation: $\mu_a(u_1)$ is extensional with respect to the fuzzy equivalence relation E as

$$\begin{aligned} &\mu_a(u_1) * E(u_1, u_2) \\ &\quad = E(u_1, a) * E(u_1, u_2) \\ &\quad = E(a, u_1) * E(u_1, u_2) \leq E(a, u_2) \\ &\quad = E(u_2, a) = \mu_a(u_2). \end{aligned}$$

Thus, given a fuzzy equivalence relation and a crisp point 'a', we can define (generate) a fuzzy set about the point 'a'. This is called fuzzification and plays an important role in the design of fuzzy systems. We illustrates this fuzzification with the following algorithm.

Algorithm FUZZ: fuzzification

Step 1. Given n fuzzy sets $A_1, A_2, \ldots, A_n$ defined over some universe of discourse U and $a \in U$.

Step 2. Construct a fuzzy equivalence relation $E(u_1, u_2)$ from $A_1, A_2, \ldots, A_n$ using (2).

Step 3. Set $\delta > 0$. Define a fuzzy set about the point $a \in U$ from the fuzzy equivalence relation $E(u_1, u_2)$ by

$$\mu_a(u_1) = E(u_1, a), \text{ in the interval } a - \delta \leq u_1 \leq a + \delta, \delta > 0$$
$$= 0, \quad \text{otherwise.}$$

Example 1. Let the domain set be $U = \{0.0, 0.1, 0.2, 0.3, 0.4, \ldots, 9.9, 10.0\}$.

Let A_r be the fuzzy sets corresponding the points $r = 0.0, 0.5, 1.0, 1.5, \ldots, 10.0$, i.e. the set of fuzzy sets $\{\mu_{A_i} | i \in R, \text{ a finite index set}\}$ where

$$\mu_{A_i}(u) = 1 - \min\{|u - i|, 1\} \quad [1].$$

Now, a fuzzy equivalence relation induced by the fuzzy sets as given in Figure 1 and using the algorithm *FUZZ* is given in Figure 2.

Choosing $\delta = 5.0$, we find the extensional hulls of the crisp values 5.0 and 7.5 with respect to this fuzzy equivalence relation are the fuzzy sets

$$\mu_{5.0}(u) = E(u, 5.0) \text{ if } 5.0 - 5 \leq u \leq 5.0 + 5, \text{ otherwise } \mu_{5.0} = 0 \text{ and}$$

$$\mu_{7.5}(u) = E(u, 7.5) \text{ if } 7.5 - 5 \leq u \leq 7.5 + 5, \text{ otherwise } \mu_{7.5} = 0.$$

Similarity measure - inference

The similarity between two objects suggests the degree to which the properties of one may be inferred from those of the other. In this section, we present some similarity measures that exist in the literature and their performances are studied. At the end, we investigate similarity-based fuzzy reasoning techniques with the best of these measures.

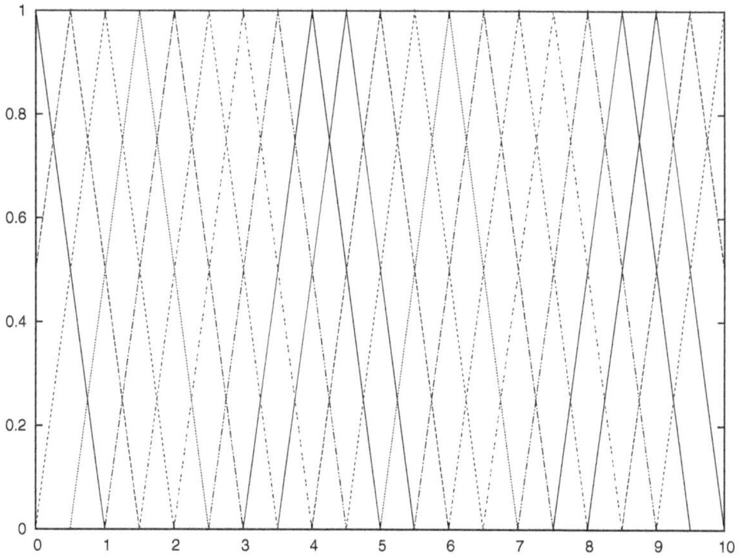

Figure 1 A typical fuzzy partition. A number of fuzzy sets on a domain draw a fuzzy partition.

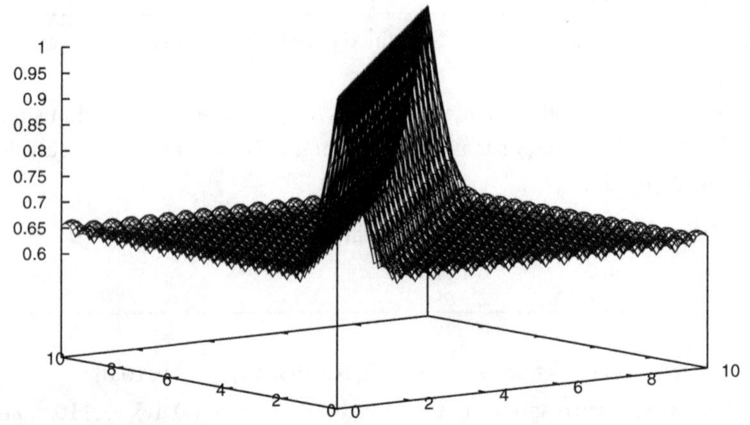

Figure 2 An equivalence relation. An equivalence relation from the fuzzy partition.

Similarity measure

A similarity matching degree can be defined from the distance functions according to the following: $S(\bullet, \bullet) = 1 - d(\bullet, \bullet)$. The most important class of distance function is the Minkowski's r-metric. Another important class of distance functions is given by the Hausdorff metric. It is a generalisation of the distance between two points in a metric space to two compact non-empty subsets of the space [3]. A set theoretic approach to a family of similarity functions can be given by $S(A, B) = \theta f(A \cap B) - \alpha f(A - B) - \beta f(B - A)$ for some function f and parameters $\theta, \alpha, \beta \geq 0$ [4].

Let us assume the universe of discourse U to be a finite set and $A = \sum_{u \in U} \{\mu_A(u)/u\}$, $B = \sum_{u \in U} \{\mu_B(u)/u\}$ be two fuzzy sets defined over U. A similarity index between the pair $\{A, B\}$ is denoted as $S(A, B; U)$ or simply $S(A, B)$. In the following, a number of existing similarity measures are listed from the literature.

1. $\text{sim1} = 1 - \left(\frac{\Sigma_u | \mu_A(u) - \mu_B(u) |^q}{n} \right)^{\frac{1}{q}}$ [5,6], where n is the cardinality of the universe of discourse and q is the family parameter.
2. $\text{sim2} = 1 - \max_{u \in U} (|\mu_A(u) - \mu_B(u)|)$ [7].
3. $\text{sim3} = 1 - \frac{\Sigma_{u \in U} |(\mu_A(u) - \mu_B(u))|}{\Sigma_{u \in U} (\mu_A(u) + \mu_B(u))}$ [7].
4. $\text{sim4} = 1 - \frac{1}{n} \Sigma_{u \in U} |\mu_A(u) - \mu_B(u)|$ [7].
5. $\text{sim5} = \frac{\max_{u \in U} \{\mu_A(u) \wedge \mu_B(u)\}}{\max_{u \in U} \{\max(\mu_A(u)), \max(\mu_B(u))\}}$ [6].
6. $\text{sim6} = \frac{\max\{\mu_A(u) \cdot \mu_B(u)\}}{\max\left\{\Sigma_{u \in U} \mu_A^2(u), \Sigma_{u \in U} \mu_B^2(u)\right\}}$ [7].
7. $\text{sim7} = \frac{\min(\mu_A(u), \mu_B(u))}{\max(\mu_A(u), \mu_B(u))}$ [7].
8. $\text{sim8} = \frac{1}{n} \Sigma_{u \in U} \frac{\min(\mu_A(u), \mu_B(u))}{\max(\mu_A(u), \mu_B(u))}$ [8].
9. $\text{sim9} = \max_{u \in U} \{\min(\mu_A(u), \mu_B(u))\}$ [9].
10. $\text{sim10} = \frac{C(A,B)}{\sqrt{T(A), T(B)}}$ [8], where $T(A) = \Sigma_{i=1}^n \left(\mu_A(u_i)^2 . (1 - \mu_A(u_i))^2\right)$,
$C(A, B) = \Sigma_{i=1}^n \{\mu_A(u_i) . \mu_B(u_i) + (1 - \mu_A(u_i)) . (1 - \mu_B(u_i))\}$.
11. $\text{sim11} = \min\{\alpha(A, B), \alpha(A^c, B^c)\}$ [6],
where $\alpha(A, B) = \sqrt{\frac{\Sigma_{u \in U} \mu_A(u) . \mu_B(u)}{\Sigma_{u \in U} (\max(\mu_A(u), \mu_B(u)))^2}}$.
12. $\text{sim12} = \frac{1}{2}\left[(A \leftrightarrow B) + (A' \leftrightarrow B')\right]$ [10], where
$A \leftrightarrow B = (A \to B) \wedge (B \to A)$, $A' = 1 - A$. $\wedge$ being a conjunction operator and $\to$ an implication operator. The above formula is equivalent to

$S(A,B) = \frac{1}{n}\Sigma_{u\in U}\frac{1}{2}\,[(\mu_A(u) \leftrightarrow \mu_B(u)) + (\mu_{A'}(u) \leftrightarrow \mu_{B'}(u))]$, where $(\mu_A(u) \leftrightarrow \mu_B(u)) = (\mu_A(u) \rightarrow \mu_B(u)) \wedge (\mu_B(u) \rightarrow \mu_A(u))$, n being the cardinality of the universal set U, and $\wedge$ and $\rightarrow$ are defined logically. Now, if we interpret the $\leftrightarrow$ implication operator as $\alpha \leftrightarrow \beta = 1 - |\alpha - \beta|$, then $S(A,B) = 1 - \frac{1}{n}\Sigma_{i=1}^{n}|\mu_A(u) - \mu_B(u)|$.

13. $\text{sim13} = \frac{\text{card}(\mu_A(u)\wedge\mu_B(u))}{\max(\text{card}(A),\text{card}(B)))}$
14. $\text{sim14} = \frac{\text{card}(\mu_A(u)\wedge\mu_B(u))}{\text{card}(A))}$
15. $\text{sim15} = \frac{\text{card}(\mu_A(u)\wedge\mu_B(u))}{\text{card}(B))}$
16. $\text{sim16} = \min\left\{\frac{\text{card}(A\cap B)}{\text{card}(A)}, \frac{\text{card}(A\cap B)}{\text{card}(B)}\right\}$

A good working measure of similarity between two countable infinite fuzzy sets can be given as in the following:

$$S(A,B) = 1 - \sup_{u\in U} \mid \mu_A(u) - \mu_B(u) \mid . \qquad (4)$$

It is easy to see that this can be an effective measure if we consider an infinite fuzzy set. But the problem in working with such a measure is that it gives importance to the sup operation only. The work in this regard can be found in [11].

Now, once a similarity index is defined - How can we compare this with other existing indices? How should we judge the goodness of such an index?

In this regard, the authors in [12,13] have reviewed different similarity measures, as suggested in the literature in the general case and as adapted to fuzzy sets. They have also presented an experimental design for linguistic approximation and discussed at length the suitability of application of different measures of similarity [14].

In [8], the authors presented a comparative study on the basis of a set of axioms. They have also investigated some similarities and dissimilarities in performance.

All the similarity measures listed above satisfy the reflexivity, symmetry and boundedness property. These three properties are indispensable for any similarity measures [9]. In this regard, all measures are equally useful. Besides these three properties, similarity measures should also satisfy properties like computational simplicity, monotonicity and non-dissimilarity. These are some desirable properties. Similarity measures based on the computation of overall sup as well as max between elements are such that they give more importance to a particular value and ignore the presence of others. Thus, two fuzzy sets are often found to be similar when they have the same sup and/or max. Of course, one can define two fuzzy sets to be similar as and when they have the same cardinality or they have the same support. This may work for mathematical theory construction. But in order to assist the decision maker in a real-life situation, the practical meaning of similarity concept is of vital importance. We are considering those indices that play a crucial role in the theory of fuzzy reasoning. This demands similarity measures based on separate membership degrees of each concept.

Next, let us consider the similarity measure defined in (4). In order to illustrate the drawback underlying it, let us consider a simple case as in the following:

$$\begin{aligned}\mu_A(u) &= 1\ \forall u \in U \text{ and}\\ \mu_B(u) &= 0 \text{ for a particular } u_0 \in U \text{ and}\\ &= 1 \text{ otherwise.}\end{aligned}$$

Even in such an almost similar pair of fuzzy sets, it is found that the similarity index is 0, showing thereby that they are completely dissimilar. Now, it can safely be concluded that it is practically impossible to single out one possible similarity measure that works well for all purpose.

In the following, we present an axiomatic definition of similarity between fuzzy sets, defined over the same universe of discourse. Some important deductions are also provided to illustrate the proposed measure's soundness. In order to provide a definition for similarity index, a number of factors must be considered. A primary consideration is that whatever way we choose to define such an index, it must satisfy the properties as already mentioned. Similarity measures are, in general, found to be non-transitive.

Under these circumstances, a similarity measure $S(A,B)$ should satisfy the following properties:

For all fuzzy sets A , B:

P1. $S(B,A) = S(A,B)$.

P2. $S(A^c,B^c) = S(A,B)$, A^c being some negation of A.

P3. $0 \leq S(A,B) \leq 1$.

P4. $A = B$ if and only if $S(A,B) = 1$.

P5. If $S(A,B) = 0$, then either $A \cap B = \Phi$ (null) or $A^c \cap B^c = \Phi$, or $B = 1 - A$.

For $0 \leq \epsilon \leq 1$, if $S(A,B) \geq \epsilon$, we say that the two fuzzy sets A and B are ϵ-similar. Thus, the case for $\epsilon = 1$ correspond to equality of fuzzy sets. There could be many functions satisfying properties P1 through P5. One such measure of similarity satisfying properties P1 through P5 is given next.

Definition 4. Let $A = \sum_{u \in U} \mu_A(u)/u$ and $B = \sum_{u \in U} \mu_B(u)/u$ be two fuzzy sets defined over the same universe of discourse U. The similarity index of the pair $\{A,B\}$ is denoted by $S(A,B)$ and is defined by

$$S(A,B) = 1 - \left(\frac{\sum_{u} | \mu_A(u) - \mu_B(u) |^q}{n} \right)^{\frac{1}{q}} \tag{5}$$

where n is the cardinality of the universe of discourse and $q \geq 1$ is the family parameter.

Theorem 4. *If* $S(A,B) = 1$ *and* $S(B,C) = 1$, *then* $S(A,C) = 1$.

Theorem 5. *For all fuzzy sets A, B, C, if either* $A \subseteq B \subseteq C$ *or* $A \supseteq B \supseteq C$, *then* $S(A,C) \leq \min\{S(A,B), S(B,C)\}$.

Theorem 5 motivates us to consider the property of monotonicity of similarity between fuzzy sets to satisfy another axiom for some kind of monotonicity. So, we are now in a position to rewrite the axioms for similarity measure as in the following.

For all fuzzy sets A , B and C defined over the universe of discourse U, we have:

A1. $S(B,A) = S(A,B)$.

A2. $S(A^c,B^c) = S(A,B)$, A^c being some negation of A.

A3. $0 \leq S(A,B) \leq 1$.

A4. $A = B$ if and only if $S(A, B) = 1$.

A5. $S(A, B) = 0$ if and only if $A \cap B = \Phi$.

A6. If $A \supseteq B \supseteq C$, then $S(A, B) \geq S(A, C)$.

Here, we note that A^c, the complement of a fuzzy set A, is to be defined first. We used the idea of '1-' as the complementation. On the basis of the above axioms, it is easy to see that the family of similarity measures defined in *Definition 4* is a valid choice. All the measures satisfy axioms A1, A3, A4 and A5 for either identical or non-overlapping fuzzy sets.

Thus, we find that the similarity between fuzzy sets can be captured by aggregating the distinguishability between membership values of each element in the corresponding fuzzy sets. The similarity index between two fuzzy sets is a pure number and does not give any information about the inclusion. This explains why these measures are not transitive, in general. Let us tabulate the performance of different measures in the following.

Performance of different similarity measures

In this sub-section, we observe six cases (Figures 3, 4, 5, 6, 7 and 8), where each case compares two fuzzy sets in consideration. Also, for each case, 16 similarity measures mentioned earlier are calculated. A performance chart of various similarity measures are given in Table 1.

Approximate reasoning

Approximate reasoning is defined as the process or processes by which an approximate conclusion can be deduced from a set of possibly imprecise information using some inexact rule for the derivation. Since its inception in 1973, significant theoretical advances have established approximate reasoning as an important field of research. Different techniques of approximate reasoning have been proposed and discussed in the literature.

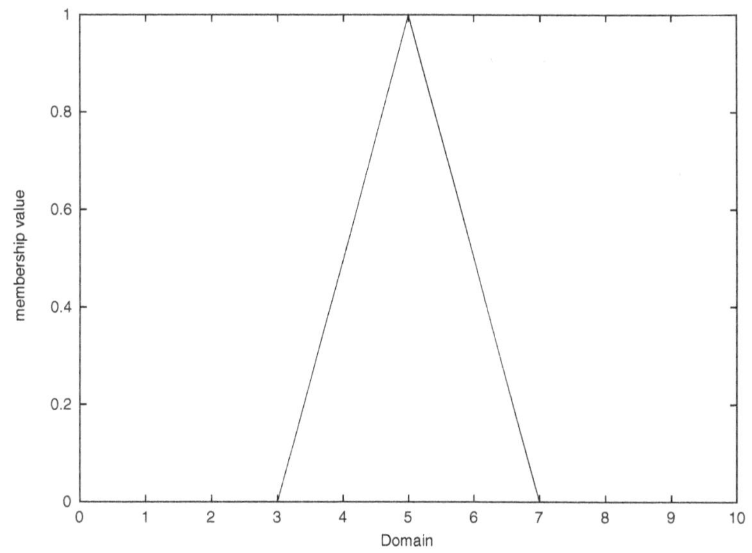

Figure 3 Case 1: two identical fuzzy sets. Fuzzy sets for measuring similarity.

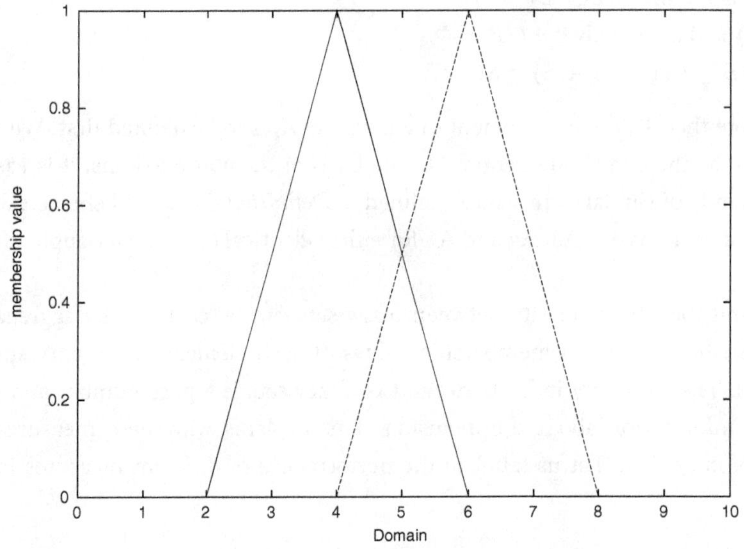

Figure 4 Case 2: two fuzzy sets almost close to each other. Fuzzy sets for measuring similarity.

Zadeh's [15] concept of approximate reasoning is based on fuzzy logic and the theory of fuzzy sets. In order to have an adequate understanding of the theory of approximate reasoning in this paper, some basic concepts are considered. The concept of linguistic variable plays an essential role in the theory of approximate reasoning. It is a tool for approximate characterisation of the values of the variables and their interrelations. For example, the *height* of a person may be short, the *volume* of a container may be huge, the *code section* of some programme may be tiny, two *numbers* may be approximately equal and so on. Zadeh [16] called such variables - linguistic variables.

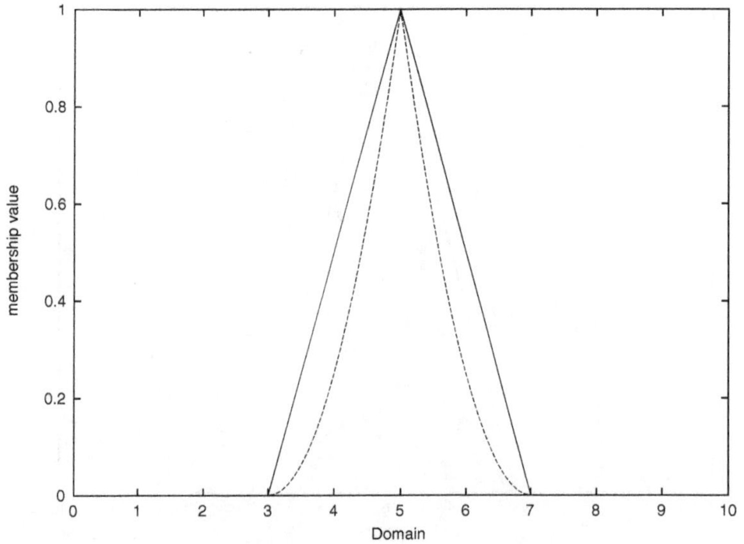

Figure 5 Case 3: two fairly close fuzzy sets. Fuzzy sets for measuring similarity.

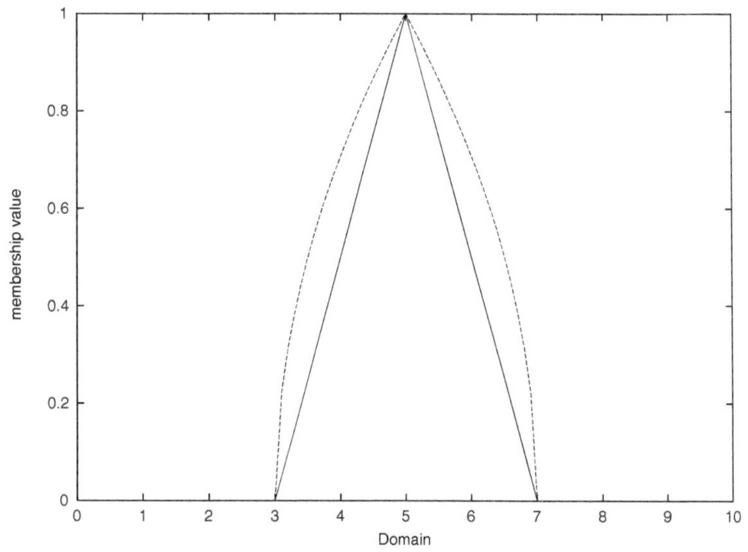

Figure 6 Case 4: two overlapping fuzzy sets. Fuzzy sets for measuring similarity.

The two basic inference rules considered for approximate reasoning based on fuzzy sets and fuzzy relations are the compositional rule of inference and the generalised *modus ponens*.

1. max-min *composition.* From 'X is A' and '(X, Y) is R' infer 'Y is B', where $\mu_B(y) = \max_x \min(\mu_A(x), \mu_R(x,y))$.
2. *Generalised modus ponens.* From 'X is A^*' and 'if X is A, then Y is B' infer 'Y is B^*', where $\mu_{B^*}(y) = \sup_x(\mu_{A^*}(x) \circ (\mu_A(x) \rightarrow \mu_B(y)))$, $\circ$ and $\rightarrow$ may have different interpretation.

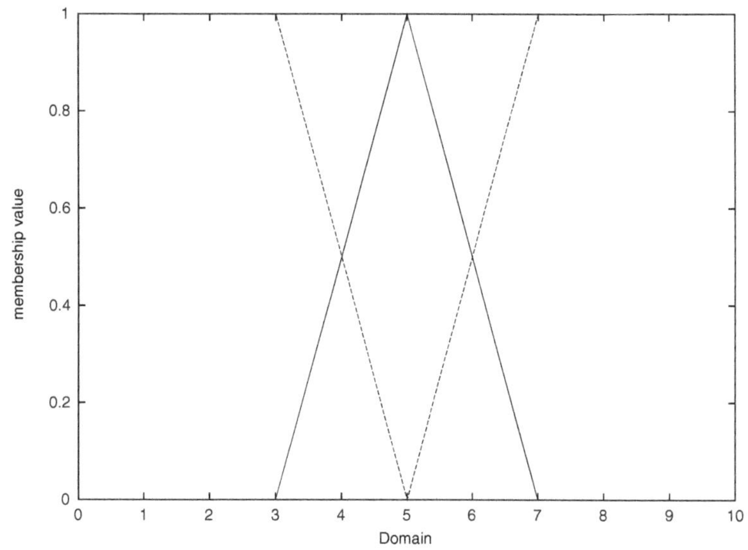

Figure 7 Case 5: two complementary fuzzy sets. Fuzzy sets for measuring similarity.

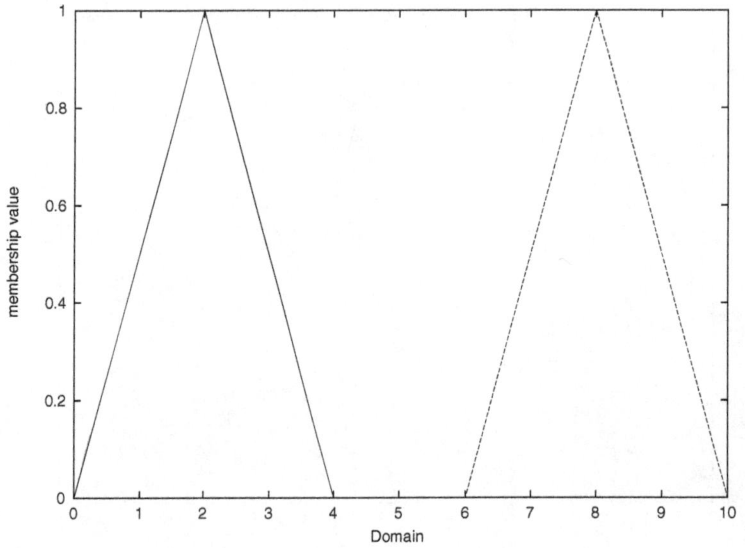

Figure 8 Case 6: two non-overlapping fuzzy sets. Fuzzy sets for measuring similarity.

Many fuzzy systems are based on Zadeh's compositional rule of inference [17]. Despite their success in various systems, researchers have indicated certain drawbacks [18] in the technique.

As for example, let $U = \{u_1, u_2, u_3, u_4\}$ and $V = \{v_1, v_2, v_3, v_4\}$ be the universes of discourse, $A = 1.0/u_1 + 0.75/u_2 + 0.5/u_3 + 0.25/u_4$ and

$$R = \begin{array}{c|cccc} & u_1 & u_2 & u_3 & u_4 \\ \hline v_1 & 1.00 & 0.75 & 0.50 & 0.25 \\ v_2 & 0.75 & 1.00 & 0.75 & 0.50 \\ v_3 & 0.50 & 0.75 & 1.00 & 0.75 \\ v_4 & 0.25 & 0.50 & 0.75 & 1.00 \end{array}.$$

Then, taking $T = \min$ and using compositional rule of inference (CRI), we find $B = 1.00/v_1, 0.75/v_2, 0.75/v_3 + 0.50/v_4$. This shows that the linguistic variables X

Table 1 Performance chart of various similarity measures

	Similarity measures							
	sim1	sim2	sim3	sim4	sim5	sim6	sim7	sim8
Case 1	1.00	1.00	1.00	1.00	1.00	1.00	1.00	1.00
Case 2	0.82	0.75	0.80	0.83	1.00	0.75	0.67	0.51
Case 3	0.82	0.75	0.86	0.83	1.00	0.80	0.75	0.68
Case 4	0.78	0.75	0.76	0.79	0.85	0.92	0.62	0.46
Case 5	0.14	0.00	0.20	0.20	0.50	0.09	0.11	0.15
Case 6	0.41	0.00	0.00	0.48	0.00	0.00	0.00	0.00
	sim9	sim10	sim11	sim12	sim13	sim14	sim15	sim16
Case 1	1.00	1.00	1.00	1.00	1.00	1.00	1.00	1.00
Case 2	1.00	0.79	0.87	0.87	0.67	0.67	1.00	0.67
Case 3	1.00	0.93	0.89	0.89	0.75	1.00	0.75	0.75
Case 4	0.46	0.85	0.79	0.83	0.76	0.76	0.76	0.76
Case 5	0.15	0.50	1.00	0.28	0.12	0.50	0.12	0.12
Case 6	0.00	0.00	0.00	0.00	0.00	0.00	0.00	0.00

and Y are approximately equal. A careful scrutiny of the relation also says so. The conclusion B will remain the same if we choose $A = 1.0/u_1 + 0.75/u_3 + 0.50/u_4$, which is highly dissimilar to A. Next, if we take $A = 1.0/u_1$, then from R we have $B = 1.00/v_1 + 0.75/v_2 + 0.50/v_3 + 0.25/v_4$; again, if we take $A = 1.0/u_4$, then $B = 0.25/u_1 + 0.50/u_2 + 0.75/u_3 + 1.00/u_4$. This shows that even if the input values are strongly complementary to each other, significant conclusions can be drawn using Zadeh's CRI.

This motivates the introduction of similarity-based reasoning techniques as proposed in [18-23].

Similarity-based approximate reasoning

To begin with, in this section, we will look at the different methods of inference based on a similarity measure. In [18,24-26], the authors proposed a similarity-based method called 'approximate analogical reasoning schema'. The method is applicable to both point-valued and interval-valued fuzzy sets. In [19], the author proposed two similar methods for medical diagnosis problems. Two other methods based on different modification procedures have been proposed in [27]. In the framework of existing approaches to similarity-based inference methodology, recently, in [21], the authors proposed two other similarity-based methods for reasoning and made a comparative study of the above similarity-based fuzzy reasoning methods.

In all these studies, it is proposed that similarity-based fuzzy reasoning methods do not require the construction of a fuzzy relation. Accordingly, they are based on the computation of the degree of similarity between the fact and the antecedent of a rule in a rule-based system. Then, based on the similarity value between the membership values of the elements of the fuzzy set representation of the fact and the corresponding fuzzy set in the antecedent of the rule, the membership value of each element of the consequent fuzzy set of the rule is modified to obtain a conclusion. This is the same for all existing similarity-based reasoning schemes. The modification procedure is different for different schemes.

We proposed two similarity-based approximate reasoning methods. One such method is a modification of the method presented in [18] and the other is a modification of Zadeh's compositional rule of inference. In the proposed methods, for inference in a rule-based system, the conditional rule is first expressed as a fuzzy binary relation. In translation, we prefer to use triangular norms for a better understanding. New facts are then used to compute the similarity between the fact and the antecedent of the rule to modify the above fuzzy binary relation and not the consequence of the rule as applied in the existing similarity-based reasoning techniques. The modification is based on a measure of similarity following some scheme to be presented. The result can be interpreted as the induced fuzzy binary relation. The inference is computed from the induced fuzzy binary relation using the well-known sup operation.

The above scheme is used in formulating different models (rule-based and resolution-based). We will provide simple numerical examples for a better understanding of the proposed schemes.

In similarity-based approximate reasoning scheme, we see that from a given fact, the desired conclusion is derived using only a measure of similarity between the fact and the antecedent in a rule-based system. In some cases, a threshold value τ is associated with a rule. If the degree of similarity, between the antecedent of the rule and the given fact, exceeds the real value of τ, associated with the rule under consideration, only then is that rule assumed to be fired. The conclusion is derived using some modification procedure.

As an illustration, let us consider the two premises as in Table 2.

Here, A and A' are fuzzy sets defined over the same universe of discourse $U = \{u_1, u_2, \ldots, u_m\}$ and B, B' are defined over the universe of discourse $V = \{v_1, v_2, \ldots, v_n\}$. Let $S(A, A')$ denote some measure of similarity between two fuzzy sets A and A'. In the existing techniques, if $S(A, A') > \tau$, then the rule will be fired and the consequent of the rule is modified to produce the desired conclusion. Based on the change of membership grade of the consequent, two types of modification procedures can be proposed as in [18,28] - expansion-type inference and reduction-type inference.

Let $B' = \sum_{i=1}^{n} \{\mu_{B'}(v_i)/v_i\}$ and $s = s(S(A, B), \tau)$.

Expansion form: $\mu_{B'}(v_i) = \min(1, \mu_B(v_i)/s)$. Reduction form: $\mu_{B'}(v_i) = (\mu_B(v_i).s)$.

The methods proposed in [19,29] use the threshold value, a confidence factor and the reduction form of inference without providing any argument as to the choice of modification procedure. In one of them [19], each fuzzy set is first conceived as an m-component vector, and then the concept of vector dot product is used for finding the similarity. If $S(A, A') \geq \tau$, the predefined threshold value, then the rule will be fired and strength of confirmation is calculated by $S(A, A').\mu$, where μ is the membership value associated with the rule. In the other method [29], the author used weights with each propositions for the calculation of similarity. The procedure for the computation of the conclusion remains the same.

In [27], the authors used the value of certainty factor associated with the rules in the modification procedure. The inference is based on the number of propositions in the antecedent of the rule(s) as also the operator(s) connecting them. In each case, the inference is one of expansion type. In [21], they have also presented two more modification procedures and claimed for two new fuzzy reasoning methods. One modification is based on Zadeh's inclusion and cardinality measure and the other on equality and cardinality measure. Other operations remain almost identical.

Proposed method

In this section, we show how conclusions can be obtained from the given premises with the help of such a similarity measure. Let X, Y be two linguistic variables and let U, V respectively denote the universes of discourse. Two typical propositions p and q are given, and we like to derive a conclusion according to similarity-based inference. The scheme can be best described in Table 2.

Table 2 Ordinary approximate reasoning

p:	X is A	then	Y is B
q:	X is A'		
r:			Y is B'.

Let $U = \{u_1, u_2, \ldots, u_l\}$, $V = \{v_1, v_2, \ldots, v_m\}$ denote the respective universe of discourse of the linguistic variables X and Y. Let fuzzy sets A, A' and B in Table 2 be defined as

$$A = \sum_{i=1}^{l} \{\mu_A(u_i)/u_i\};\ A' = \sum_{i=1}^{l} \{\mu_{A'}(u_i)/u_i\};$$

$$B = \sum_{i=1}^{m} \{\mu_B(v_i)/v_i\};\ B' = \sum_{i=1}^{m} \{\mu_{B'}(v_i)/v_i\}.$$

All the existing methods [18,20,21,27] use the similarity measure for a direct computation of inference without considering the induced relation, i.e. how the underlying relation (a condition) is modified in the presence of the given fact. This is important in deriving a consequence of the fact from the rule. Consequently, those methods provide the same conclusion, if A and A' are interchanged in the propositions concerned. Thus, if p, q and p', q' be defined as in the following:

(i) p : if X is A, then Y is B, τ and $q : X$ is A';
(ii) p : if X is A', then Y is B, τ and $q : X$ is A,

then both *(i)* and *(ii)* will produce the same conclusion which is not appealing. This happens because the conclusion is derived by a modification of the consequent of the rule. It should be noted here that this is not the case with Zadeh's compositional rule of inference. Another notable fact is that we need to consider the threshold or certainty factor in order to tackle the problem of rule misfiring.

The first drawback can be eliminated if we consider the interpretation of the relational operator present in the conditional premise, as is done in executing compositional rule of inference. It is easy to verify that for a class of nested fuzzy sets, each different from the other, the consequence of a rule using CRI becomes the same. We seek a reasoning system which should be such that every change in the concept(s) as appears in the antecedent of the rule and that in the fact should be incorporated in the induced relation between the variables defining a rule, in this case, X and Y. Only then the inference will be influenced by the change concerned.

In order to avoid the use of certainty factor for rule misfiring, we modify the inference scheme in such a way that a significant change will make the conclusion less specific. This is done if an expansion type of inference scheme be chosen. Here, the 'UNKNOWN' case, i.e. the fuzzy set $B' = V$, can be taken as the limit. Explicitly, when the similarity value becomes low, i.e when A and A' differ significantly, the reasoning process should be such that the only inference be $B' = V$. As $A' = A$, we expect that $B' = B$. This, in turn, implies that nothing better than what the rule says should be allowed as a valid conclusion.

Schema

In view of the above observations, we propose a similarity-based inference method for deriving the consequence r. We first generate the fuzzy relation between the antecedent variable(s) and the consequent variable as done in executing CRI. We then compute the absolute change in linguistic labels, represented as fuzzy sets, and systematically propagate the same into the conditional relation in order to obtain the induced modified conditional relation. From this induced modified relation, a possible conclusion can be drawn using the sup operation. The scheme for computation can be presented in the following algorithm.

Algorithm SAR: similarity-based ordinary approximate reasoning

Step 1. Translate premise p into a relation $R(A, B)$ and compute it using any suitable translating rule possibly, a t-norm operator.

Step 2. Compute similarity $S(A, A')$ by *Definition 4.*

Step 3. Modify $R(A, B)$ with $S(A, A')$ to obtain the modified conditional relation $R(A' \mid A, B)$ according to some scheme C.

Step 4. Use sup-projection operation on $R(A' \mid A, B)$ to obtain B' as

$$\mu_{B'}(v) = \sup_{u} \mu_{R(A'|A,B)}(u, v). \tag{6}$$

Now, for a given fact $q : X$ is A' and from the condition p : if X is A, then Y is B, we propose two schemes $C1$ and $C2$ for computation of the modified conditional relation $R(A' \mid A, B)$ as given in Step 3.

Scheme C1. The first scheme C1 is based on a concept similar (but NOT identical) to the method proposed in [18]. We may recall here that the authors computed the conclusion $B' = \min(1, B/s)$, where s is the measure of similarity between fuzzy sets A and A' without considering the information suggested by the conditional rule. Here, we propose to modify the conditional relation according to (7).

$$R(A' \mid A, B) = \left[r'_{u,v}\right]_{l \times m} = \begin{bmatrix} r'_{u,v} & = \min(1, r_{u,v}/s) & \text{if } s > 0 \\ & = 1 & \text{otherwise} \end{bmatrix}. \tag{7}$$

The difference between the proposed scheme from the one presented in [18] can be noted easily. It is clear that the proposed scheme, unlike the schemes in [18,20], does not produce the same conclusion when A and A' are interchanged. It is not difficult to see that in (7), if $s \le r_{u,v}$ for some $v \in V$, then $r'_{u,v}$ becomes equal to 1. This makes the membership of that v in the resultant fuzzy set equal to one. This scheme, although a heuristic one, is intuitively a plausible scheme. Our next scheme C2 for computation of $R(A' \mid A, B)$ is based on a set of axioms.

Scheme C2. We believe that in a similarity-based reasoning methodology, a scheme for computation of the induced relation, when a fact and a conditional statement is given, should satisfy the following axioms:

A1. If $S\left(A, A'\right) = 1$, i.e. if $A' = A$, then $\mu_{R(A'|A,B)}(u, v) = \mu_{R(A,B)}(u, v), \forall (u, v) \in U \times V$.

A2. If $S(A, A') = 0$, i.e. if $A' \cap A = \Phi$, then $\mu_{R(A'|A,B)} = 1 \ \forall (u, v) \in U \times V$.

A3. As $S(A, A')$ increase from 0 to 1, $\mu_{R(A'|A,B)}(u, v)$ decreases uniformly from 1 to $\mu_{R(A,B)}(u, v); \forall (u, v) \in U \times V$.

Axiom A1 asserts that we should not modify the conditional relation as and when A' and A remain equal. Axiom A2 asserts that when A' is completely dissimilar to A, i.e. A' and A have disjoint support, we should not conclude specifically. In such a situation, anything is possible. A3 says that as the fact A' changes from the most dissimilar case (similarity value 0) to the most similar one (similarity value 1), the inferred conclusion should change from the most non-specific case, i.e. the UNKNOWN case ($B' = V$) to the most specific case, i.e. $B' = B$. This, in turn, means that whatever A' be, $R(A' \mid A, B) \supseteq R(A, B)$, i.e. the induced relation should not be more specific than what is given as a condition.

For notational simplicity, let us denote $S(A, A')$ by s and $R_{A'|A,B}$ by r'. Now, axiom A3 *uniquely* suggest a function of the form

$$\frac{dr'}{ds} = k \text{ (a constant) } \Rightarrow r' = k\,s + c, \quad c \text{ is a constant.}$$

These two constants can be determined from the conditions already prescribed in axiom A1 and axiom A2. More explicitly, when $s = 1$, we know that $r' = r$ (from axiom A1), and when $s = 0$, we know that $r' = 1$ (from axiom A2). This gives, $r' = 1 - (1 - r).s$ as our new scheme for the modification of the conditional relational.

Therefore, axiom A1 through axiom A3 *uniquely* suggest the scheme C2 as

$$\mu_{R(A'|A,B)}(u, v) = 1 - \left(1 - \mu_{R(A,B)}(u, v)\right).S\left(A, A'\right). \tag{8}$$

From (7) and (8), we observe that when $S(A, A') = 0$ we have $B' = V$. In other words, it is impossible to conclude anything when $\{A, A'\}$ are completely dissimilar. Again, when $S(A, A')$ is close to unity, then $R(A' \mid A, B)$ is close to $R(A, B)$ and the inferred fuzzy set B' will be close to B, i.e. $S(B, B')$ is close to unity. Axiom A3 also suggests that a small change in the input produces a small change in the output. In this sense, the above mechanism of inference is stable. As in the previous case, in (8), if either $S(A, A') = 0$ or $\mu_{R(A,B)} = 1$, then $r'_{u,v}$ becomes equal to 1.

Let us see how the above scheme can be modified to handle the concept of threshold associated with a rule in a natural manner. Let τ be the threshold associated with the rule. For that, we are to modify axiom A2 according to the following:

A4. If $S(A, A') \leq \tau$, then $\mu_{R(A'|A,B)} = 1 \ \forall (u, v) \in U \times V$. Accordingly, simple calculations as before resulted in the following $\mu_{R(A'|A,B)} = \min\left[1, \left(1 - \left(1 - \mu_{R(A,B)}\right).\frac{s - \tau}{1 - \tau}\right)\right]$ as the general scheme for relation membership modification. The case $\tau = 0$ corresponds to the scheme presented in (8). This scheme ensures that with all fuzzy sets A' having similarity value $S(A, A')$ less or equal to the threshold value τ, the inference B' using (7) will be 'UNKNOWN'.

Let A be a normal fuzzy set. If we assume that the translating rule used in generating the conditional relation is one of the t-norm types, then, as is already proposed, a basic and desirable result of the inferred proposition *nothing better than what the rule says can be concluded* can be established as in the following. For that, let us consider the model as in Table 2. For all A, A', the following proposition is valid.

Theorem 6. $B' \supseteq B$. *[5,30]*

A few translation rules are presented in Table 3, and the corresponding rules for modification of relations are presented in Table 4, based on the interpretation of the $\rightarrow$ operators in the formula $(s \rightarrow \mu_R)$, where s is the similarity measure between A and A', i.e. $s = S(A, A')$. They can be categorised into groups - expansion-type modifier and reduction-type modifier. For the first class, $\mu_{R'} \geq \mu_R \ \forall (u, v) \in U \times V$, and that for the second class, $\mu_{R'} \leq \mu_R \ \forall (u, v) \in U \times V$.

Table 3 For translation of relations

R	Conditional relations	Translation ($\mu_R(u,v)$)
R_1	Reichenbach s-implication	$1-\mu_A(u)+\mu_A(u)*\mu_B(v)$
R_2	Kleen-Dienes s-implication	$\max(1-\mu_A(u),\mu_B(v))$
R_3	Lukasiewicz r- and s-implications	$\min(1-\mu_A(u)+\mu_B(v),1)$
R_4	Rescher-Gaines r-implication	$\begin{cases}1, & \text{if } \mu_A(u)\le\mu_B(v))\\ 0, & \text{otherwise}\end{cases}$
R_5	Godel r-implication	$\begin{cases}1, & \text{if } \mu_A(u)\le\mu_B(v))\\ \mu_B(v), & \text{otherwise}\end{cases}$
R_6	Goguen r-implication	$\begin{cases}\min(\mu_B(v)/\mu_A(u),1), & \text{if } \mu_A(u)>0\\ 1, & \text{otherwise}\end{cases}$
R_7	Standard t-norm	$\min(\mu_A(u),\mu_B(v))$
R_8	Bounded sum t-norm	$\max(\mu_A(u)+\mu_B(v)-1,0)$
R_9	Product t-norm	$\mu_A(u)*\mu_B(v)$

A few examples of translating rules for a simple conditional statement is presented in Table 3. Each rule actually identifies a fuzzy relation. With input, these relations are modified using schemes presented in Table 4. The effects of different translating rules and modification procedures are presented in Table 5, which shows that for identical matching as in Figure 9, the output of rule firing is consistent if we choose t-norm and r-implication for translation.

For distinct A, A' as in Figure 10, the result of rule firing is shown in Table 6. In this case, also, the performance of translating rules is satisfactory for t-norm (R_7, R_8, R_9) as well as r-implication (R_4, R_5, R_6), and modification rules mr1, mr2 and mr3, respectively. Simulation study reveals the modification procedure mr2 is consistent. An output fuzzy set B' is shown in Figure 11, using the modification scheme mr1 in Table 6.

With the above understanding of similarity-based reasoning methodology, let us now propose another module of a fuzzy system - defuzzification.

Specificity measure - defuzzification

The result of rule firing, using any of the above-mentioned approaches to inference, is a fuzzy set. This is interpreted at the semantic level as the desired output. Often, we need to determine a precise action as output. The purpose of defuzzification is to obtain a scalar value $u \in U$, from the said output fuzzy set, as the action. Then, if necessary, denormalisation is performed on the output so as to obtain the corresponding action on its physical domain.

Table 4 For modification of relations

Scheme	Modified relations $\mu_{R'}(u,v)$	Type
mr1	$\begin{cases}\min(\mu_R(u,v)/s,1), & \text{if } s>0\\ 1, & \text{otherwise}\end{cases}$	Expansion
mr2	$\min((1-s)+s.\mu_R(u,v),1)$	Reduction
mr3	$\min(1-s+\mu_R(u,v),1)$	Expansion
mr4	$\max(1-s.\mu_A(u),\mu_B(v))$	Reduction
mr5	$\max(1-s.\mu_A(u),\min(\mu_A(u),\mu_B(v)))$	
mr6	$\max(1-s,\min(s,\max(1-\mu_A(u),\min(\mu_A(u),\mu_B(v)))))$	
mr7	$\min(1-\mu_A(u).s+s.\mu_A(u).\mu_B(v),1)$	

Table 5 Results for different reasoning schemes for $A' = A$

Modified relation	Relation data (A')	Input	$S(A, A')$	Output (B')	$S(B, B')$
mr1, mr2, mr3	R_1	A	1.0	$B' \supset B$	0.524941
	R_2	A	1.0	$B' \supset B$	0.526477
	R_3	A	1.0	$B' \supset B$	0.522559
	R_4	A	1.0	$B' = B$	1.0
	R_5	A	1.0	$B' = B$	1.0
	R_6	A	1.0	$B' \supset B$	0.945352
	R_7	A	1.0	$B' = B$	1.0
	R_8	A	1.0	$B' = B$	1.0
	R_9	A	1.0	$B' = B$	1.0
mr4		A	1.0	$B' \supset B$	0.526477
mr5		A	1.0	$B' \supset B$	0.526477
mr6		A	1.0	$B' \supset B$	0.526477
mr7		A	1.0	$B' \supset B$	0.524941

Specificity measure of fuzzy set estimates the precision of an information represented by the fuzzy set rather than an estimate of its fuzziness which is measured by the entropy of the fuzzy set. In order to provide a definition for any specificity index, a number of factors must be considered. A fuzzy set with maximum specificity value corresponds to a precise assessment of the values of a variable. In trying to capture the form of the specificity index, a number of properties are required or desirable.

According to Dubois and Prade, a specificity measure $\mathrm{Sp}(A)$ [31] should satisfy the following properties. Let X be a linguistic variable defined on a universe of discourse U. A and B are normalised fuzzy subsets of U.

P1. $\forall A \subseteq U, \mathrm{Sp}(A) \in [0, 1]$.

P2. $\mathrm{Sp}(A) = 1$ if and only if A is a singleton of S.

P3. If $A \subseteq B$, then $\mathrm{Sp}(A) \geq \mathrm{Sp}(B)$.

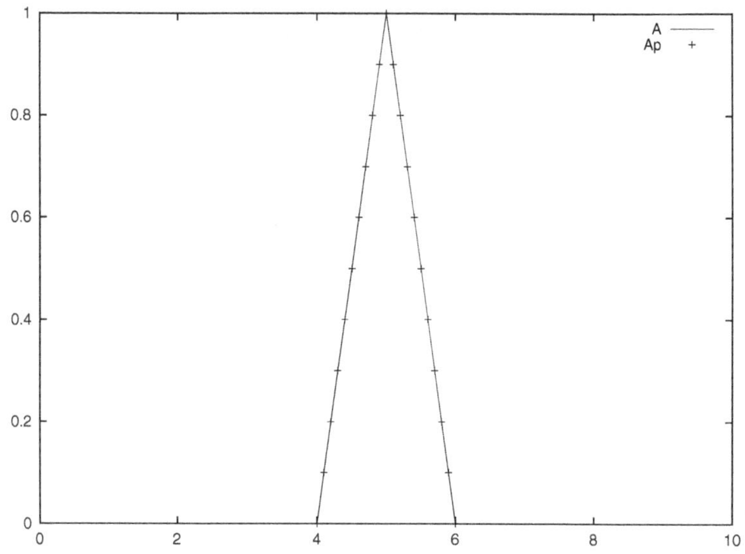

Figure 9 The output when $A = A'$.

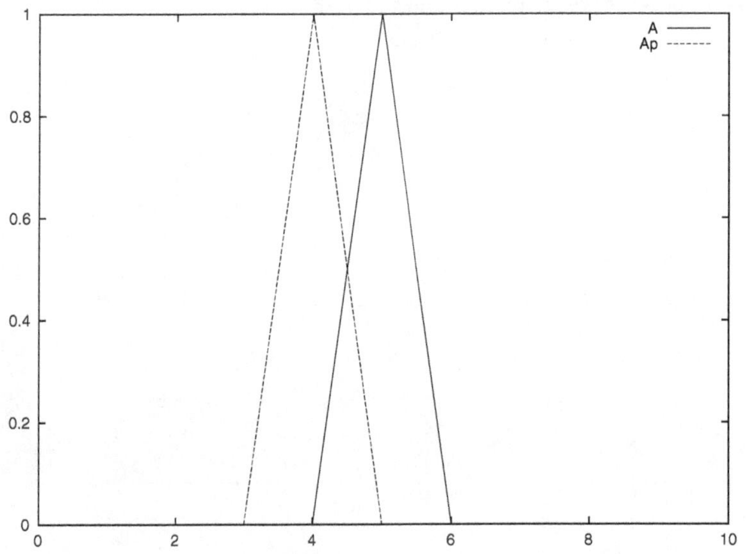

Figure 10 The ouput when *A* and *A′* are different.

Yager [32] introduced one such measure of specificity that satisfies the above properties. When U is finite, Yager proposed an expression for defining the specificity. Let us assume that A be a fuzzy set defined over the universal set U and A_α be the α-level set of A. The specificity associated with A is denoted as Sp(A) and is defined as

$$\text{Sp}(A) = \int_0^{\alpha_{\max}} \frac{1}{\text{card}A_\alpha} d\alpha \tag{9}$$

Let us now list some properties [32] associated with the above definition.

For all A, Sp(A) assumes its maximum value 1, when $A = \{1/u\}$ for some particular $u \in U$.

For all A, Sp(A) $\in [0, 1]$ and it assumes its minimum value 0, when $A = \Phi$.

If, for all A, $\mu_A(u) = k$ for all $u \in U$, then $\text{Sp}(A) = \frac{k}{n}$ where n is the cardinality of the ordinary set U.

Defuzzification is a procedure applied to reduce the anxiety in a decision. Accordingly, we propose a new technique for defuzzification based on a measure of precision. Let there be m clipped fuzzy sets $\{A^{(k)}; k = 1, 2, \ldots, m\}$ and let $\{s_{(k)}, p^{(k)}; k = 1, 2, \ldots, m\}$ be the specificity associated with $A^{(k)}$ as well as the peak of the consequent fuzzy set of the kth rule. Then, the defuzzified value u^* will be given by

$$u^* = \frac{\sum_{k=1}^{m} p^{(k)}.s_{(k)}}{\sum_{k=1}^{m} s_{(k)}} \tag{10}$$

The *height* method of defuzzification demands strictly convex fuzzy sets. The individual peak values of consequent fuzzy sets of the fired rules are used to generate the weighted average of these peak values. It is a simple method and works faster than the centre of

Table 6 Results for different reasoning schemes for $A' \neq A$

Modified relation	Relation	Input data (A')	$S(A, A')$	Output (B')	$S(B, B')$
mr1	R_1	A'	0.683772	$B' \supset B$	0.438764
	R_2	A'	0.683772	$B' \supset B$	0.441981
	R_3	A'	0.683772	$B' \supset B$	0.437005
	R_4	A'	0.683772	$B' = B$	1.0
	R_5	A'	0.683772	$B' \supset B$	0.935526
	R_6	A'	0.683772	$B' \supset B$	0.906437
	R_7	A'	0.683772	$B' \supset B$	0.935526
	R_8	A'	0.683772	$B' \supset B$	0.935526
	R_9	A'	0.683772	$B' \supset B$	0.935526
mr2	R_1	A'	0.683772	$B' \supset B$	0.437789
	R_2	A'	0.683772	$B' \supset B$	0.439834
	R_3	A'	0.683772	$B' \supset B$	0.436249
	R_4	A'	0.683772	$B' \supset B$	0.707154
	R_5	A'	0.683772	$B' \supset B$	0.697115
	R_6	A'	0.683772	$B' \supset B$	0.691508
	R_7	A'	0.683772	$B' \supset B$	0.697115
	R_8	A'	0.683772	$B' \supset B$	0.697115
	R_9	A'	0.683772	$B' \supset B$	0.697115
mr3	R_1	A'	0.683772	$B' \supset B$	0.407110
	R_2	A'	0.683772	$B' \supset B$	0.410883
	R_3	A'	0.683772	$B' \supset B$	0.405919
	R_4	A'	0.683772	$B' \supset B$	0.701754
	R_5	A'	0.683772	$B' \supset B$	0.687924
	R_6	A'	0.683772	$B' \supset B$	0.682404
	R_7	A'	0.683772	$B' \supset B$	0.687924
	R_8	A'	0.683772	$B' \supset B$	0.687924
	R_9	A'	0.683772	$B' \supset B$	0.687924
mr4		A'	0.683772	$B' \supset B$	0.440355
mr5		A'	0.683772	$B' \supset B$	0.440355
mr6		A'	0.683772	$B' \supset B$	0.525030
mr7		A'	0.683772	$B' \supset B$	0.437789

sums method. Let $p^{(k)}$ be the peak value of $A^{(k)}$ and $h^{(k)}$ be the corresponding height of the clipped version of $A^{(k)}$. Then, the defuzzified value will be given by

$$u^* = \frac{\sum_{k=1}^{m} p^{(k)}.h^{(k)}}{\sum_{k=1}^{m} h^{(k)}}. \tag{11}$$

Specificity, height and peak values are sometimes used simultaneously to compute defuzzified value. Here, we propose a method for the defuzzification of fuzzy sets as in the following:

$$u^* = \frac{\sum_{k=1}^{m} p^{(k)}.h^{(k)}.s_{(k)}}{\sum_{k=1}^{m} h^{(k)}.s_{(k)}}. \tag{12}$$

Example 2. In this example, we consider two fuzzy sets as given in Figure 12 and corresponding clipped fuzzy sets in Figure 13. From these sets, we compute the defuzzified

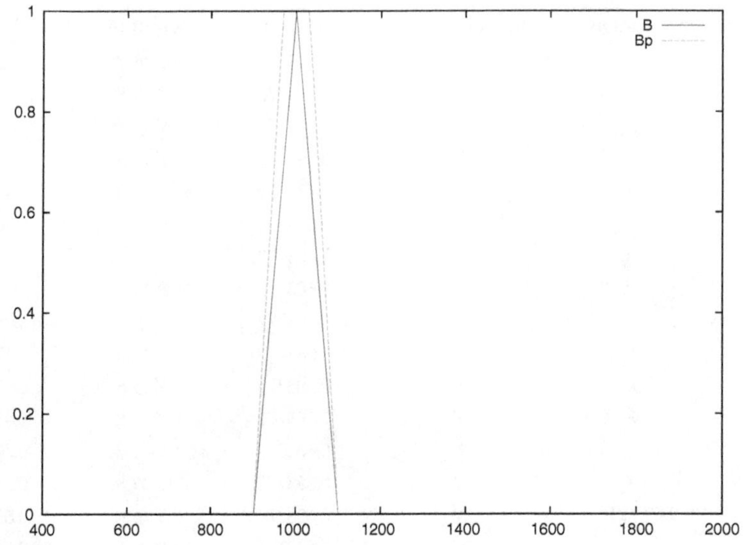

Figure 11 *B* and *B′* for mr1. Comparison of output with *B*.

value of the fuzzy sets using the methods of defuzzification as given in (10), (11) and (12), respectively.

$\mathrm{Sp}(A^{(1)}) = 0.175$, $\mathrm{Sp}(A^{(2)}) = 0.275$; $p^{(1)} = 3$, $p^{(2)} = 4$ and $h^{(1)} = 0.6$, $h^{(2)} = 0.8$.
From (10), (11) and (12), we get $u^* = 3.61, u^* = 3.57$ and $u^* = 3.67$.
We apply these three types of defuzzification methods in our fuzzy systems.

Application in different models

Let us consider a generalised model as presented in Table 7. This form of reasoning is used in rule-based fuzzy systems. In particular, it is used in pattern classification and fuzzy control. Let there be n linguistic variables associated with another linguistic variable Y according to the following m fuzzy rules. The problem is to find the linguistic value of the variable Y as suggested by the rules, when the values of the n variables are given.

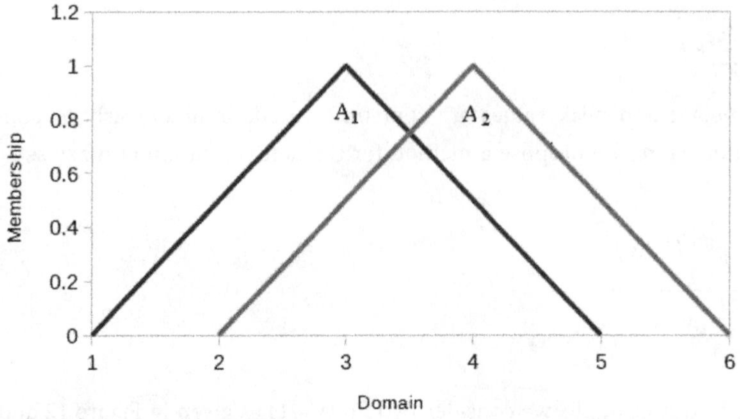

Figure 12 Fuzzy sets for defuzzification.

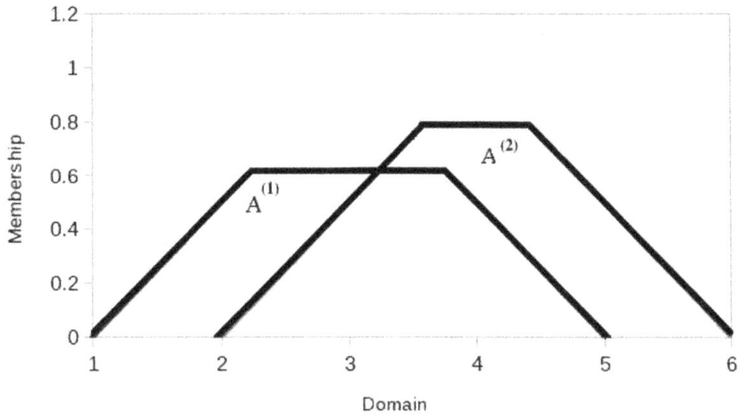

Figure 13 Clipped fuzzy sets for defuzzification.

Under the conventional technique, for each rule, the consequent fuzzy set is calculated according to existing method of inference as already described, and then the union of all consequent fuzzy sets is taken as the conclusion which is then defuzzified, if necessary, using some defuzzification scheme. In similarity-based reasoning, this cannot be done. Here, the membership values computed from the modified induced relation becomes less and less specific as the similarity between the facts and antecedent of a rule decreases. In conventional practice, the membership values of various elements become equal to the maximum, making it an ambiguous one (more alternatives with similar membership values at the positive level) with the reduction of the firing strength (used in deriving a conclusion), but the membership values at which the ambiguity occurs become less than one. For example, in case of Mamdani type of reasoning, if the firing strength of a rule is say 0.3, then all alternatives which have membership values greater than or equal to 0.3 take membership values of 0.3. On the other hand, in the present case, if the similarity value is 0.3, then the membership values of elements in the inferred fuzzy set will be at least 0.3. Moreover, the elements having membership value greater than or equal to 0.3 in the consequent of the rule will be equal to one in the consequent fuzzy set. This means that with decrease in similarity the computed membership values increase and ultimately move close to the least specific case (with membership values of 1 for all alternatives). For this reason, we propose a new scheme, for computing the final conclusion, based on a measure of similarity. Our method is based on rule selection and then rule execution. In both cases, we use the concept of similarity between fuzzy sets as a basis of the task. For that, first of all, we compute $S(A_{ij}, A_i); i = 1, 2, \ldots, m$. Then, we perform the same

Table 7 Applicable form of approximate reasoning

if	X_1 is A_{11}	and	X_2 is A_{12}	$\cdots$	X_n is A_{1n}	then	Y is B_1
else if	X_1 is A_{21}	and	X_2 is A_{22}	$\cdots$	X_n is A_{2n}	then	Y is B_2
			$\vdots$		$\vdots$		
else if	X_1 is A_{m1}	and	X_2 is A_{m2}	$\cdots$	X_n is A_{mn}	then	Y is B_m
	X_1 is A_1	and	X_2 is A_2	$\cdots$	X_n is A_n		
Conclusion							Y is B

operation for different $j = 1, 2, \ldots, n$. Let s_{ij} denote the different similarity values. We next compute the overall rule matching index from the above data as

$$s^i = \min_j s_{ij} \tag{13}$$

From among the m distinct rules, we choose those rules for which $s^i > \epsilon$. This ϵ can be interpreted as a threshold in our case. We then apply the algorithm *SEAR* to generate a conclusion from each rule conformal for firing. The output can be generated using the intersection of fuzzy sets. It is important to note that the intersection operation is chosen in order to justify the rule selection procedure. Here, fewer rules are fired and the output of each rule is significant.

To compute B' as in Table 7, we apply the following algorithm.

Algorithm AFSAR: applicable form of similarity-based approximate reasoning

Step 1. Compute s_{ij} for $i = 1, 2, \ldots, m; j = 1, 2, \cdots, n$ and then s^i according to (13).

Step 2. Define ϵ and find the rules conformal for firing.

Step 3. Translate the ith rule, provided $s^i > \epsilon$, and compute the relation R_i using any suitable translating rule possibly, a t-norm operator as given in Table 3.

Step 4. Modify R_i with s^i to obtain the modified conditional relation R_i' according to some modification scheme as given in Table 4.

Step 5. Use sup-projection operation on R_i' to obtain B_i' as given in (14).

$$\mu_{B_i'}(v) = \sup_{u_1, u_2, \ldots, u_k} \mu_{R(A_1|A_{i1}, A_2|A_{i2}, \ldots, A_k'|A_{ik}, B)}(u_1, u_2, \ldots, u_k, v). \tag{14}$$

Step 6. Compute the specificity measure of B_i' denoted by $\mathrm{Sp}(B_i')$, for all i for which a rule is fired and set $\mathrm{Sp} = \max_i \mathrm{Sp}(B_i')$. If $\mathrm{Sp} > 1 - \epsilon$ $(\epsilon > 0)$, the predefined threshold, then output $B = \left\{ \bigcap_j B_j' | \mathrm{Sp}(B_j') > 1 - \epsilon \right\}$.

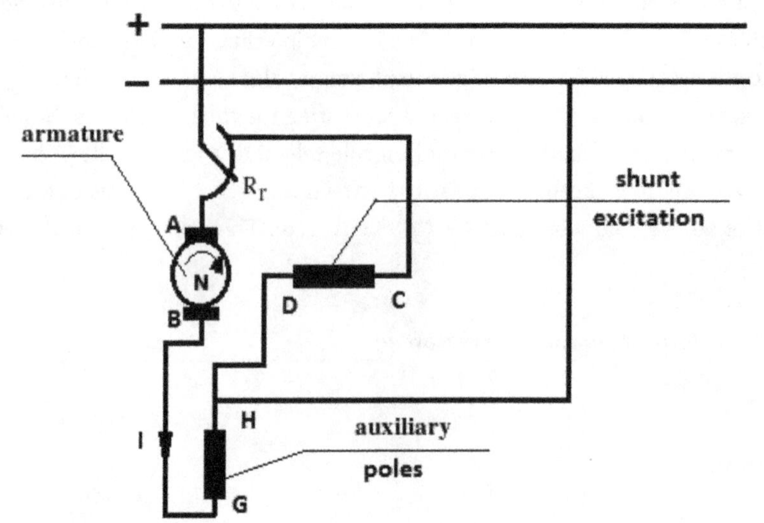

Figure 14 DC shunt motor.

Table 8 Rule base of DC shunt motor

If $I = A_1$,	then $N = B_{17}$	also	If $I = A_{12}$,	then $N = B_7$	also
If $I = A_2$,	then $N = B_{15}$	also	If $I = A_{13}$,	then $N = B_9$	also
If $I = A_3$,	then $N = B_{13}$	also	If $I = A_{14}$,	then $N = B_{11}$	also
If $I = A_4$,	then $N = B_{11}$	also	If $I = A_{15}$,	then $N = B_{13}$	also
If $I = A_5$,	then $N = B_9$	also	If $I = A_{16}$,	then $N = B_{15}$	also
If $I = A_6$,	then $N = B_7$	also	If $I = A_{17}$,	then $N = B_{17}$	also
If $I = A_7$,	then $N = B_5$	also	If $I = A_{18}$,	then $N = B_{15}$	also
If $I = A_8$,	then $N = B_3$	also	If $I = A_{19}$,	then $N = B_{13}$	also
If $I = A_9$,	then $N = B_1$	also	If $I = A_{20}$,	then $N = B_{11}$	also
If $I = A_{10}$,	then $N = B_3$	also	If $I = A_{21}$,	then $N = B_9$	also
If $I = A_{11}$,	then $N = B_5$	also			

In the process, we find a conceptual change in similarity-based inference mechanism. A closer look at the connection between the proposed schemes and the existing schemes allows us to conclude that our schemes can be thought of as an integration of Zadeh's compositional rule of inference and similarity-based inference schemes. Such a scheme is expected to produce efficiency in inference mechanisms.

A case study on DC shunt motor

In this section, let us consider the DC shunt motor in Figure 14 as investigated in [33]. From the measurement of the current value I, the rotating speed value N in steady states was determined by $N = f(I)$.

Human experts observed the behaviour of the DC shunt motor and described the relation between current and speed in the form of fuzzy conditional statements as in the following, where I is the linguistic variable representing the motor current and N is the linguistic variable representing the motor rotations.

Let the domain set for the variable I be $U = \{0.0, 0.1, 0.2, 0.3, 0.4, \ldots, 9.9, 10.0\}$. Let $A_1, A_2, A_3, A_4, \ldots, A_{21}$ be the fuzzy sets corresponding the points $0.0, 0.5, 1.0, 1.5, \ldots, 10.0$. Let $V = \{400, 410, 420, \ldots, 1,990, 2,000\}$ be the universe of discourse of the linguistic variable N. Let $B_1, B_2, \ldots, B_{17}$ be the fuzzy sets corresponding the points $400, 500, 600, \ldots, 2,000$. Now, we describe the behaviour of the motor (the specific relation between current and speed) using fuzzy rules as in Table 8.

The data for the fuzzy model is given in Table 9, and the corresponding real static curve is given in Figure 15.

For a particular observed value of current expressed in natural language, we first translate the inexact concepts into fuzzy sets (the simple observation) or fuzzy relations (the complex rule) over the specified universe of discourse using triangular membership functions. We then perform approximate reasoning to obtain the corresponding speed of the DC motor using the algorithm SAR.

Table 9 Real data for DC shunt motor

I	N	I	N	I	N	I	N	I	N	I	N	I	N
0.0	2,000	1.5	1,400	3.0	800	4.5	600	6.0	1,200	7.5	1,800	9.0	1,600
0.5	1,800	2.0	1,200	3.5	600	5.0	800	6.5	1,400	8.0	2,000	9.5	1,400
1.00	1,600	2.5	1,000	4.0	400	5.5	1,000	7.0	1,600	8.5	1,800	10.0	1,200

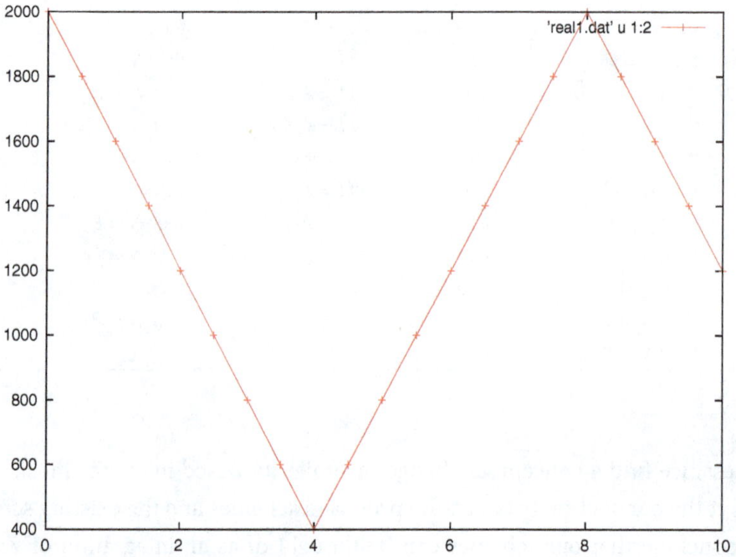

Figure 15 Real static characteristic of a DC motor.

The defuzzified input/output are plotted for a comparative assessment of the utility of the proposed similarity-based approximate reasoning methodology. The simulation results are presented in the following self-explanatory diagrams given in Figure 16.

Conclusions

Developing intelligent systems becomes necessary to handle modern computer-based technologies managing different kinds of information and knowledge. This paper discusses a theory to help provide solutions to difficult problems in the construction of intelligent systems in which the available information is supplied by human experts which

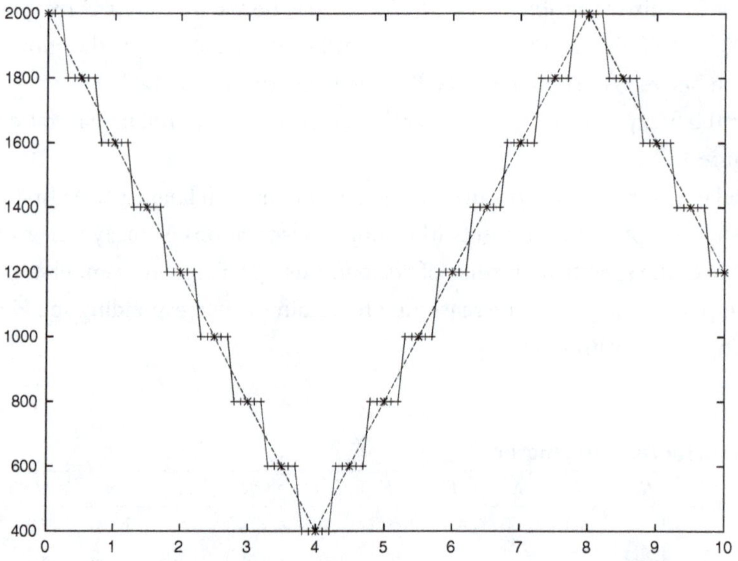

Figure 16 Comparison: real and proposed inference.

at times are found incomplete, imprecise or even uncertain in nature and therefore inherently ambiguous. It is hoped that by upgrading existing methodologies through addition of concepts and techniques drawn from the fuzzy set theory opens the door to a substantial enhancement of our ability to model reality. In the process, we have developed a mechanism to compute the matching degree of two fuzzy sets - representation of imprecise concepts (e.g. *low* speed and *very low* speed of a DC motor).

It has been shown that the concept of similarity is inherent in approximate reasoning methodology. Different problems arising out of the use of existing compositional rule of inference as well as similarity-based reasoning have been discussed with suitable examples. Different functions used to measure the similarity between two inexact concepts are reviewed. We have proposed axioms to compute the similarity between two fuzzy sets and appropriately introduced the concept in approximate reasoning methodology. It may be argued that the proposed similarity-based approximate reasoning technique is a combination of Zadeh's compositional rule of inference and Turksen's similarity-based reasoning. It is shown that this method is a more general characterisation of similarity-based approximate reasoning, and Turksen's method is a special case of the proposed method.

We have suggested relevant issues involved in the design of fuzzy systems - introduced similarity in reasoning, similarity relation in fuzzification and the concept of specificity measure in defuzzification. The concept of similarity is effectively used in system control. It is hoped that the introduction of the specificity-based defuzzification technique will prove to be a powerful technique in qualitative control. Further research on the use of similarity-based approximate reasoning is necessary for a better understanding of the effect of the same on the cognitive process involved in qualitative modelling and simulation. Similarity-based reasoning is a basic mode of inference in fuzzy logic in a wide sense that provides a basis for formalisation of commonsense reasoning and a non-*ad hoc* computational framework for combining and deriving evidence in expert systems. Similarity-based approximate reasoning may be popularised because of the scope of its applications in wide and challenging fields of investigation.

Acknowledgements

This research has been partially supported by the UGC SAP (DRS) Phase-II project under the Department of Mathematics, Visva-Bharati and the UGC funded major research project titled 'Role of similarity in approximate reasoning'.

References

1. Klawonn, F, Castro, JL: Similarity in fuzzy reasoning. Mathware Soft Comput. **2**, 336–350 (1995)
2. Valverde, L: On the structure of F-indistinguishability operators. Fuzzy Set. Syst. **17**(3), 313–328 (1985). doi:10.1016/0165-0114(85)90096-X
3. Bouchon-Meunier, B, Rifqi, M, Bothorel, S: Towards general measures of comparison of objects. Fuzzy Set. Syst. **84**(2), 143–153 (1996)
4. Tversky, A: Features of similarity. Psychol. Rev. **84**, 327–352 (1977)
5. Mondal, B, Mazumdar, D, Raha, S: Similarity in approximate reasoning. Int. J. Comput. Cogn. **4**(3), 46–56 (2006)
6. Raha, S, Pal, NR, Ray, KS: Similarity based approximate reasoning: methodology and application. IEEE Trans. Syst. Man Cybern. Syst. Hum. **32**(4), 541–547 (2002)
7. Pappis, C, Karacapilidis, N: A comparative assessment of measures of similarity of fuzzy values. Fuzzy Set. Syst. **56**, 171–174 (1993)
8. Wang, WJ: New similarity measures on fuzzy sets and on elements. Fuzzy Set. Syst. **85**, 305–309 (1997)
9. Kwang, HL, Song, YS, Lee, KM: Similarity measure between fuzzy sets and between elements. Fuzzy Set. Syst. **62**, 291–293 (1994)
10. Pedrycz, W: Neurocomputations in relational systems. IEEE Trans. Pattern Anal. Mach. Intell. **13**(1), 287–297 (1991)
11. Hong, DH, Hwang, SY: A note on the value similarity of fuzzy systems variables. Fuzzy Set. Syst. **66**, 383–386 (1994)

12. Zwick, R, Carlstein, E, Budescu, DV: Measures of similarity among fuzzy concepts: a comparative analysis. Inter. J. App. Reas. **1**, 221–242 (1987)
13. Lesot, M-J, Rifqi, M, Benhada, H: Similarity measures for binary and numerical data: a survey. Int. J. Knowl. Eng. Soft Data Paradigms **1**(1), 63–84 (2009)
14. Bouchon-Menuier, B, Rifqi, M, Lesot, M-J: Similarities in fuzzy data mining: from a cognitive view to real-world applications. In: Zurada, JM, Yen, GG, Wang, J (eds.) *Computational Intelligence: Research Frontiers. Lecture Notes in Computer Science, vol. 5050*, pp. 349–367. Springer, Berlin, (2008)
15. Zadeh, LA: A theory of approximate reasoning. In: Hayes, J, Michie, D, Mikulich, LI (eds.) *Machine Intelligence vol. 9*, pp. 149–194. Elsevier, New York, (1979)
16. Zadeh, LA: The concept of a linguistic variable and its application to approximate reasoning I, II. Inf. Sci. **8**, 199–249301357 (1975)
17. Zadeh, LA: Outline of a new approach to the analysis of complex systems and decision processes. IEEE Trans. SMC **3**, 28–44 (1973)
18. Turksen, IB, Zhong, Z: An approximate analogical reasoning approach based on similarity measures. IEEE Trans. SMC **18**(6), 1049–1056 (1988)
19. Chen, SM: A new approach to handling fuzzy decision making problems. IEEE Trans. SMC **18**, 1012–1016 (1988)
20. Tsang, FCC, Lee, JWT, Yeung, DS: Similarity-based fuzzy reasoning methods for fuzzy production rules. In: *Proc. Sixth IFSA World Congress*, pp. 157–160, Sao Paolo, (22–28 July 1995)
21. Yeung, DS, Tsang, ECC: A comparative study on similarity-based fuzzy reasoning methods. IEEE Trans. SMC - Part B: Cyber. **27**(2), 216–227 (1997)
22. Esteva, F, Garcia, P, Godo, L, Rodrigue, R: A modal account of similarity-based reasoning. Int. J. Approximate Reas. **16**(3–4), 235–260 (1997)
23. Esteva, F, Garcia, P, Godo, L, Rupsini, E, Valverde, L: On similarity logic and the generalised modus ponens. In: *Proceedings of the Third IEEE Conference on Fuzzy Systems, vol. 2*, pp. 1423–1427, Orlando, (26–29 June 1994)
24. Bouchon-Menuier, B, Valverde, L: Analogy relations and inference. In: *Proceedings of the Second IEEE International Conference on Fuzzy Systems, vol. 2*, pp. 1140–1144, San Francisco, (28 Mar–1 Apr 1993)
25. Bouchon-Menuier, B, Valverde, L: A fuzzy approach to analogical reasoning. Soft Comput. **3**(3), 141–147 (1999)
26. Bouchon-Menuier, B, Delechamp, J, Marsala, C, Rifqi, M: Several forms of fuzzy analogical reasoning. In: *Proceedings of 6th IEEE International Conference on Fuzzy Systems, FUZZ-IEEE'97, vol. 1*, pp. 45–50, Barcelona, (1–5 July 1997)
27. Yeung, DS, Tsang, ECC: Improved fuzzy knowledge representation and rule evaluation using fuzzy petri nets and degree of subsethoods. Intell. Syst. **9**(12), 1083–1100 (1994)
28. Turksen, IB, Tian, Y: Combination of rules or their consequences in fuzzy expert systems. Fuzzy set. Syst. **58**(1), 3–40 (1993)
29. Chen, SM: A weighted fuzzy reasoning algorithm for medical diagnosis. Decis. Support Syst. **11**, 37–43 (1994)
30. Raha, S, Hossain, A, Ghosh, S: Similarity based approximate reasoning: fuzzy control. J. Appl. Logic. **6**(1), 47–71 (2008)
31. Dubois, D, Prade, H: A note on measures of specificity for fuzzy sets. Int. J. Gen Syst. **10**, 279–283 (1985)
32. Yager, RR: Measuring tranquility and anxiety in decision making: an application of fuzzy sets. Int. J. Gen Syst. **8**, 139–146 (1982)
33. Goralczany, MB, Kiszka, JB, Stachowicz, MS: Some problems of studying adequacy of fuzzy models. In: Yager, RR (ed.) *Fuzzy Set and Possibility Theory*, pp. 14–31. Pergamon Press, New York, (1982)

6

Three-layer supply chain in an imperfect production inventory model with two storage facilities under fuzzy rough environment

Amalesh Kumar Manna[1], Jayanta Kumar Dey[2*] and Shyamal Kumar Mondal[1]

*Correspondence:
dey_jaykum@yahoo.com
[2]Department of Mathematics, Mahishadal Raj College, Mahishadal, West Bengal, 721628, India
Full list of author information is available at the end of the article

Abstract

This article focuses on an imperfect production inventory model considering product reliability and reworking of imperfect items in three-layer supply chain under fuzzy rough environment. In the model, the supplier receives the raw materials, all are not of perfect quality, in a lot and delivers the items of superior quality to the manufacturer and the inferior quality items are sold at a reduced price in a single batch by the end of the cent percent screening process. The manufacturer produces a mixture of perfect and imperfect quality items. A portion of the imperfect items is transformed into perfect quality items after rework. Another portion of imperfect items, termed as 'less perfect quality items,' is sold at a reduced price to the retailer, and the portion which cannot be either transformed to the perfect quality items or sold at a reduce price is being rejected. Here, retailer purchases both the perfect and imperfect quality items from the manufacturer to sell the items to the customers through his/her respective showrooms of finite capacities. A secondary warehouse of infinite capacity is hired by the retailer on rental basis to store the excess quantity of perfect quality items. This model considers the impact of business strategies such as optimal order size of raw materials, production rate, and unit production cost in different sectors in a collaborating marketing system that can be used in the industry, like textile, footwear, and electronics goods. An analytical method has been used to optimize the production rate and raw material order size for maximization of the average profit of the integrated model. Finally, a numerical example is given to illustrate the model.

Keywords: Inventory; Three-layer supply chain; Imperfect production; Two-warehouse; Stock-dependent demand; Fuzzy rough parameters

Introduction

Business organizations all over the world are striving hard to evolve strategies to survive in the era of competition ushered in by globalization. Supply chain management (SCM) is one such strategy. It is an effective methodology and presents an integrated approach to resolve issues in sourcing customer service, demand flow, and distribution. The focus is on the customer. The results are in the form of reduced operational costs, improved flow of supplies, reduction in delays of production, and increased customer satisfaction. While the goal of supply chain management is to reduce cost of producing and reaching the finished products to the customers, inventory control is the means to achieve the goal. Researchers as well as practitioners in manufacturing industries have given importance to develop inventory control problems in supply chain management. All steps from supply of raw materials to finished products can be included into a supply chain, connecting raw materials supplier, manufacturer, retailer, and finally, customers. Recent reviews on supply chain management are provided by Weng [1], Munson and Rosenblatt [2], Yang and Wee [3], Khouja [4], Yao et al. [5], Chaharsooghi et al. [6], Wang et al. [7], and others.

Nowadays, it is common to all industries that a certain percentage of produced or ordered items are a mixture of perfect and imperfect quality. It is also important to a supply manager of any organization to control and maintain the inventories of perfect and imperfect quality items. Salameh and Jaber [8] developed an inventory model for imperfect quality items using the economic production quantity (EPQ)/economic order quantity (EOQ) formulae and assumed that inferior quality items are sold as a single batch at the end of the total screening process. Thereafter, Goyal and Cardenas-Barron [9] extended the idea of Salameh and Jaber's [8] model and proposed a practical approach to determine EPQ for items with imperfect quality. Yu et al. [10] generalized the models of Salameh and Jaber [8], incorporating deterioration and partial back ordering. Liu and Yang [11] investigated a single-stage production system with imperfect process delivering two types of defects: reworkable and non-reworkable items. The reworkable items are sent for reworking, whereas non-reworkable items are immediately discarded from the system. They determined the optimal lot size that maximized the expected total profit over the expected time length of the production cycle. Panda and Maiti [12] represented a geometric programming approach for multi-item inventory models with price-dependent demand under flexibility and reliability with imprecise space constraint. Ma et al. [13] considered the effects of imperfect production processes and the decision on whether and when to implement a screening process for defective items generated during a production run. Sana [14] develops two inventory models in an imperfect production system and showed that the inferior quality items could be reworked at a cost where overall production inventory costs could be reduced significantly. Sana [15] extended the idea of imperfect production process in three-layer supply chain management system.

Inventory management is generally attracted for large stock for several reasons: an attractive price discount for bulk purchase, the replenishment cost including transportation cost is higher than the inventory related cost, the demand of an item is very high, and so on. Therefore, due to space limitation of showroom, one (or sometimes more than one) warehouse(s) is hired on rental basis to store the excess items. The secondary warehouse (SW) may be located away from the showroom or nearer to the showroom. The actual service to the customer is done at the showroom only. Hartely [16] first introduced the basic two warehouses problem in his book *Operations Research - A Managerial Emphasis.* After Hartely [16], a number of research papers have been published by different authors. Among them, the work done by Sarma [17], Dave [18], Goswami and Chaudhuri [19], Pakkala and Achary [20], Bhunia and Maiti [21], Benkherouf [22], Zhou [23], Kar et al. [24], and Chung and Huang [25] are worth mentioning. Dey et al. [26] considered a finite-time horizon inventory problem for a deteriorating item having two separate warehouses with interval-valued lead time under inflation and a time value of money. Liang and Zhou [27] investigated a two-warehouse inventory model for deteriorating items under conditionally permissible delay in payments. Hariga [28] proposed an EOQ model with multiple storage facilities where both owned and rented warehouses had limited stock capacity. They assumed the rented warehouse had higher unit holding costs than the own warehouse but offered better preservation resulting in a lower rate of deterioration for the goods than in the own warehouse.

Dubois and Prade [29] first studied the fuzzification problem of rough sets. Furthermore, Morsi and Yakout [30] defined the upper and lower approximations of the fuzzy sets with respect to a fuzzy min-similarity relation. Additionally, Radzikowska and Kerre [31], Xu and Zhou [32], Liu and Sai [33], Chen [34], and others generalized the above definitions of the fuzzy rough set to a more general case. Different types of uncertainty such as randomness, fuzziness, and roughness are common factors in any production inventory problem. But in some problems in production inventory system, both fuzziness and roughness occur simultaneously. In many cases, it is found that some inventory parameters involve both the fuzzy and rough uncertainties. For example, the inventory related costs holding cost, set-up cost, idle costs, etc. depend on several factors such as bank interest, inflation, etc. which are uncertain in fuzzy rough sense. To be more specific, inventory holding cost is sometimes represented by a fuzzy number and it depends on the storage amount which may be imprecise and range within an interval due to several factors such as scarcity of storage space, market fluctuation, and human estimation thought process, i.e., it may be represented by a rough set.

In this paper, a supply chain model consisting of supplier, manufacturer, and retailer has been considered. Here supplier receives the raw materials in a lot and then the superior quality items of the raw materials are sold at a higher price to the manufacturer after the screening the imperfect raw materials as well as inferior quality items of the raw materials are also sold to another manufacturer at a reduced price in a single batch by the end of cent percent screening process. A mixture of perfect and imperfect quality items is produced by the manufacturer. After some rework, some repairable portion of imperfect quality items are transformed into perfect quality items and some of non-repairable portion of imperfect items are sold with reduced price to the retailer. Retailer purchases both perfect and imperfect quality items and sells both items to the customers through his/her respective showrooms of finite capacities at a market place.

It is assumed that the customers' demand is stock dependent and selling price dependent for the perfect quality items and less perfect items, respectively. Since the storage space of the showroom for perfect quality items is limited due to space problem and the demand of the corresponding items is stock dependent, hence a secondary warehouse is hired by the retailer on rental basis to store the excess amount of perfect quality items and these items are continuously transferred to the showroom concerned. The literature suggests that the holding cost of secondary warehouse per unit item per unit time is more than the holding cost of the showroom due to the preservation cost for maintaining the quality of the product and other costs related to holding large quantity of the product in the secondary warehouse. But in this paper, it has been considered that the holding cost of perfect quality items in the secondary warehouse is less than the holding cost of the showroom as the nature of the items are non-deteriorating and so having no preservation cost. The perfect quality items are transported to the showroom via secondary warehouse, and the less perfect quality items are directly transported to the related showroom. For this purpose, transportation cost is incurred to transport both quality items at the respective showrooms from the production center. Due to complexity of environment, inventory holding costs, idle costs, set-up costs, and transportation costs are considered as fuzzy rough type and these are reduced to crisp ones using fuzzy rough expectation. In order to optimize the production rate and raw material order size (the decision variables), the average profit function of the manufacturer is maximized as the manufacturer acts as a leader (Stakelberg approach), and the supplier as well as retailer are the followers of that chain. The decision variables are also optimized by maximizing the integrated average profit function of the chain. Finally, a comparative study has been made between both Stakelberg and integrated approaches. A numerical example is provided to illustrate the feasibility of the model.

Necessary knowledge about fuzzy-rough (Fu-Ro) set

In this section, we discuss some basic concepts, theorems, and lemmas on fuzzy rough theory by Xu and Zhou [32]. These results are crucial for the remainder of this paper.

Definition 1. Xu and Zhou [32] proposed some definitions and discussed some important properties of fuzzy rough variable. Let U be a universe, and X be a set representing a concept. Then, its lower and upper approximation is defined by

$$\underline{X} = \{x \in U \mid R(x) \subset X\} \text{ and } \overline{X} = \bigcup_{x \in X} R(x) \text{ respectively.}$$

Definition 2. The collection of all sets having the same lower and upper approximations is called a *rough set,* denoted by $(\underline{X}, \overline{X})$. The figure of the rough set is depicted in Figure 1.

Example 1. Let ξ focus on the continuous set in the one dimension real space R. There are still some vague sets which cannot be directly fixed and need to be described by the rough approximation. Let set R be the universe, a similarity relation ! is defined as $a \simeq b$ if and only if $|a - b| \leq 10$. Let us define for the set $[20, 50]$, its lower approximation $\underline{[20, 50]} = [30, 40]$ and its upper approximation $\overline{[20, 50]} = [10, 60]$. Then the upper and lower approximations of the set $[20, 50]$ make up a rough set $([30, 40], [10, 60])$ which is the collection of all sets having the same lower approximation $[30, 40]$ and upper approximation $[10, 60]$.

Definition 3. A *fuzzy rough variable* ξ is a fuzzy variable with uncertain parameter $\rho \in X$, where X is approximated by $(\underline{X}, \overline{X})$ according to the similarity relation R, namely, $\underline{X} \subseteq X \subseteq \overline{X}$.

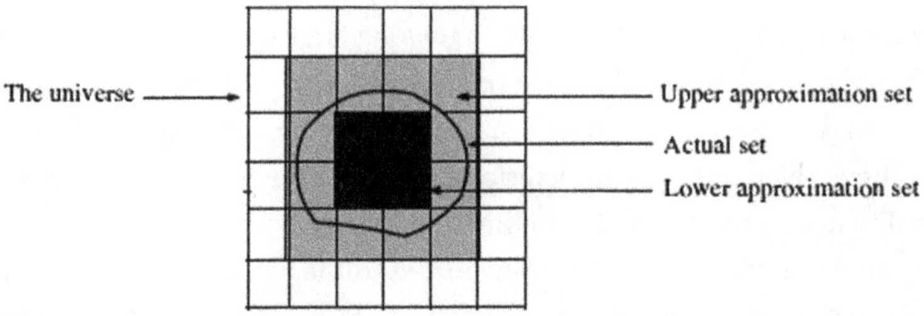

Figure 1 A rough set.

For convenience, we usually denote $\rho \vdash (\underline{X}, \overline{X})_R$ expressing that ρ is in some set A which is approximated by $(\underline{X}, \overline{X})$ according to the similarity relation R, namely, $\underline{X} \subseteq A \subseteq \overline{X}$.

Example 2. Let us consider the LR fuzzy variable ξ with the following membership function:

$$\mu_\xi(x) = \begin{cases} L\left(\frac{\rho - x}{\alpha}\right) & \text{if } \rho - \alpha < x < \rho \\ 1 & \text{if } x = \rho \\ L\left(\frac{x-\rho}{\beta}\right) & \text{if } \rho < x < \rho + \beta \end{cases}$$

where $L(x) = 1 - x$ and $\rho \vdash ([1,2], [0,3])$, then ξ is a fuzzy rough variable.

Theorem 1. *If fuzzy rough variables $\tilde{\bar{c}}_{ij}$ are defined as $\tilde{\bar{c}}_{ij}(\lambda) = (\bar{c}_{ij1}, \bar{c}_{ij2}, \bar{c}_{ij3}, \bar{c}_{ij4})$ with $\bar{c}_{ijt} \vdash ([c_{ijt2}, c_{ijt3}], [c_{ijt1}, c_{ijt4}])$, for $i = 1, 2, \ldots, m$, $j = 1, 2, \ldots, n$, $t = 1, 2, 3, 4$, $x = (x_1, x_2, \ldots, x_m)$, $0 \le c_{ijt1} \le c_{ijt2} < c_{ijt3} \le c_{ijt4}$, then $E\left[\tilde{\bar{c}}_1^T x\right], E\left[\tilde{\bar{c}}_2^T x\right], \ldots, E\left[\tilde{\bar{c}}_n^T x\right]$ is respectively equivalent to*

$$\frac{1}{16}\sum_{j=1}^{n}\sum_{t=1}^{4}\sum_{k=1}^{4} c_{1jtk}x_j, \frac{1}{16}\sum_{j=1}^{n}\sum_{t=1}^{4}\sum_{k=1}^{4} c_{2jtk}x_j, \ldots, \frac{1}{16}\sum_{j=1}^{n}\sum_{t=1}^{4}\sum_{k=1}^{4} c_{njtk}x_j.$$

Proof. The proof of the theorem is in reference Xu and Zhou [32] in page 308. □

Theorem 2. *If fuzzy rough variables $\tilde{\bar{a}}_{rj}, \tilde{\bar{b}}_r$ defined as follows, $\tilde{\bar{a}}_{rj}(\lambda) = (\bar{a}_{rj1}, \bar{a}_{rj2}, \bar{a}_{rj3}, \bar{a}_{rj4})$ with $\bar{a}_{rjt} \vdash ([a_{rjt2}, a_{rjt3}], [a_{rjt1}, a_{rjt4}])$, $\tilde{\bar{b}}_r(\lambda) = (\bar{b}_{r1}, \bar{b}_{r2}, \bar{b}_{r3}, \bar{b}_{r4})$ with $\bar{b}_{rt} \vdash ([b_{rt2}, b_{rjt3}], [b_{rt1}, b_{rt4}])$, for $r = 1, 2, \ldots, p$, $j = 1, 2, \ldots, n$, $t = 1, 2, 3, 4$, $0 \le a_{rt1} \le a_{rt2} < a_{rt3} \le a_{rt4}$, $0 \le b_{rt1} \le b_{rt2} < b_{rt3} \le b_{rt4}$. Then $E\left[\tilde{\bar{a}}_{rj}^T x\right] \le E\left[\tilde{\bar{b}}_{rj}\right]$, $r = 1, 2, \ldots, p$ is equivalent to*

$$\frac{1}{16}\sum_{j=1}^{n}\sum_{t=1}^{4}\sum_{k=1}^{4} a_{rjtk}x_j \le \frac{1}{16}\sum_{t=1}^{4}\sum_{k=1}^{4} b_{rtk}, \ r = 1, 2, \ldots, p$$

Proof. The proof of the theorem is in reference Xu and Zhou [32] in page 316. □

Lemma 1. *Assume that ξ and η are the introduction of variables with finite expected values. Then for any real numbers a and b, we have*

$$E[a\xi + b\eta] = aE[\xi] + bE[\eta].$$

Proof. The proof of the Lemma is in reference Xu and Zhou [32] in page 313. □

Single-objective Fu-Ro model

Let us consider the following single-objective decision-making model with fuzzy rough coefficients:

$$\begin{cases} \max \ \{f(x, \xi)\} \\ \text{s.t.} \ \begin{cases} g_r(x, \xi) \le 0, \ r = 1, 2, \ldots, p \\ x \in X \end{cases} \end{cases} \tag{1}$$

where x is a n-dimensional decision vector, $\xi = (\xi_1, \xi_2, \xi_3, \ldots, \xi_n)$ is a Fu-Ro vector, and $f(x, \xi)$ is objective function. Because of the existence of Fu-Ro vector ξ, problem (3) is not well-defined, that is, the meaning of maximizing $f(x, \xi)$ is not clear and constraints $g_r(x, \xi) \leq 0, r = 1, 2, \ldots, p$ do not define a deterministic feasible set.

Equivalent crisp model for single-objective problem with Fu-Ro parameters

For the single-objective model with Fu-Ro parameters, we cannot deal with it directly, we should use some tools to make it have mathematical meaning, we then can solve it. In this subsection, we employ the expected value operator to transform the fuzzy rough model into Fu-Ro EVM, i.e., crisp model. Based on the definition of the expected value of fuzzy rough events $f(x, \xi)$, $g_r(x, \xi)$ and Theorems 1 and 2, the Fu-Ro EVM is proposed as follows:

$$\begin{cases} \max \ \mathrm{E}\left[f\left(x, \xi\right)\right] \\ \text{s.t.} \quad \begin{cases} E\left[g_r\left(x, \xi\right)\right] \leq 0, \ r = 1, 2, \ldots, p \\ x \in X \end{cases} \end{cases} \tag{2}$$

where x is the n-dimensional decision vector and ξ is the n-dimensional fuzzy rough variable.

Assumptions and notations

The following notations and assumptions has been used in develop the proposed models.

Notations

For convenience, the following notations are used throughout the entire paper:

R Replenishment lot size of the supplier
P Production rate for the manufacturer which is also the demand rate of supplier
x Screening rate of supplier
θ Percentage of inferior quality items in each lot received by the supplier
t_1 Cycle length of supplier
t' Total screening time of R units order per cycle
A_s Set-up cost of supplier
c_s Purchase cost per unit item of supplier
h_s Holding cost per unit item for per unit time for supplier
I_{cs} Cost per unit idle time of supplier
s_c Screening cost per unit item
w_s Selling price per unit of superior quality item for supplier
w'_s Selling price per unit of inferior quality item for supplier
β Percentage of imperfect quality items suitable for rework to make perfect items
γ Percentage of imperfect items which are suitable for sale through reduction
$q_{1m}(t)$ Inventory level of perfect quality items for the manufacturer at any time t
$q_{2m}(t)$ Inventory level of less perfect quality items for the manufacturer at any time t
D_r Demand rate of the retailer for perfect quality items
D'_r Demand rate of the retailer for less perfect quality items
A_m Set-up cost of manufacturer
h_m Holding cost per unit item for per unit time for perfect item in manufacturer
h'_m Holding cost per unit item for per unit time for imperfect item in manufacturer

I_{cm} Cost per unit idle time of manufacturer
I_{sm} Inspection cost per unit item for manufacturer
r_{cm} Reworking cost per unit item for manufacturer
$C(P)$ Production cost per unit item
s_{m} Selling price per unit of perfect quality items for manufacturer
s'_{m} Selling price per unit less perfect quality items for manufacturer
$q_{1r}(t)$ Inventory level of perfect quality items for the retailer at any time t
$q_{2r}(t)$ Inventory level of less perfect quality items for the retailer at any time t
A_{r} Set-up cost for perfect quality items of retailer
A'_{r} Set-up cost for less perfect quality items of retailer
h_{r} Holding cost per unit item for per unit time of perfect quality items in PW_1 for retailer
h_{rs} Holding cost per unit item for per unit time of perfect quality items for the retailer in secondary ware house
h'_{r} Holding cost per unit item for per unit time of less perfect quality items in PW_2 for retailer
s_{r} Selling price per unit of perfect quality items for retailer PW_1
s'_{r} Selling price per unit of less perfect quality items for retailer PW_2
c_{tr} Transportation cost perfect item for retailer
c'_{tr} Transportation cost of less perfect quality items for retailer
S Capacity of secondary warehouse
W_j Capacity of $\text{PW}_j(j = 1, 2)$.
D_{c} Demand rate of customer for perfect quality items PW_1
D'_{c} Demand rate of customer for less perfect quality items PW_2
$\simeq$ Denotes the fuzzy rough parameters.

Assumptions

The following assumptions are used throughout the entire paper.

(*i*) Joint effect of supplier, manufacturer, and retailer is considered in a supply chain management.
(*ii*) Model is developed for single-item products and lead time is negligible.
(*iii*) Production rate is a decision variable.
(*iv*) Demand for perfect quality items is deterministic and function of current stock level.
(*v*) Replenishment rate of manufacture is instantaneously infinite but its size is finite.
(*vi*) Unit production cost is a function of production rate.
(*vii*) The manufacturer has ignored the machine breakdown.
(*viii*) Cost of idle times of supplier and manufacturer is taken into account.
(*ix*) Showrooms PW_1 and PW_2 of the retailer are adjacent.

Formulation of three-layer supply chain production inventory model

Block diagram and pictorial representation of the proposed supply chain production inventory model are respectively depicted in Figures 2 and 3. Formulation of the model for supplier, manufacturer, and retailer are given in the subsections 'Formulation of the supplier,' 'Formulation of the manufacturer,' and 'Formulation of the retailer,' respectively.

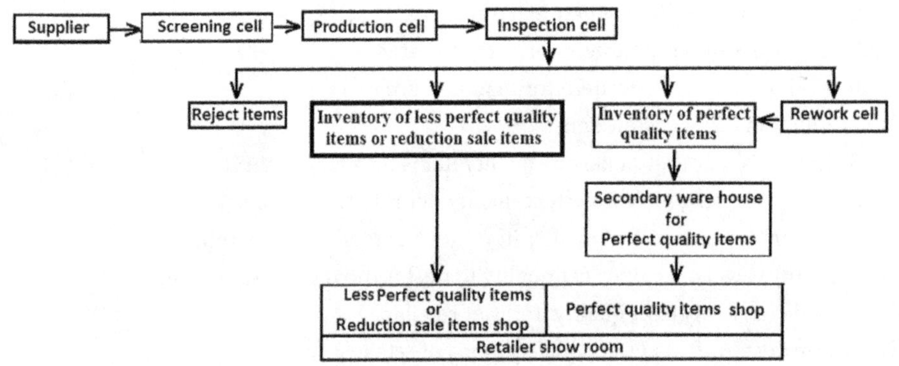

Figure 2 Block diagram of the model.

Formulation of the supplier

Here, let R be the lot-size received by the supplier at $t = 0$. A screening process of the lot is conducted at a rate of x units per unit time and t' be the total screening time of R units. Defective items are kept in stock and sold prior to receiving the next shipment as a single batch at a discounted price of w_s' per unit. $R\theta$ is the number of inferior quality items withdrawn from inventory, and t_1 is the cycle length of the supplier. The number of superior quality items in each lot, denoted by $N(R, \theta)$, is given by

$$N(R, \theta) = (R - R\theta) \tag{3}$$

Supplier supplies the superior quality items as raw materials to the manufacturer at a rate P up to the time t_1 and to avoid shortages; it is assumed that the number of superior quality items $N(r, \theta)$ is at least equal to the demand during screening time t', i.e.,

$$N(R, \theta) \geq Pt' \tag{4}$$

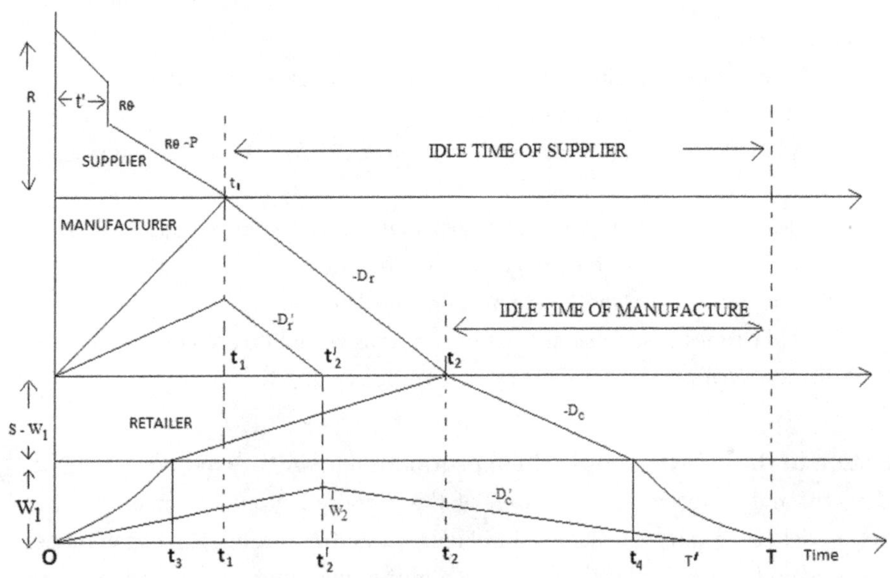

Figure 3 Pictorial representation of inventory situation of the integrated model.

Substituting Equation 3 in Equation 4 and replacing t' by $\frac{R}{x}$, the value of R is restricted to $R \le 1 - \frac{P}{x}$.

Sales revenue from superior quality items per cycle $= w_s(R - R\theta)$.

Sales revenue from inferior quality items per cycle $= w'_s R\theta$.

Procurement cost for the supplier per cycle = (Set-up cost + Purchasing cost) $= A_s + c_s R$.

Screening cost per cycle $= s_c R$.

Holding cost during $(0, t_1) = h_s \left[\frac{(R-R\theta)t_1}{2} + \frac{R^2\theta}{x}\right]$.

Idle time cost per cycle $= I_{cs}(T - t_1)$.

Therefore, the average profit (APS) of the supplier during $(0, T)$ is given by

$$\begin{aligned}
\text{APS} &= \frac{1}{T}\big[(\text{Sales revenue of superior quality items}) + (\text{Sales revenue of inferior quality items}) \\
&\quad -(\text{Procurement cost}) - (\text{screening cost}) - (\text{Holding cost}) - (\text{Idle time cost})\big] \\
&= \frac{1}{T}\left[w_s(R - R\theta) + w'_s R\theta - (A_s + c_s R) - s_c R - h_s\left\{\frac{(R-R\theta)t_1}{2} + \frac{R^2\theta}{x}\right\} - I_{cs}(T - t_1)\right] \\
&= \frac{1}{T}\left[-Z_{0s} + Z_{1s}R + Z_{2s}\frac{R}{P} - Z_{3s}\frac{R^2}{2P} - Z_{4s}R^2\right]. \qquad (5)
\end{aligned}$$

where $t_1 = \frac{(R-R\theta)}{P}$ and Z_{is}, $i = 0, 1, 2, 3, 4$ are independent of R and P. (See Appendix 1).

Formulation of the manufacturer

It is considered that a manufacturer produces some perfect and imperfect quality items at the rate of P units during the period $(0, t_1)$, receiving the raw material from the supplier at the same rate P during the period $(0, t_1)$. $Pe^{-\alpha t}$ and $P\left(1 - e^{-\alpha t}\right)$ are respectively the expected quantity of perfect and imperfect quality items at any time t, where α be the reliability parameter given by $\alpha = \frac{\text{Number of failures}}{\text{Total units of operating hours}}$. Among the imperfect quality items, only $\beta P\left(1 - e^{-\alpha t}\right)$ units per unit time become perfect quality after reworking and the portion $\gamma\,(1 - \beta)\,P\left(1 - e^{-\alpha t}\right)$ are less perfect quality which are sold at a reduced price to the retailer. Here D_r and D'_r denote the demand rates of a retailer for perfect quality and less perfect quality items which are met by manufacturer during $(0, t_2)$ and $\left(0, t'_2\right)$, respectively. The production cost $C(P)$ per unit item is considered as $C(P) = L + \frac{G}{P} + HP$, where G be the total labor cost for manufacturing the items, L and H are respectively the material cost and tool/die cost per unit item.

For perfect quality items of manufacturer

The rate of change of inventory level of the manufacturer for perfect quality items can be represented by the following differential equations:

$$\frac{dq_{1m}}{dt} = \begin{cases} Pe^{-\alpha t} + \beta P\left(1 - e^{-\alpha t}\right) - D_r, & 0 \le t \le t_1 \\ -D_r, & t_1 \le t \le t_2 \end{cases} \qquad (6)$$

with boundary conditions $q_{1m}(t) = 0$ at $t = 0$ and $t = t_2$.

The solution of above differential equations are given by

$$q_{1m}(t) = \begin{cases} \frac{P}{\alpha}(1 - \beta)\left(1 - e^{-\alpha t}\right) + (P\beta - D_r)\,t, & 0 \le t \le t_1 \\ D_r\,(t_2 - t), & t_1 \le t \le t_2 \end{cases} \qquad (7)$$

From continuity at $t = t_1$, following condition is obtained:

$$\frac{P}{\alpha}(1-\beta)\left(1-e^{-\alpha t_1}\right) + (P\beta - D_r)\,t_1 = D_r\,(t_2 - t_1)$$

$$\text{which implies } t_2 = \frac{P}{D_r}\left[\frac{1}{\alpha}(1-\beta)\left(1-e^{-\alpha t_1}\right) + \beta t_1\right]. \tag{8}$$

Now holding cost (HCM_1) for perfect quality items for manufacture is given by

$$\begin{aligned} HCM_1 &= h_m \int_0^{t_1} q_{1m}(t)\,dt + h_m \int_{t_1}^{t_2} q_{1m}(t)\,dt \\ &= h_m \int_0^{t_1} \frac{P}{\alpha}(1-\beta)\left(1-e^{-\alpha t}\right) dt + h_m \int_0^{t_1} (P\beta - D_r)\,t dt - D_r h_m \int_{t_1}^{t_2} t\,dt \\ &= \frac{h_m P}{\alpha}(1-\beta)t_1 - \frac{P h_m}{\alpha^2}(1-\beta)\left(1-e^{-\alpha t_1}\right) + h_m P\beta\frac{t_1^2}{2} - D_r h_m \frac{t_2^2}{2}. \end{aligned}$$

and reworking cost (RCM) for manufacture:

$$\begin{aligned} RCM &= r_{cm} \int_0^{t_1} P\beta\left(1-e^{-\alpha t}\right) dt \\ &= r_{cm} P\beta\left[t_1 - \frac{1}{\alpha}\left\{1-e^{-\alpha t_1}\right\}\right]. \end{aligned}$$

For less perfect quality items of manufacturer

The rate of change of inventory level of less perfect quality items for manufacturer can be represented by the following differential equations:

$$\frac{dq_{2m}}{dt} = \begin{cases} \gamma(1-\beta)P\left(1-e^{-\alpha t}\right) - D'_r, & 0 \le t \le t_1 \\ -D'_r, & t_1 \le t \le t'_2 \end{cases} \tag{9}$$

with boundary conditions $q_{2m}(t) = 0$ at $t = 0$ and $t = t'_2$.

The solution of above differential equations are given by

$$q_{2m}(t) = \begin{cases} -\frac{P}{\alpha}\gamma(1-\beta)\left(1-e^{-t\alpha}\right) + \left[\gamma(1-\beta)P - D'_r\right]t, & 0 \le t \le t_1 \\ D'_r\left(t'_2 - t\right), & t_1 \le t \le t'_2 \end{cases} \tag{10}$$

From continuity at $t = t_1$, following condition is obtained:

$$t'_2 = \frac{1}{D'_r}\left[-\frac{P}{\alpha}\gamma(1-\beta)\left(1-e^{-\alpha t_1}\right) + \gamma(1-\beta)Pt_1\right]. \tag{11}$$

Now holding cost (HCM_2) for less perfect quality items for manufacture

$$\begin{aligned} HCM_2 &= h'_m \int_0^{t_1} q_{2m}(t)\,dt + h'_m \int_{t_1}^{t'_2} q_{2m}(t)\,dt \\ &= -h'_m \int_0^{t_1} \frac{P}{\alpha}\gamma(1-\beta)\left(1-e^{-\alpha t}\right) dt + h'_m \int_0^{t_1} \gamma\{P(1-\beta) - D'_r\}\,t\,dt + D'_r h'_m \int_{t_1}^{t'_2} (t'_2 - t)\,dt \\ &= h'_m \frac{P\gamma}{\alpha^2}(1-\beta)\left[1-e^{-t_1\alpha}\right] - h'_{m\frac{P\gamma}{\alpha}}(1-\beta)t_1 + h'_m\left[P\gamma(1-\beta) - D'_r\right]\frac{{t_1}^2}{2} + h'_m \frac{D'_r}{2}\left(t'_2 - t_1\right)^2. \end{aligned}$$

Production cost for the manufacturer $= C(P)Pt_1$.

Inspection cost $= I_{sm}Pt_1$.

Holding cost for the manufacturer $= [HCM_1 + HCM_2]$.

Set-up cost of the manufacturer $= A_m$.

Idle time cost for the manufacturer $= I_{cm}(T - t_2)$.

Revenue of perfect quality items for the manufacturer $= s_m \int_0^{t_2} D_r dt = s_m D_r t_2$.

Revenue of less perfect quality items for the manufacturer $= s'_m \int_0^{t'_2} D'_r dt, = s'_m D'_r t'_2$.

Average profit of manufacturer

Average profit (APM) of the manufacturer during the period $(0, T)$ is given by

$$
\begin{aligned}
\text{APM} &= \frac{1}{T}\big[(\text{Revenue of perfect and less perfect quality items for the manufacturer}) - w_s P t_1 \\
&\quad -(\text{Production cost} + \text{Inspection cost} + \text{Holding cost} + \text{Reworking cost} + \text{Idle time cost}) \\
&\quad - \text{Set-up cost}\big] \\
&= \frac{1}{T}\big[(s_{\text{m}} D_{\text{r}} t_2 + s'_{\text{m}} D'_{\text{r}} t'_2) - w_s P t_1 - C(P) P t_1 - I_{\text{sm}} P t_1 - \text{HCM}_1 - \text{HCM}_2 - \text{RCM} \\
&\quad - A_{\text{m}} - I_{\text{cm}}(T - t_2)\big] \\
&= \frac{1}{T}\Bigg[(s_{\text{m}} D_{\text{r}} t_2 + s'_{\text{m}} D'_{\text{r}} t'_2) - \{w_s + C(P) + I_{\text{sm}}\} P t_1 - r_{\text{cm}}\left\{-\frac{P\beta}{\alpha}\left(1 - e^{-\alpha t_1}\right) + P\beta t_1\right\} \\
&\quad - A_{\text{m}} - h_{\text{m}}\left\{\frac{P}{\alpha}(1-\beta)t_1 - \frac{P}{\alpha^2}(1-\beta)\left(1 - e^{-\alpha t_1}\right) + P\beta\frac{t_1^2}{2} - D_{\text{r}}\frac{t_2^2}{2}\right\} - I_{\text{cm}}(T - t_2) \\
&\quad - h'_{\text{m}}\left\{\frac{P}{\alpha^2}\gamma(1-\beta)\left(1 - e^{-\alpha t_1}\right) - \frac{\gamma}{\alpha}(1-\beta)P t_1 + \left(\gamma(1-\beta)P - D'_{\text{r}}\right)\frac{t_1^2}{2} + \frac{D'_{\text{r}}}{2}\left(t_1 - t'_2\right)^2\right\}\Bigg] \\
&= \frac{1}{T}\left[-Z_{0m} - Z_{1m}\frac{R}{P} + Z_{2m} R + Z_{3m}\frac{R^2}{P^2} + Z_{4m}\frac{R^2}{P} + Z_{5m} R^2 - Z_{6m} P R^2\right].
\end{aligned}
$$

where $Z_{im},\ i = 0, 1, 2, \ldots, 6,$ are independent of R and P. (See Appendix 1).

Formulation of the retailer

The customers' demand for both perfect and less perfect quality items are met by the retailer through the adjacent showrooms PW_1 and PW_2, respectively. Retailer has a secondary warehouse SW to store the excess perfect quality items which are continuously transferred to the showroom PW_1. Less perfect quality items are directly transferred to the showroom PW_2. Transportation cost is taken into account to transfer each items from production center to the showrooms.

For perfect quality items of retailer

In this case the demand rate (D_{c}) of customers at PW_1 has consider as stock dependent as the following form:

$$
D_{\text{c}} = \begin{cases} \alpha_1 + \beta_1 q_{1r}, & 0 \le t \le t_3 \\ \alpha_1 + \beta_1 W_1, & t_3 \le t \le t_4 \\ \alpha_1 + \beta_1 q_{1r}, & t_4 \le t \le T \end{cases}
$$

Now, the corresponding rate of change of on hand inventory of perfect quality items are given by

$$
\frac{dq_{1r}}{dt} = \begin{cases} D_{\text{r}} - D_{\text{c}}, & 0 \le t \le t_2 \\ -D_{\text{c}}, & t_2 \le t \le T \end{cases}
$$

with boundary conditions

$$
q_{1r}(t) = \begin{cases} 0, & \text{at } t = 0 \\ W_1, & \text{at } t = t_3 \\ S, & \text{at } t = t_2 \\ W_1, & \text{at } t = t_4 \\ 0, & \text{at } t = T \end{cases}
$$

Therefore, the solutions of above differential equations are given by

$$q_{1r}(t) = \begin{cases} \frac{D_r-\alpha_1}{\beta_1}\left(1-e^{-\beta_1 t}\right), & 0 \le t \le t_3 \\ W_1 + [D_r - (\alpha_1 + \beta_1 W_1)](t - t_3), & t_3 \le t \le t_2 \\ S - [(\alpha_1 + \beta_1 W_1)](t - t_2), & t_2 \le t \le t_4 \\ -\frac{\alpha_1}{\beta_1}\left[1 - e^{-\beta_1(t-T)}\right], & t_4 \le t \le T \end{cases} \tag{12}$$

$$\text{Now, } q_{1r}(t_3) = W_1 \text{ implies } t_3 = -\frac{1}{\beta_1}\log\left(1 - \frac{W_1\beta_1}{D_r - \alpha_1}\right) \tag{12a}$$

$$q_{1r}(t_2) = S \text{ gives } t_4 = t_2 + \frac{S - W_1}{\alpha_1 + \beta_1 W_1} \text{ and } W_1 + [D_r - (\alpha_1 + \beta_1 W_1)](t_2 - t_3) = S \tag{12b}$$

$$q_{1r}(t_4) = W_1 \text{ implies } T = t_4 + \frac{1}{\beta_1}\log\left[\left(1 - \frac{D_r}{\alpha_1}\right)\left(1 - e^{-\beta_1 t_3}\right)\right] \tag{12c}$$

Holding cost (HCRW) of the secondary warehouse SW is given by

$$\begin{aligned} \text{HCRW} &= h_{rs}\int_{t_3}^{t_2}\{q_{1r}(t) - W_1\}\,dt + h_{rs}\int_{t_2}^{t_4}\{q_{1r}(t) - W_1\}\,dt \\ &= h_{rs}\int_{t_3}^{t_2}[D_r - (\alpha_1 + \beta_1 W_1)](t - t_3)\,dt - h_{rs}\int_{t_2}^{t_4}[\alpha_1 + \beta_1 W_1](t - t_4)\,dt \\ &= \frac{h_{rs}}{2}\left[(D_r - (\alpha_1 + \beta_1 W_1))(t_2 - t_3)^2 - (\alpha_1 + \beta_1 W_1)(t_2 - t_4)^2\right]. \end{aligned}$$

Holding cost (HCRS_1) of the showroom PW_1 is given by

$$\begin{aligned} \text{HCRS}_1 &= h_r\int_0^{t_3} q_{1r}(t)\,dt + h_r\int_{t_3}^{t_4} W_1\,dt + \int_{t_4}^{T} q_{1r}(t)\,dt \\ &= h_r\int_0^{t_3}\frac{D_r - \alpha_1}{\beta_1}(1 - e^{-\beta_1 t})\,dt + h_r\int_{t_3}^{t_4} W_1\,dt + \int_{t_4}^{T} -\frac{\alpha_1}{\beta_1}\left[1 - e^{-\beta_1(t-T)}\right]dt \\ &= h_r\frac{D_r-\alpha_1}{\beta_1}\left[t_3 + \frac{e^{-\beta_1 t_3}}{\beta_1} - \frac{1}{\beta_1}\right] + W_1 h\Big]_r (t_4 - t_3) - h_r\frac{\alpha_1}{\beta_1}\left[(T - t_4) + \frac{1}{\beta_1} - \frac{e^{-\beta_1(t_4-T)}}{\beta_1}\right]. \end{aligned}$$

Transportation cost (TCPR) for perfect quality items of retailer is given by

$$\begin{aligned} \text{TCPR} &= c_{tr}\int_0^T D_C\,dt \\ &= c_{tr}\left[\int_0^{t_3}(\alpha_1 + \beta_1 q_{1r})\,dt + \int_{t_3}^{t_4}(\alpha_1 + \beta_1 W_1)\,dt + \int_{t_4}^{T}(\alpha_1 + \beta_1 q_{1r})\,dt\right] \\ &= c_{tr}\left[(\alpha_1 + \beta_1 W_1)(t_4 - t_3) + D_r t_3 - \frac{D_r}{\beta_1} + \frac{D_r - \alpha_1}{\beta_1}e^{-\beta_1 t_3} + \frac{\alpha_1}{\beta_1}e^{-\beta_1(t_4-T)}\right]. \end{aligned}$$

Revenue from selling perfect quality items is given by

$$\begin{aligned}\mathrm{REPR} &= s_{\mathrm{r}}\int_0^T D_{\mathrm{c}}\,dt\\ &= s_{\mathrm{r}}\left[(\alpha_1+\beta_1 W_1)(t_4-t_3)+D_{\mathrm{r}}t_3-\frac{D_{\mathrm{r}}}{\beta_1}+\frac{D_{\mathrm{r}}-\alpha_1}{\beta_1}e^{-\beta_1 t_3}+\frac{\alpha_1}{\beta_1}e^{-\beta_1(t_4-T)}\right].\end{aligned}$$

Hence, the total profit of the retailer from the perfect quality items is given by

$$\begin{aligned}\mathrm{TPR}_1 &= \text{Revenue from selling perfect quality items} - \text{Transportation cost for perfect quality items}\\ &\quad - \text{Holding cost of SW} - \text{Holding cost of PW}_1 - \text{Set-up cost} - s_{\mathrm{m}}(PA)_m\\ &= s_{\mathrm{r}}\left[(\alpha_1+\beta_1 W_1)(t_4-t_3)+D_{\mathrm{r}}t_3-\frac{D_{\mathrm{r}}}{\beta_1}+\frac{D_{\mathrm{r}}-\alpha_1}{\beta_1}e^{-\beta_1 t_3}+\frac{\alpha_1}{\beta_1}e^{-\beta_1(t_4-T)}\right]\\ &\quad -c_{\mathrm{tr}}\left[(\alpha_1+\beta_1 W_1)(t_4-t_3)+D_{\mathrm{r}}t_3-\frac{D_{\mathrm{r}}}{\beta_1}+\frac{D_{\mathrm{r}}-\alpha_1}{\beta_1}e^{-\beta_1 t_3}+\frac{\alpha_1}{\beta_1}e^{-\beta_1(t_4-T)}\right]\\ &\quad -\frac{h_{\mathrm{rs}}}{2}\left[(D_{\mathrm{r}}-(\alpha_1+\beta_1 W_1))(t_2-t_3)^2-(\alpha_1+\beta_1 W_1)(t_2-t_4)^2\right]\\ &\quad -h_{\mathrm{r}}\left[\frac{D_{\mathrm{r}}-\alpha_1}{\beta_1}\left(t_3+\frac{e^{-\beta_1 t_3}}{\beta_1}-\frac{1}{\beta_1}\right)+W_1(t_4-t_3)-\frac{\alpha_1}{\beta_1}\left\{(T-t_4)+\frac{1}{\beta_1}-\frac{e^{-\beta_1(t_4-T)}}{\beta_1}\right\}\right]\\ &\quad -A_{\mathrm{r}}-s_{\mathrm{m}}\left[\frac{P}{\alpha}(1-\beta)t_1-\frac{P}{\alpha^2}(1-\beta)\left(1-e^{-\alpha t_1}\right)+P\beta\frac{t_1^2}{2}-D_{\mathrm{r}}\frac{t_2^2}{2}\right].\end{aligned}$$

For less perfect quality items of retailer

In this case demand rate (D'_{c}) of customers at PW_2 is assumed as selling price dependent defined as $D'_{\mathrm{c}}=(a-bs'_{\mathrm{r}})$ and corresponding change of on hand inventory of less perfect quality items are given by

$$\frac{dq_{2r}}{dt}=\begin{cases}D'_{\mathrm{r}}-D'_{\mathrm{c}}, & 0\le t\le t'_2\\ -D'_{\mathrm{c}}, & t'_2\le t\le T'\end{cases}$$

with boundary conditions $q_{2r}(t)=0$ at $t=0$ and $t=T'$.

The solution of above differential equations is given by

$$q_{2r}(t)=\begin{cases}\left(D'_{\mathrm{r}}-D'_{\mathrm{c}}\right)t, & 0\le t\le t'_2\\ -D'_{\mathrm{c}}\left(t-T'\right), & t'_2\le t\le T'\end{cases}\tag{13}$$

From continuity condition at $t=t'_2$, $T'=\frac{1}{D'_{\mathrm{c}}}\left[\gamma(1-\beta)Pt_1-\frac{P}{\alpha^2}\gamma(1-\beta)\left(1-e^{-\alpha t_1}\right)\right]$.

Also $q_{2r}\left(t'_2\right)=W_2$, i.e., $W_2=\left(1-\frac{D'_{\mathrm{c}}}{D'_{\mathrm{r}}}\right)\left[-\frac{P}{\alpha}\gamma(1-\beta)\left(1-e^{-\alpha t_1}\right)+\gamma(1-\beta)Pt_1\right]$.

Holding cost (HCRS_2) of the showroom PW_2 is given by

$$\begin{aligned}\mathrm{HCRS}_2 &= h'_{\mathrm{r}}\int_0^{t'_2}q_{2r}(t)\,dt+h'_{\mathrm{r}}\int_{t'_2}^{T'}q_{2r}(t)\,dt\\ &= h'_{\mathrm{r}}\int_0^{t'_2}\left\{D'_{\mathrm{r}}-D'_{\mathrm{c}}\right\}t\,dt+h'_{\mathrm{r}}\int_{t'_2}^{T'}\left\{-D'_{\mathrm{c}}\left(t-T'\right)\right\}dt\\ &= \frac{h'_{\mathrm{r}}}{2}\left[\left(D'_{\mathrm{r}}-D'_{\mathrm{c}}\right)t_2'^{2}+D'_{\mathrm{c}}\left(t'_2-T'\right)^2\right].\end{aligned}$$

Transportation cost (TCIR) for less perfect quality items of retailer is given by

$$\text{TCIR} = c'_{\text{tr}} \int_0^{T'} D'_{\text{c}}\, dt = c'_{\text{tr}} D'_{\text{c}} T'.$$

Revenue (REIR) from selling less perfect quality items of retailer is given by

$$\text{REIR} = s'_{\text{r}} \int_0^{T'} D'_{\text{c}}\, dt = s'_{\text{r}} D'_{\text{c}} T'.$$

Total profit (TPR_2) of retailer for less perfect quality items is given by

$$\begin{aligned}
\text{TPR}_2 &= \text{Revenue from selling less perfect quality items} - (\text{Transportation cost for} \\
&\quad \text{less perfect quality items}) - (\text{Holding cost of PW}_2) - \text{Set-up cost-}s'_{\text{m}}(DA)_{\text{m}} \\
&= \left\{s'_{\text{r}} - c'_{\text{tr}}\right\} D'_{\text{c}} T' - \frac{h'_{\text{r}}}{2}\left[\left(D'_{\text{r}} - D'_{\text{c}}\right) {t'_2}^2 + D'_{\text{c}}\left(t'_2 - T'\right)^2\right] - s'_{\text{m}}\left[\frac{P\gamma}{\alpha^2}(1-\beta)\left(1 - e^{-t_1\alpha}\right)\right. \\
&\quad \left. - \frac{P\gamma}{\alpha}(1-\beta)t_1 + \left(P\gamma(1-\beta) - D'_{\text{r}}\right)\frac{{t_1}^2}{2} + \frac{D'_{\text{r}}}{2}\left(t_1 - t'_2\right)^2\right] - A'_{\text{r}}.
\end{aligned}$$

Average profit of retailer

Average profit (APR) of the retailer during $(0, T)$ is

$$\begin{aligned}
\text{APR} &= \frac{1}{T}\left[\text{TPR}_1 + \text{TPR}_2\right] \\
&= \frac{1}{T}\left[s_{\text{r}}\left\{(\alpha_1 + \beta_1 W_1)(t_4 - t_3) + D_{\text{r}} t_3 - \frac{D_{\text{r}}}{\beta_1} + \frac{D_{\text{r}} - \alpha_1}{\beta_1} e^{-\beta_1 t_3} + \frac{\alpha_1}{\beta_1} e^{-\beta_1(t_4 - T)}\right\}\right. \\
&\quad - c_{\text{tr}}\left\{(\alpha_1 + \beta_1 W_1)(t_4 - t_3) + D_{\text{r}} t_3 - \frac{D_{\text{r}}}{\beta_1} + \frac{D_{\text{r}} - \alpha_1}{\beta_1} e^{-\beta_1 t_3} + \frac{\alpha_1}{\beta_1} e^{-\beta_1(t_4 - T)}\right\} \\
&\quad - \frac{h_{\text{rs}}}{2}\left\{(D_{\text{r}} - (\alpha_1 + \beta_1 W_1))(t_2 - t_3)^2 - (\alpha_1 + \beta_1 W_1)(t_2 - t_4)^2\right\} + s'_{\text{r}} D'_{\text{c}} T' \\
&\quad - c'_{\text{tr}} D'_{\text{c}} T' - \frac{h'_{\text{r}}}{2}\left\{\left(D'_{\text{r}} - D'_{\text{c}}\right){t'_2}^2 + D'_{\text{c}}\left(t'_2 - T'\right)^2\right\} - h_{\text{r}}\left\{\frac{D_{\text{r}} - \alpha_1}{\beta_1}\left(t_3 + \frac{e^{-\beta_1 t_3}}{\beta_1}\right.\right. \\
&\quad \left.\left. - \frac{1}{\beta_1}\right) + W_1(t_4 - t_3) - \frac{\alpha_1}{\beta_1}\left\{(T - t_4) + \frac{1}{\beta_1} - \frac{e^{-\beta_1(t_4 - T)}}{\beta_1}\right\}\right\} - s_{\text{m}}\left\{\frac{P}{\alpha}(1-\beta)t_1\right. \\
&\quad \left. - \frac{P}{\alpha^2}(1-\beta)\left(1 - e^{-\alpha t_1}\right) + P\beta\frac{t_1^2}{2} - D_{\text{r}}\frac{t_2^2}{2}\right\} - s'_{\text{m}}\left\{\frac{P\gamma}{\alpha^2}(1-\beta)\left(1 - e^{-t_1\alpha}\right)\right. \\
&\quad \left.\left. - \frac{P\gamma}{\alpha}(1-\beta)t_1 + \left(P\gamma(1-\beta) - D'_{\text{r}}\right)\frac{{t_1}^2}{2} + \frac{D'_{\text{r}}}{2}\left(t_1 - t'_2\right)^2\right\}\right] - A_{\text{r}} - A'_{\text{r}} \\
&= \frac{1}{T}\left[Z_{0r} + Z_{1r}R + Z_{2r}\frac{R^2}{2P} + Z_{3r}R^2 + Z_{4r}\frac{R^3}{P} + Z_{5r}\frac{R^4}{P^2}\right].
\end{aligned}$$

where $Z_{ir},\ i = 0, 1, \ldots, 5$ are independent of R and P. (See Appendix 1).

Integrated average profit

Average profit (IAP) for the Integrated Model during $(0, T)$ is

$$\begin{aligned}\text{IAP} &= [\text{APS} + \text{APM} + \text{APR}]\\ &= \frac{1}{T}\left[Z_0 + Z_1\frac{R}{P} + Z_2R - Z_3\frac{R^2}{2P} + Z_4\frac{R^2}{P^2} - Z_5PR^2 - Z_6R^2 + Z_7\frac{R^3}{P} + Z_8\frac{R^4}{P^2}\right]\end{aligned}$$

where $Z_i,\ i = 0, 1, \ldots, 8$ are independent of R and P. (See Appendix 1).

In fuzzy rough environment

In this environment, all holding cost, idle cost, set-up cost, and transportation cost have been considered fuzzy-rough parameters. Then the corresponding fuzzy-rough objective functions for supplier, manufacturer, and retailer are given by

$$\begin{aligned}\tilde{\tilde{\text{APS}}} &= \frac{1}{T}\left[-\tilde{\tilde{Z}}_{0s} + \tilde{\tilde{Z}}_{1s}R + \tilde{\tilde{Z}}_{2s}\frac{R}{P} - \tilde{\tilde{Z}}_{3s}\frac{R^2}{2P} - \tilde{\tilde{Z}}_{4s}R^2\right]\\ \tilde{\tilde{\text{APM}}} &= \frac{1}{T}\left[-\tilde{\tilde{Z}}_{0m} - \tilde{\tilde{Z}}_{1m}\frac{R}{P} + \tilde{\tilde{Z}}_{2m}R + \tilde{\tilde{Z}}_{3m}\frac{R^2}{P^2} + \tilde{\tilde{Z}}_{4m}\frac{R^2}{P} + \tilde{\tilde{Z}}_{5m}R^2 - \tilde{\tilde{Z}}_{6m}PR^2\right]\\ \text{and } \tilde{\tilde{\text{APR}}} &= \frac{1}{T}\left[\tilde{\tilde{Z}}_{0r} + \tilde{\tilde{Z}}_{1r}R + \tilde{\tilde{Z}}_{2r}\frac{R^2}{2P} + \tilde{\tilde{Z}}_{3r}R^2 + \tilde{\tilde{Z}}_{4r}\frac{R^3}{P} + \tilde{\tilde{Z}}_{5r}\frac{R^4}{P^2}\right]\end{aligned}$$

Also, the fuzzy-rough objective functions for integrated model is given by

$$\begin{aligned}\tilde{\tilde{\text{IAP}}} &= \left[\tilde{\tilde{\text{APS}}} + \tilde{\tilde{\text{APM}}} + \tilde{\tilde{\text{APR}}}\right]\\ &= \frac{1}{T}\left[\tilde{\tilde{Z}}_0 + \tilde{\tilde{Z}}_1\frac{R}{P} + \tilde{\tilde{Z}}_2R - \tilde{\tilde{Z}}_3\frac{R^2}{2P} + \tilde{\tilde{Z}}_4\frac{R^2}{P^2} - \tilde{\tilde{Z}}_5PR^2 - \tilde{\tilde{Z}}_6R^2 + \tilde{\tilde{Z}}_7\frac{R^3}{P} + \tilde{\tilde{Z}}_8\frac{R^4}{P^2}\right]\end{aligned}$$

where fuzzy rough parameters $\tilde{\tilde{h}}_s, \tilde{\tilde{h}}_m, \tilde{\tilde{h}}'_m, \tilde{\tilde{h}}_r, \tilde{\tilde{h}}'_r, \tilde{\tilde{h}}_{rs}, \tilde{\tilde{A}}_s, \tilde{\tilde{A}}_m, \tilde{\tilde{A}}_r, \tilde{\tilde{I}}_{cs}, \tilde{\tilde{I}}_{cm}, \tilde{\tilde{c}}_{tp}, \tilde{\tilde{c}}'_{tp}$ are defined as follows:

$\tilde{\tilde{h}}_s = \left(\overline{h}_{s1}, \overline{h}_{s2}, \overline{h}_{s3}, \overline{h}_{s4}\right)$ with $\overline{h}_{st} \vdash ([h_{st2}, h_{st3}], [h_{st1}, h_{st4}])$, $0 \le h_{st1} \le h_{st2} < h_{st3} \le h_{st4}$, $\tilde{\tilde{h}}_m = \left(\overline{h}_{m1}, \overline{h}_{m2}, \overline{h}_{m3}, \overline{h}_{m4}\right)$ with $\overline{h}_{mt} \vdash ([h_{mt2}, h_{mt3}], [h_{mt1}, h_{mt4}])$, $0 \le h_{mt1} \le h_{mt2} < h_{mt3} \le h_{mt4}$. $\tilde{\tilde{h}}'_m = \left(\overline{h'}_{m1}, \overline{h'}_{m2}, \overline{h'}_{m3}, \overline{h'}_{m4}\right)$ with $\overline{h'}_{mt} \vdash \left(\left[h'_{mt2}, h'_{mt3}\right], \left[h'_{mt1}, h'_{mt4}\right]\right)$, $0 \le h'_{mt1} \le h'_{mt2} < h'_{mt3} \le h'_{mt4}$. $\tilde{\tilde{h}}_r = \left(\overline{h}_{r1}, \overline{h}_{r2}, \overline{h}_{r3}, \overline{h}_{r4}\right)$ with $\overline{h}_{rt} \vdash ([h_{rt2}, h_{rt3}], [h_{rt1}, h_{rt4}])$, $0 \le h_{rt1} \le h_{rt2} < h_{rt3} \le h_{rt4}$. $\tilde{\tilde{h}}'_r = \left(\overline{h'}_{r1}, \overline{h'}_{r2}, \overline{h'}_{r3}, \overline{h'}_{r4}\right)$ with $\overline{h'}_{rt} \vdash \left(\left[h'_{rt2}, h'_{rt3}\right], \left[h'_{rt1}, h'_{rt4}\right]\right)$, $0 \le h'_{rt1} \le h'_{rt2} < h'_{rt3} \le h'_{rt4}$. $\tilde{\tilde{h}}_{rs} = \left(\overline{h}_{rs1}, \overline{h}_{rs2}, \overline{h}_{rs3}, \overline{h}_{rs4}\right)$ with $\overline{h}_{rst} \vdash ([h_{rst2}, h_{rst3}], [h_{rst1}, h_{rst4}])$, $0 \le h_{rst1} \le h_{rst2} < h_{rst3} \le h_{rst4}$. $\tilde{\tilde{A}}_s = \left(\overline{A}_{s1}, \overline{A}_{s2}, \overline{A}_{s3}, \overline{A}_{s4}\right)$ with $\overline{A}_{st} \vdash ([A_{st2}, A_{st3}], [A_{st1}, A_{st4}])$, $0 \le A_{st1} \le A_{st2} < A_{st3} \le A_{st4}$. $\tilde{\tilde{A}}_m = \left(\overline{A}_{m1}, \overline{A}_{m2}, \overline{A}_{m3}, \overline{A}_{m4}\right)$ with $\overline{A}_{mt} \vdash ([A_{mt2}, A_{mt3}], [A_{mt1}, A_{mt4}])$, $0 \le A_{mt1} \le A_{mt2} < A_{mt3} \le A_{mt4}$. $\tilde{\tilde{A}}_r = \left(\overline{A}_{r1}, \overline{A}_{r2}, \overline{A}_{r3}, \overline{A}_{r4}\right)$ with $\overline{A}_{rt} \vdash ([A_{rt2}, A_{rt3}], [A_{rt1}, A_{rt4}])$, $0 \le A_{rt1} \le A_{rt2} < A_{rt3} \le A_{rt4}$. $\tilde{\tilde{I}}_{cs} = \left(\overline{I}_{cs1}, \overline{I}_{cs2}, \overline{I}_{cs3}, \overline{I}_{cs4}\right)$ with $\overline{I}_{cst} \vdash ([I_{cst2}, I_{cst3}], [I_{cst1}, I_{cst4}])$, $0 \le I_{cst1} \le I_{cst2} < I_{cst3} \le I_{cst4}$. $\tilde{\tilde{I}}_{cm} = \left(\overline{I}_{cm1}, \overline{I}_{cm2}, \overline{I}_{cm3}, \overline{I}_{cm4}\right)$ with $\overline{I}_{cmt} \vdash ([I_{cmt2}, I_{cst3}], [I_{cmt1}, I_{cmt4}])$, $0 \le I_{cmt1} \le I_{cmt2} < I_{cmt3} \le I_{cmt4}$. $\tilde{\tilde{c}}_{tp} = \left(\overline{c}_{tp1}, \overline{c}_{tp2}, \overline{c}_{tp3}, \overline{c}_{tp4}\right)$ with $\overline{c}_{tpt} \vdash \left([c_{tpt2}, c_{tpt3}], [c_{tpt1}, c_{tpt4}]\right)$, $0 \le c_{tpt1} \le c_{tpt2} < c_{tpt3} \le c_{tpt4}$. $\tilde{\tilde{c}}'_{tp} = \left(\overline{c'}_{tp1}, \overline{c'}_{tp2}, \overline{c'}_{tp3}, \overline{c'}_{tp4}\right)$ with $\overline{c'}_{tpt} \vdash \left(\left[c'_{tpt2}, c'_{tpt3}\right], \left[c'_{tpt1}, c'_{tpt4}\right]\right)$, $0 \le c'_{tpt1} \le c'_{tpt2} < c'_{tpt3} \le c'_{tpt4}$.

In equivalent crisp environment

In this environment, using Lemma 1 and Theorems 1 and 2, the fuzzy rough objective functions for supplier, manufacturer, and retailer are given by

$$\text{EAPS} = E\left[\tilde{\tilde{\text{APS}}}\right] = \frac{1}{T}\left[-E\left[\tilde{\tilde{Z}}_{0s}\right] + E\left[\tilde{\tilde{Z}}_{1s}\right]R + E\left[\tilde{\tilde{Z}}_{2s}\right]\frac{R}{P} - E\left[\tilde{\tilde{Z}}_{3s}\right]\frac{R^2}{2P} - E\left[\tilde{\tilde{Z}}_{4s}\right]R^2\right]$$

$$\begin{aligned}\text{EAPM} &= E\left[\tilde{\tilde{\text{APM}}}\right] \\ &= \frac{1}{T}\left[-E\left[\tilde{\tilde{Z}}_{0m}\right] - E\left[\tilde{\tilde{Z}}_{1m}\right]\frac{R}{P} + E\left[\tilde{\tilde{Z}}_{2m}\right]R + E\left[\tilde{\tilde{Z}}_{3m}\right]\frac{R^2}{P^2} + E\left[\tilde{\tilde{Z}}_{4m}\right]\frac{R^2}{P}\right. \\ &\quad \left. + E\left[\tilde{\tilde{Z}}_{5m}\right]R^2 - E\left[\tilde{\tilde{Z}}_{6m}\right]PR^2\right]\end{aligned}$$

$$\begin{aligned}\text{and EAPR} &= E\left[\tilde{\tilde{\text{APR}}}\right] = \frac{1}{T}\left[E\left[\tilde{\tilde{Z}}_{0r}\right] + E\left[\tilde{\tilde{Z}}_{1r}\right]R + E\left[\tilde{\tilde{Z}}_{2r}\right]\frac{R^2}{2P} + E\left[\tilde{\tilde{Z}}_{3r}\right]R^2 + E\left[\tilde{\tilde{Z}}_{4r}\right]\frac{R^3}{P}\right. \\ &\quad \left. + E\left[\tilde{\tilde{Z}}_{5r}\right]\frac{R^4}{P^2}\right]\end{aligned}$$

Also, the objective functions for integrated model is given by

$$\begin{aligned}\text{EIAP} &= E\left[\tilde{\tilde{\text{IAP}}}\right] = \left[E\left[\tilde{\tilde{\text{APS}}}\right] + E\left[\tilde{\tilde{\text{APM}}}\right] + E\left[\tilde{\tilde{\text{APR}}}\right]\right] \\ &= \frac{1}{T}\left[E\left[\tilde{\tilde{Z}}_0\right] + E\left[\tilde{\tilde{Z}}_1\right]\frac{R}{P} + E\left[\tilde{\tilde{Z}}_2\right]R - E\left[\tilde{\tilde{Z}}_3\right]\frac{R^2}{2P} + \tilde{\tilde{Z}}_4\frac{R^2}{P^2} - E\left[\tilde{\tilde{Z}}_5\right]PR^2\right. \\ &\quad \left. - \tilde{\tilde{Z}}_6 R^2 + E\left[\tilde{\tilde{Z}}_7\right]\frac{R^3}{P} + E\left[\tilde{\tilde{Z}}_8\right]\frac{R^4}{P^2}\right]\end{aligned}$$

where

$$E\left[\tilde{\tilde{h}}_{\text{s}}\right] = \frac{1}{16}\sum_{t=1}^{4}\sum_{k=1}^{4} h_{\text{stk}}, \quad E\left[\tilde{\tilde{h}}_{\text{m}}\right] = \frac{1}{16}\sum_{t=1}^{4}\sum_{k=1}^{4} h_{\text{mtk}}, \quad E\left[\tilde{\tilde{h}}'_{\text{m}}\right] = \frac{1}{16}\sum_{t=1}^{4}\sum_{k=1}^{4} h'_{\text{mtk}},$$

$$E\left[\tilde{\tilde{h}}_{\text{r}}\right] = \frac{1}{16}\sum_{t=1}^{4}\sum_{k=1}^{4} h_{\text{rtk}}, \quad E\left[\tilde{\tilde{h}}'_{\text{r}}\right] = \frac{1}{16}\sum_{t=1}^{4}\sum_{k=1}^{4} h'_{\text{rtk}}, \quad E\left[\tilde{\tilde{h}}_{\text{rs}}\right] = \frac{1}{16}\sum_{t=1}^{4}\sum_{k=1}^{4} h_{\text{rstk}},$$

$$E\left[\tilde{\tilde{A}}_{\text{s}}\right] = \frac{1}{16}\sum_{t=1}^{4}\sum_{k=1}^{4} A_{\text{stk}}, \quad E\left[\tilde{\tilde{A}}_{\text{m}}\right] = \frac{1}{16}\sum_{t=1}^{4}\sum_{k=1}^{4} A_{\text{mtk}}, \quad E\left[\tilde{\tilde{A}}_{\text{r}}\right] = \frac{1}{16}\sum_{t=1}^{4}\sum_{k=1}^{4} A_{\text{rtk}},$$

$$E\left[\tilde{\tilde{I}}_{\text{cs}}\right] = \frac{1}{16}\sum_{t=1}^{4}\sum_{k=1}^{4} I_{\text{cstk}}, \quad E\left[\tilde{\tilde{I}}_{\text{cm}}\right] = \frac{1}{16}\sum_{t=1}^{4}\sum_{k=1}^{4} I_{\text{cmtk}}, \quad E\left[\tilde{\tilde{c}}_{\text{tp}}\right] = \frac{1}{16}\sum_{t=1}^{4}\sum_{k=1}^{4} c_{\text{tptk}},$$

$E\left[\tilde{\tilde{c}}'_{\text{tp}}\right] = \frac{1}{16}\sum_{t=1}^{4}\sum_{k=1}^{4} c'_{\text{tptk}}$.

Stakelberg approach (leader-follower relationship)

In this case, the manufacturer is the leader, and the supplier and retailer are the followers. Also, the optimum values of the average profit of supplier and retailer are obtained by putting the optimum value of the decision variables, which are obtained by optimizing the average profit of manufacturer. Using Equations 6, 12a, 12b, and $\left(1 - e^{-\alpha t_1}\right) \approx \alpha t_1$, in Equation 12c, the relation $T = \frac{1-\theta}{D_{\text{r}}}R + t_0$ is obtained where t_0 is given by $t_0 = \frac{S-W_1}{\alpha_1+\beta_1 W_1} + \frac{1}{\beta_1}\log\left[\left(1 - \frac{D_{\text{r}}}{\alpha_1}\right)\left(1 - e^{-\beta_1 t_3}\right)\right]$ and it is independent of R and P.

When both R and P are decision variables

The average profit of manufacturer is given by

$$\mathrm{EAPM}(R,P) = \frac{1}{T}\Big[-E\Big[\tilde{\tilde{Z}}_{0m}\Big] - E\Big[\tilde{\tilde{Z}}_{1m}\Big]\frac{R}{P} + E\Big[\tilde{\tilde{Z}}_{2m}\Big]R + E\Big[\tilde{\tilde{Z}}_{3m}\Big]\frac{R^2}{P^2} + E\Big[\tilde{\tilde{Z}}_{4m}\Big]\frac{R^2}{P}$$
$$+E\Big[\tilde{\tilde{Z}}_{5m}\Big]R^2 - E\Big[\tilde{\tilde{Z}}_{6m}\Big]PR^2\Big]$$

where $E\Big[\tilde{\tilde{Z}}_{im}\Big], i = 0, 1, 2, \ldots, 6,$ are independent of R and P. (See Appendix 1).

The necessary conditions for maximum value of EAPM(R,P) are $\frac{\partial}{\partial R}(\mathrm{EAPM}) = 0$ and $\frac{\partial}{\partial P}(\mathrm{EAPM}) = 0$.

Now, $\frac{\partial}{\partial R}(\mathrm{EAPM}) = 0$

$$\text{i.e., } \frac{(1-\theta)}{D_r T}E\Big[\tilde{\tilde{Z}}_{0m}\Big] + \left\{\frac{(1-\theta)R}{D_r PT} - \frac{1}{P}\right\}E\Big[\tilde{\tilde{Z}}_{1m}\Big] + \left\{1 - \frac{(1-\theta)R}{D_r T}\right\}E\Big[\tilde{\tilde{Z}}_{2m}\Big]$$
$$+\left\{\frac{2R}{P^2} - \frac{(1-\theta)R^2}{D_r TP^2}\right\}E\Big[\tilde{\tilde{Z}}_{3m}\Big] + \left\{\frac{2R}{P} - \frac{(1-\theta)R^2}{D_r TP}\right\}E\Big[\tilde{\tilde{Z}}_{4m}\Big]$$
$$+\left\{2R - \frac{(1-\theta)R^2}{D_r T}\right\}E\Big[\tilde{\tilde{Z}}_{5m}\Big] + \left\{\frac{(1-\theta)R^2 P}{D_r T} - 2PR\right\}E\Big[\tilde{\tilde{Z}}_{6m}\Big] = 0 \tag{14}$$

and $\frac{\partial}{\partial P}(\mathrm{EAPM}) = 0$

$$\text{i.e., } E\Big[\tilde{\tilde{Z}}_{1m}\Big]\frac{R}{P^2} - 2E\Big[\tilde{\tilde{Z}}_{3m}\Big]\frac{R^2}{P^3} - E\Big[\tilde{\tilde{Z}}_{4m}\Big]\frac{R^2}{P^2} - E\Big[\tilde{\tilde{Z}}_{6m}\Big]R^2 = 0 \tag{15}$$

Solving (14) and (15), we can obtain the optimum value of R and P, say R^* and P^*.

If $\frac{\partial^2}{\partial R^2}(\mathrm{EAPM}) < 0$, $\frac{\partial^2}{\partial P^2}(\mathrm{EAPM}) < 0$ and $\left\{\frac{\partial^2}{\partial R^2}(\mathrm{EAPM})\right\}\left\{\frac{\partial^2}{\partial P^2}(\mathrm{EAPM})\right\} - \left\{\frac{\partial^2}{\partial P \partial R}(\mathrm{EAPM})\right\}^2 > 0$ holds for $R = R^*$ and $P = P^*$, then EAPM (R^*, P^*) is maximum.

Now, $\frac{\partial^2}{\partial R^2}(\mathrm{EAPM})\Big]_{\text{at }(R^*,P^*)} < 0$

$$\text{i.e., } -\frac{(1-\theta)^2}{D_r^2 T^{*2}}E\Big[\tilde{\tilde{Z}}_{0m}\Big] + \left\{\frac{(1-\theta)}{D_r T^* P^*} - \frac{(1-\theta)^2 R^*}{D_r^2 T^{*2} P^*}\right\}E\Big[\tilde{\tilde{Z}}_{1m}\Big]$$
$$+\left\{\frac{(1-\theta)^2 R^*}{D_r^2 T^{*2}} - \frac{(1-\theta)}{D_r T^*}\right\}E\Big[\tilde{\tilde{Z}}_{2m}\Big] + \left\{\frac{(1-\theta)^2 R^{*2}}{D_r^2 T^{*2} P^{*2}} + \frac{1}{P^{*2}} - \frac{2(1-\theta)R^*}{D_r T^* P^{*2}}\right\}E\Big[\tilde{\tilde{Z}}_{3m}\Big]$$
$$+\left\{\frac{(1-\theta)^2}{D_r^2 T^{*2} P^*} + \frac{1}{P^*} - \frac{2(1-\theta)R^*}{D_r T^* P^*}\right\}E\Big[\tilde{\tilde{Z}}_{4m}\Big] + \left\{1 - \frac{2(1-\theta)R^*}{D_r T^* P^*} + \frac{(1-\theta)^2 R^{*2}}{D_r^2 T^{*2}}\right\}E\Big[\tilde{\tilde{Z}}_{5m}\Big]$$
$$-\left\{P^* - \frac{2(1-\theta)P^* R^*}{D_r T^*} + \frac{(1-\theta)^2 R^{*2} P^*}{D_r^2 T^{*2}}\right\}E\Big[\tilde{\tilde{Z}}_{6m}\Big] < 0 \tag{16}$$

and $\frac{\partial^2}{\partial P^2}(\mathrm{EAPM})]_{\text{at }(R^*,P^*)} < 0$

$$\text{i.e., } -E\Big[\tilde{\tilde{Z}}_{1m}\Big]\frac{R^*}{P^{*3}} + 3E\Big[\tilde{\tilde{Z}}_{3m}\Big]\frac{R^{*2}}{P^{*4}} + E\Big[\tilde{\tilde{Z}}_{4m}\Big]\frac{R^{*2}}{P^{*3}} < 0 \tag{17}$$

$$\text{and } \left[\left\{\frac{\partial^2}{\partial R^2}(\mathrm{EAPM})\right\}\left\{\frac{\partial^2}{\partial P^2}(\mathrm{EAPM})\right\} - \left\{\frac{\partial^2}{\partial P \partial R}(\mathrm{EAPM})\right\}^2\right]_{\text{at }(R^*,P^*)} > 0 \tag{18}$$

Therefore, EAPM (R^*, P^*) is maximum if the relations (16), (17), and (18) hold.

The corresponding optimum average profit of supplier and retailer is

$$\text{EAPS}\,(R^*, P^*) = \frac{1}{T^*}\left[-E\left[\tilde{\tilde{Z}}_{0s}\right] + E\left[\tilde{\tilde{Z}}_{1s}\right] R^* + E\left[\tilde{\tilde{Z}}_{2s}\right] \frac{R^*}{P^*} - E\left[\tilde{\tilde{Z}}_{3s}\right] \frac{R^{*2}}{2P^*} - E\left[\tilde{\tilde{Z}}_{4s}\right] R^{*2}\right]$$

$$\text{EAPR}\,(R^*, P^*) = \frac{1}{T^*}\left[E\left[\tilde{\tilde{Z}}_{0r}\right] + E\left[\tilde{\tilde{Z}}_{1r}\right. [R^* + E\left[\tilde{\tilde{Z}}_{2r}\right] \frac{R^{*2}}{2P^*} + E\left[\tilde{\tilde{Z}}_{3r}\right] R^{*2} + E\left[\tilde{\tilde{Z}}_{4r}\right] \frac{R^{*3}}{P^*}\right.$$

$$\left. + E\left[\tilde{\tilde{Z}}_{5r}\right] \frac{R^{*4}}{P^{*2}}\right]$$

where $T^* = \frac{1-\theta}{D_r} R^* + t_0$.

When P is a decision variable

The necessary conditions for maximum value of $\text{EAPM}(P)$ is $\frac{d}{dP}(\text{EAPM}) = 0$

$$\text{i.e., } E\left[\tilde{\tilde{Z}}_{1m}\right] \frac{R}{P^2} - 2E\left[\tilde{\tilde{Z}}_{3m}\right] \frac{R^2}{P^3} - E\left[\tilde{\tilde{Z}}_{4m}\right] \frac{R^2}{P^2} - E\left[\tilde{\tilde{Z}}_{6m}\right] R^2 = 0 \tag{18a}$$

which gives the optimum value of P, say P^{**}. (See Appendix 2).

If $\frac{d^2}{dP^2}(\text{EAPM}) < 0$ holds for $P = P^{**}$, then EAPM (P^{**}) is maximum.

$$\text{Now } \left.\frac{d^2}{dP^2}(\text{EAPM})\right]_{\text{at } P=P^{**}} < 0 \text{ gives } -E\left[\tilde{\tilde{Z}}_{1m}\right] \frac{R}{P^{**3}} + 3E\left[\tilde{\tilde{Z}}_{3m}\right] \frac{R^2}{P^{**4}} + E\left[\tilde{\tilde{Z}}_{4m}\right] \frac{R^2}{P^{**3}} < 0 \tag{18b}$$

Therefore, EAPM(P^{**}) is maximum if the relation (18b) holds and the corresponding optimum average profits of supplier and retailer are respectively

$$\text{EAPS}\,(P^{**}) = \frac{1}{T}\left[-E\left[\tilde{\tilde{Z}}_{0s}\right] + E\left[\tilde{\tilde{Z}}_{1s}\right] R + E\left[\tilde{\tilde{Z}}_{2s}\right] \frac{R}{P^{**}} - E\left[\tilde{\tilde{Z}}_{3s}\right] \frac{R^2}{2P^{**}} - E\left[\tilde{\tilde{Z}}_{4s}\right] R^2\right]$$

$$\text{EAPR}\,(P^{**}) = \frac{1}{T}\left[E\left[\tilde{\tilde{Z}}_{0r}\right] + E\left[\tilde{\tilde{Z}}_{1r}\right] R + E\left[\tilde{\tilde{Z}}_{2r}\right] \frac{R^2}{2P^{**}} + E\left[\tilde{\tilde{Z}}_{3r}\right] R^2 + E\left[\tilde{\tilde{Z}}_{4r}\right] \frac{R^3}{P^{**}}\right.$$

$$\left. + E\left[\tilde{\tilde{Z}}_{5r}\right] \frac{R^4}{P^{**2}}\right]$$

where $T = \frac{1-\theta}{D_r} R + t_0$.

Integrated approach

When both R and P are decision variables

The necessary conditions for maximum value of $\text{EIAP}(R, P)$ are $\frac{\partial}{\partial R}(\text{EIAP}) = 0$ and $\frac{\partial}{\partial P}(\text{EIAP}) = 0$

Now, $\frac{\partial}{\partial R}(\text{EIAP}) = 0$

$$\text{i.e., } -\frac{(1-\theta)}{D_r T} E\left[\tilde{\tilde{Z}}_0\right] + \left\{\frac{1}{P} - \frac{(1-\theta)R}{D_r TP}\right\} E\left[\tilde{\tilde{Z}}_1\right] + \left\{1 - \frac{(1-\theta)R}{D_r T}\right\} E\left[\tilde{\tilde{Z}}_2\right] - \left\{\frac{R}{P} + \frac{(1-\theta)R^2}{2D_r TP}\right\}$$

$$\times E\left[\tilde{\tilde{Z}}_3\right] + \left\{\frac{2R}{P^2} - \frac{(1-\theta)R^2}{D_r TP^2}\right\} E\left[\tilde{\tilde{Z}}_4\right] - \left\{2PR + \frac{(1-\theta)PR^2}{D_r T}\right\} E\left[\tilde{\tilde{Z}}_5\right] - \left\{2R + \frac{(1-\theta)R^2}{D_r T}\right\}$$

$$\times E\left[\tilde{\tilde{Z}}_6\right] + \left\{\frac{3R^2}{P} - \frac{(1-\theta)R^3}{D_r TP}\right\} E\left[\tilde{\tilde{Z}}_7\right] + \left\{\frac{4R^3}{P^2} - \frac{(1-\theta)R^4}{D_r TP^2}\right\} E\left[\tilde{\tilde{Z}}_8\right] = 0 \tag{19}$$

and $\frac{\partial}{\partial P}(\text{EIAP}) = 0$

$$\text{i.e., } \left[-E\left[\tilde{\tilde{Z}}_1\right] \frac{R}{P^2} - E\left[\tilde{\tilde{Z}}_3\right] \frac{R^2}{2P^2} - 2E\left[\tilde{\tilde{Z}}_4\right] \frac{R^2}{P^3} - E\left[\tilde{\tilde{Z}}_5\right] R^2 - E\left[\tilde{\tilde{Z}}_7\right] \frac{R^3}{P^2} - 2E\left[\tilde{\tilde{Z}}_8\right] \frac{R^4}{P^3}\right] = 0 \tag{20}$$

Solving (19) and (20), we can obtain the optimum value of R and P, say R^* and P^*.

Table 1 Collected data for different crisp parameters

Parameters	Value	Parameter	Value	Parameter	Value
θ	0.11	C_s	40	S	335
x	20	w_s	90	W_1	45
α	0.10	w'_s	55	W_2	16
β	0.44	R_1	16	a	51
γ	0.78	H	0.02	b	0.17
α_1	20	G	10	D'_r	35
β_1	0.2	s_r	176	D_r	36
r_{cm}	2	s'_r	135	S_c	0.35
s_m	158	s'_m	108		

If $\frac{\partial^2}{\partial R^2}(\text{EIAP}) < 0$, $\frac{\partial^2}{\partial P^2}(\text{EIAP}) < 0$ and $\left\{\frac{\partial^2}{\partial R^2}(\text{EIAP})\right\}\left\{\frac{\partial^2}{\partial P^2}(\text{EIAP})\right\} - \left\{\frac{\partial^2}{\partial P \partial R}(\text{EIAP})\right\}^2 > 0$ holds for $R = R^*$ and $P = P^*$ then $\text{EIAP}(R^*, P^*)$ is maximum.

Now, $\left.\frac{\partial^2}{\partial R^2}(\text{EIAP})\right]_{\text{at}(R^*,P^*)} < 0$

$$
\begin{aligned}
\text{i.e., } & \frac{2(1-\theta)^2}{D_r{}^2 T^{*2}} E\left[\tilde{\tilde{Z}}_0\right] + \left\{\frac{2(1-\theta)^2 R^*}{D_r{}^2 T^{*2} P^*} - \frac{2(1-\theta)}{D_{rT^*} P^*}\right\} E\left[\tilde{\tilde{Z}}_1\right] + \left\{\frac{2(1-\theta)^2 R^*}{D_r{}^2 T^{*2}} - \frac{2(1-\theta)}{D_{rT^*}}\right\} E\left[\tilde{\tilde{Z}}_2\right] \\
& - \left\{1 + \frac{2(1-\theta)^2 R^{*2}}{D_r{}^2 T^{*2} P^*} - \frac{2(1-\theta) R^*}{D_{rT^*} P^*}\right\} E\left[\tilde{\tilde{Z}}_3\right] + \left\{\frac{2}{P^{*2}} + \frac{2(1-\theta)^2 R^{*2\,2}}{D_r{}^2 T^*} P^{*2} - \frac{4(1-\theta) R^*}{D_{rT^*} P^*}\right\} E\left[\tilde{\tilde{Z}}_4\right] \\
& - \left\{2P^* + \frac{2(1-\theta)^2 P^* R^{*2}}{D_r{}^2 T^{*2}} - \frac{4(1-\theta) P^* R^*}{D_{rT^*}}\right\} E\left[\tilde{\tilde{Z}}_5\right] - \left\{2 + \frac{2(1-\theta)^2 R^{*2}}{D_r{}^2 T^{*2}} - \frac{4(1-\theta) R^*}{D_{rT}}\right\} E\left[\tilde{\tilde{Z}}_6\right] \\
& + \left\{\frac{6R^*}{P^*} + \frac{2(1-\theta)^2 R^{*3}}{D_r{}^2 T^{*2} P^*} - \frac{6(1-\theta) R^{*2}}{D_{rT^*} P^*}\right\} E\left[\tilde{\tilde{Z}}_7\right] + \left\{\frac{12R^{*2}}{P^{*2}} + \frac{2(1-\theta)^2 R^{*4}}{D_r{}^2 T^{*2} P^{*2}} - \frac{8(1-\theta) R^{*3}}{D^2_{rT^* P^*}}\right\} E\left[\tilde{\tilde{Z}}_8\right] < 0
\end{aligned} \tag{21}
$$

Table 2 Collected data for different fuzzy-rough parameters

Fu-Ro parameters	Fu-Ro value	Input values	Expected value	Value
$\tilde{\tilde{h}}_s$	Near roughly (0.5)	$(\overline{0.49}, \overline{0.50}, \overline{0.51}, \overline{0.52})$ with $([-0.04, 0.04], [-0.08, 0.08])$	$E[\tilde{\tilde{h}}_s]$	5.05
$\tilde{\tilde{h}}_m$	Near roughly (1.7)	$(\overline{1.5}, \overline{1.6}, \overline{1.7}, \overline{1.8})$ with $([-0.2, 0.2], [-0.4, 0.4])$	$E[\tilde{\tilde{h}}_m]$	1.65
$\tilde{\tilde{h}}'_m$	Near roughly (1.5)	$(\overline{1.35}, \overline{1.40}, \overline{1.50}, \overline{1.55})$ with $([-0.18, 0.18], [-0.36, 0.36])$	$E[\tilde{\tilde{h}}'_m]$	1.45
$\tilde{\tilde{h}}_r$	Near roughly (1.6)	$(\overline{1.54}, \overline{1.6}, \overline{1.63}, \overline{1.67})$ with $([-0.12, 0.12], [-0.24, 0.24])$	$E[\tilde{\tilde{h}}_r]$	1.61
$\tilde{\tilde{h}}'_r$	Near roughly (1.3)	$(\overline{1.15}, \overline{1.22}, \overline{1.3}, \overline{1.37})$ with $([-0.11, 0.11], [-0.22, 0.22])$	$E[\tilde{\tilde{h}}'_r]$	1.26
$\tilde{\tilde{h}}_{rs}$	Near roughly (1.8)	$(\overline{1.67}, \overline{1.72}, \overline{1.80}, \overline{1.85})$ with $([-0.13, 0.13], [-0.26, 0.26])$	$E[\tilde{\tilde{h}}_{rs}]$	1.78
$\tilde{\tilde{A}}_s$	Near roughly (420)	$(\overline{416}, \overline{420}, \overline{425}, \overline{428})$ with $([-0.30, 0.30], [-0.60, 0.60])$	$E[\tilde{\tilde{A}}_s]$	422.25
$\tilde{\tilde{A}}_m$	Near roughly (500)	$(\overline{486}, \overline{495}, \overline{500}, \overline{510})$ with $([-0.29, 0.29], [-0.58, 0.58])$	$E[\tilde{\tilde{A}}_m]$	491.50
$\tilde{\tilde{A}}_r$	Near roughly (400)	$(\overline{395}, \overline{400}, \overline{408}, \overline{410})$ with $([-0.35, 0.35], [-0.70, 0.70])$	$E[\tilde{\tilde{A}}_r]$	403.25
$\tilde{\tilde{I}}_{cs}$	Near roughly (3)	$(\overline{2.5}, \overline{2.8}, \overline{3.0}, \overline{3.15})$ with $([-0.3, 0.3], [-0.6, 0.6])$	$E[\tilde{\tilde{I}}_{cs}]$	2.86
$\tilde{\tilde{I}}_{cm}$	Near roughly (2)	$(\overline{1.9}, \overline{1.96}, \overline{2.0}, \overline{2.04})$ with $([-0.21, 0.21], [-0.42, 0.42])$	$E[\tilde{\tilde{I}}_{cm}]$	1.97
$\tilde{\tilde{c}}_{tp}$	Near roughly (1.4)	$(\overline{1.36}, \overline{1.40}, \overline{1.43}, \overline{1.47})$ with $([-0.14, 0.14], [-0.28, 0.28])$	$E[\tilde{\tilde{c}}_{tp}]$	1.42
$\tilde{\tilde{c}}'_{tp}$	Near roughly (1.0)	$(\overline{0.90}, \overline{0.97}, \overline{1.0}, \overline{1.06})$ with $([-0.01, 0.01], [-0.02, 0.02])$	$E[\tilde{\tilde{c}}'_{tp}]$	0.98

The value of $\tilde{\tilde{h}}_s$ is near roughly $\underline{(0.5)} = (\overline{0.49}, \overline{0.50}, \overline{0.51}, \overline{.52})$ with oscillation $([-0.04, 0.04], [-0.08, 0.08])$ means that $\overline{0.49} \vdash ([0.45, 0.53], [0.41, 0.57])$, $\overline{0.50} \vdash ([0.46, 0.54], [0.42, 0.58])$, $\overline{0.51} \vdash ([0.47, 0.55], [0.43, 0.59])$ and $\overline{0.52} \vdash ([0.48, 0.56], [0.44, 0.60])$.

Table 3 Optimal result when P and R are decision variables

Approach	Total profit	EAPS	EAPM	EAPR	R	P
Stakelberg	5, 126.616	1,43.341	1,814.635	1,656.160	3,217.589	64.9821
Integrated	5, 176.606	1,596.751	1,872.695	1,707.160	3,189.629	69.8521

and $\left.\frac{\partial^2}{\partial P^2}(\text{EIAP})\right]_{\text{at}(R^*,P^*)}$

$$\text{i.e.,}\quad \left[2E\left[\tilde{\tilde{Z}}_1\right]\frac{R^*}{P^{*3}}+E\left[\tilde{\tilde{Z}}_3\right]\frac{R^{*2}}{P^{*3}}+6E\left[\tilde{\tilde{Z}}_4\right]\frac{R^{*2}}{P^{*4}}+2E\left[\tilde{\tilde{Z}}_7\right]\frac{R^{*3}}{P^{*3}}+6E\left[\tilde{\tilde{Z}}_8\right]\frac{R^{*4}}{P^{*4}}\right]<0 \quad (22)$$

$$\text{and}\quad \left[\left\{\frac{\partial^2}{\partial R^2}(\text{EIAP})\right\}\left\{\frac{\partial^2}{\partial P^2}(\text{EIAP})\right\}-\left\{\frac{\partial^2}{\partial P\partial R}(\text{EIAP})\right\}^2\right]_{\text{at}(R^*,P^*)}>0 \quad (23)$$

Therefore, IAP(R^*, P^*) is maximum if the relations (21), (22), and (23) hold and the corresponding optimum integrated average profit of the supply chain is

$$\text{EIAP}\left(R^*,P^*\right)=\frac{1}{T^*}\left[E\left[\tilde{\tilde{Z}}_0\right]+E\left[\tilde{\tilde{Z}}_1\right]\frac{R^*}{P^*}+E\left[\tilde{\tilde{Z}}_2\right]R^*-Z_3\frac{R^{*2}}{2P^*}+E\left[\tilde{\tilde{Z}}_4\right]\frac{R^{*2}}{P^{*2}}\right.$$
$$\left.-E\left[\tilde{\tilde{Z}}_5\right]P^*R^{*2}-E\left[\tilde{\tilde{Z}}_6\right]R^{*2}+E\left[\tilde{\tilde{Z}}_7\right]\frac{R^{*3}}{P^*}+E\left[\tilde{\tilde{Z}}_8\right]\frac{R^{*4}}{P^{*2}}\right].$$

where $T^*=\frac{1-\theta}{D_r}R^*+t_0$.

When P is a decision variable

The necessary conditions for maximum value of EIAP(P) is $\frac{d}{dP}(\text{EIAP})=0$ i.e., $E\left[\tilde{\tilde{Z}}_1\right]\frac{R}{P^2}-E\left[\tilde{\tilde{Z}}_3\right]\frac{R^2}{2P^2}+2E\left[\tilde{\tilde{Z}}_4\right]\frac{R^2}{P^3}+E\left[\tilde{\tilde{Z}}_5\right]R^2-E\left[\tilde{\tilde{Z}}_7\right]\frac{R^3}{P^2}+2E\left[\tilde{\tilde{Z}}_8\right]\frac{R^4}{P^3}=0$ which gives the optimum value of P, say P^{**}.

If $\frac{d^2}{dP^2}(\text{EIAP})<0$ hold for $P=P^{**}$ then EIAP (P^{**}) is maximum.

Now $\left.\frac{d^2}{dP^2}(\text{EIAP})\right]_{\text{at }P=P^{**}}<0$ gives

$$-2E\left[\tilde{\tilde{Z}}_1\right]\frac{R}{P^{**3}}+E\left[\tilde{\tilde{Z}}_3\right]\frac{R^2}{P^{**3}}-6E\left[\tilde{\tilde{Z}}_4\right]\frac{R^2}{P^{**4}}-2E\left[\tilde{\tilde{Z}}_7\right]\frac{R^3}{P^{**3}}-6E\left[\tilde{\tilde{Z}}_8\right]\frac{R^4}{P^{**4}}<0 \quad (24)$$

Therefore, EIAP(P^{**}) is maximum if the relation (24) holds and the corresponding maximum integrated average profit of the supply chain is

$$\text{EIAP}\left(P^{**}\right)=\frac{1}{T}\left[E\left[\tilde{\tilde{Z}}_0\right]+E\left[\tilde{\tilde{Z}}_1\right]\frac{R}{P^{**}}+Z_2R-Z_3\frac{R^2}{2P^{**}}+E\left[\tilde{\tilde{Z}}_4\right]\frac{R^2}{P^{**2}}\right.$$
$$\left.-Z_5P_2R^2-E\left[\tilde{\tilde{Z}}_6\right]R^2+E\left[\tilde{\tilde{Z}}_7\right]\frac{R^3}{P^{**}}+E\left[\tilde{\tilde{Z}}_8\right]\frac{R^4}{P^{**2}}\right].$$

Table 4 Optimal result when P is decision variable when $R = 3,225$

Approach	Total profit	EAPS	EAPM	EAPR	P
Stakelberg	5,012.614	1,512.752	1,714.346	1, 751.426	63.64570
Integrated	5,049.634	1,582.912	1,759.146	1, 707.576	67.37870

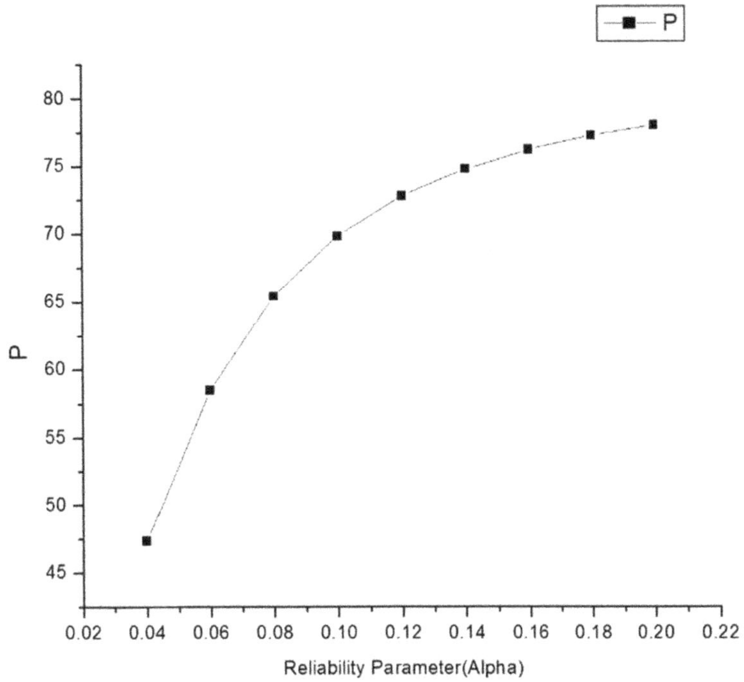

Figure 4 Graphical representation of change the optimum value of *P* with alpha.

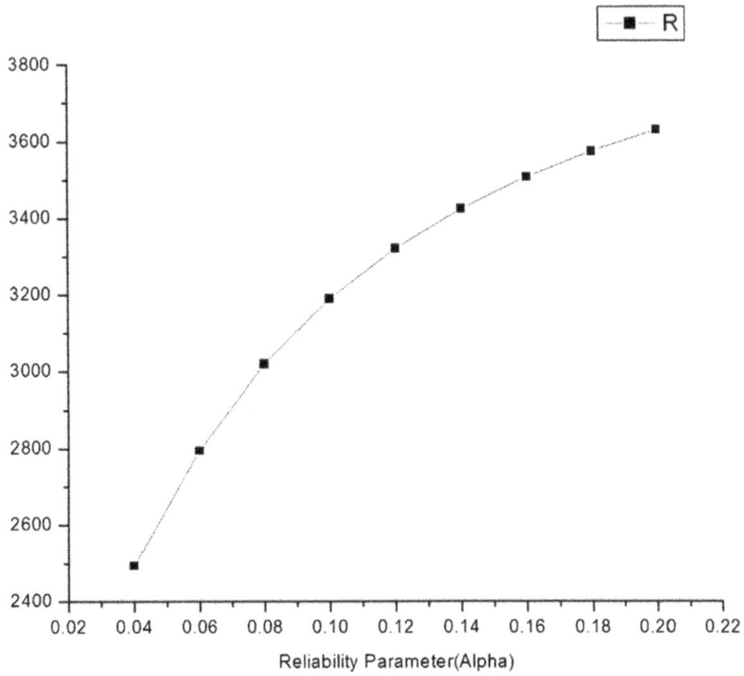

Figure 5 Graphical representation of change the optimum value of *R* with alpha.

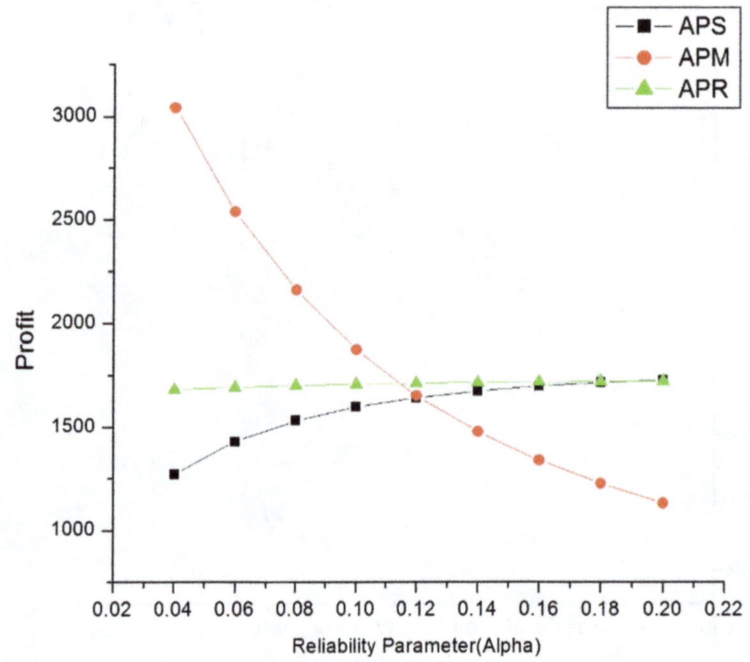

Figure 6 Graphical representation of change the optimum value of profits with alpha.

Numerical examples

To illustrate the proposed production inventory model, we consider the following numerical data in Tables 1 and 2.

The optimal values of the decision variables and corresponding profits are given in Tables 3 and 4. Also, sensitivity analysis has been performed of the profits, production

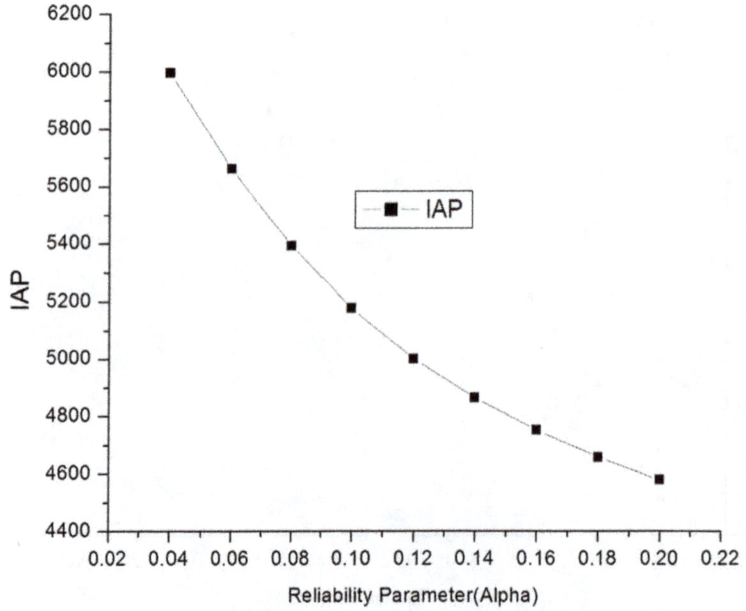

Figure 7 Graphical representation of change the optimum value of IAP with alpha.

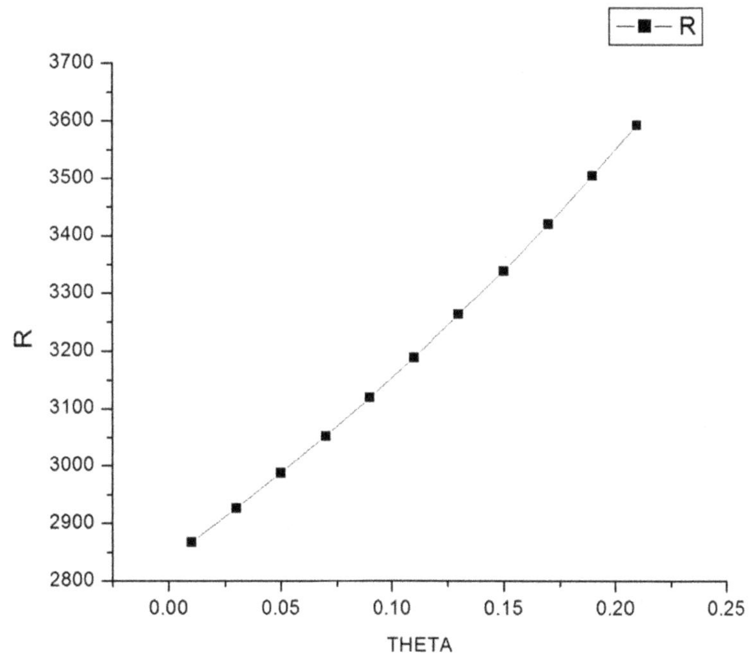

Figure 8 Graphical representation of change the optimum value of *R* with theta.

rate (*P*), and inventory level (*R*) of supplier with respect to different parameters which are shown in Figures 4, 5, 6, 7, 8, 9, 10, 11, 12, and 13.

Discussion

From Tables 1 and 2, it is observed that profits under the integrated approach is greater than the Stakelberg approach and hence the former approach is better than the later

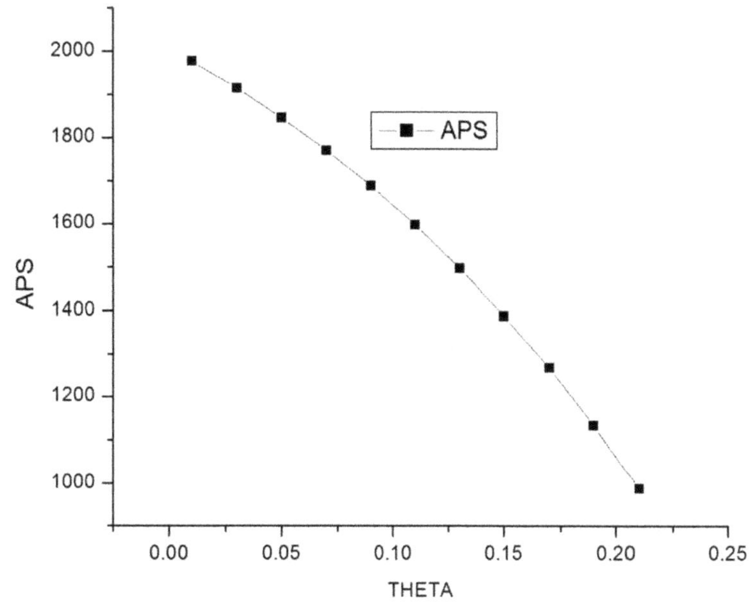

Figure 9 Graphical representation of change the optimum value of APS with theta.

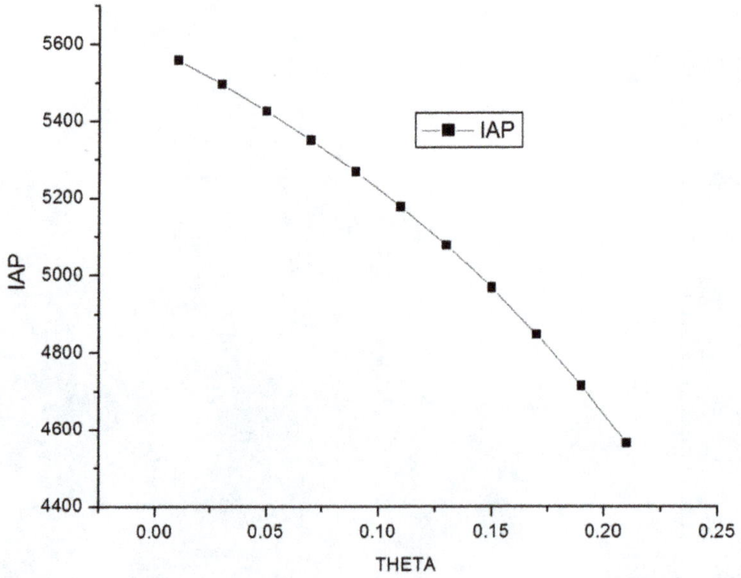

Figure 10 Graphical representation of change the optimum value of IAP with theta.

approach. The sensitivity analysis in 'Numerical examples' section shows that with increase of the reliability parameter α, *(i)* the profits of the supplier and retailer are slightly increasing (Figure 6), *(ii)* the values of P and R are gradually increasing (Figures 4 and 5) but *(iii)* both profits of the manufacturer (APM) and the integrated profit (IAP) are gradually decreasing (Figures 6 and 7). From Figures 8, 9, 10, 11, it is also seen that the values of APS and IAP are decreasing, but the initial amount of inventory level of supplier is

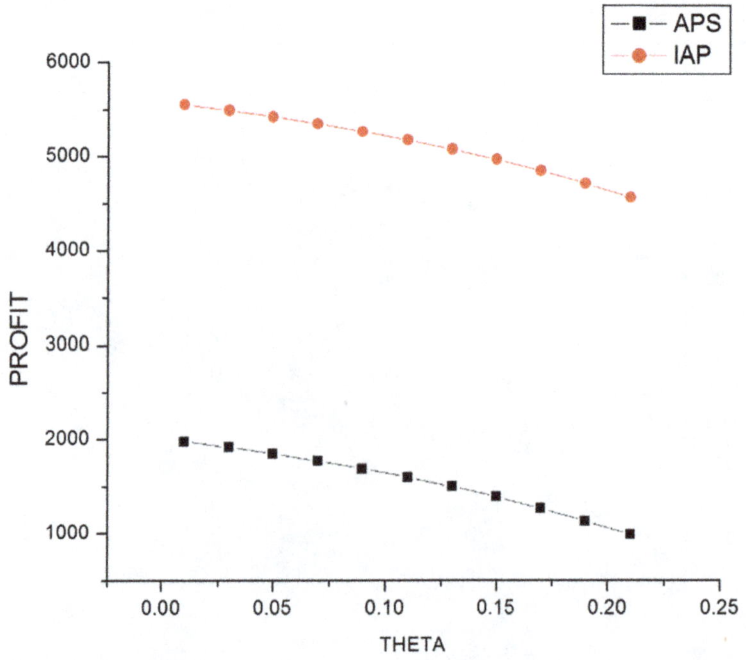

Figure 11 Graphical representation of change the optimum value of profits with theta.

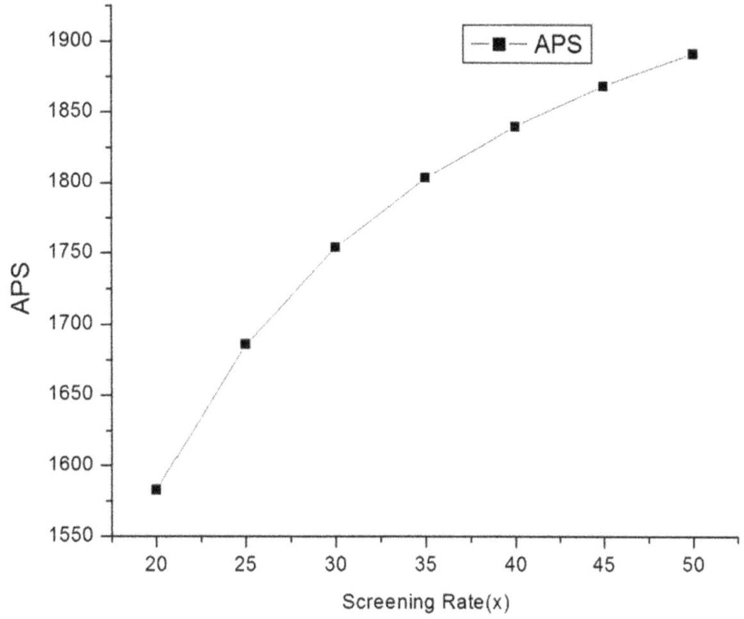

Figure 12 Graphical representation of change the optimum value of APS with *x*.

increasing with increase of θ. It is also noted that the values of APS and IAP increase with the screening rate of the supplier (x).

Conclusions

This paper develops a three-layer supply chain production inventory model involving supplier, manufacturer, and retailer as the members of the chain who are

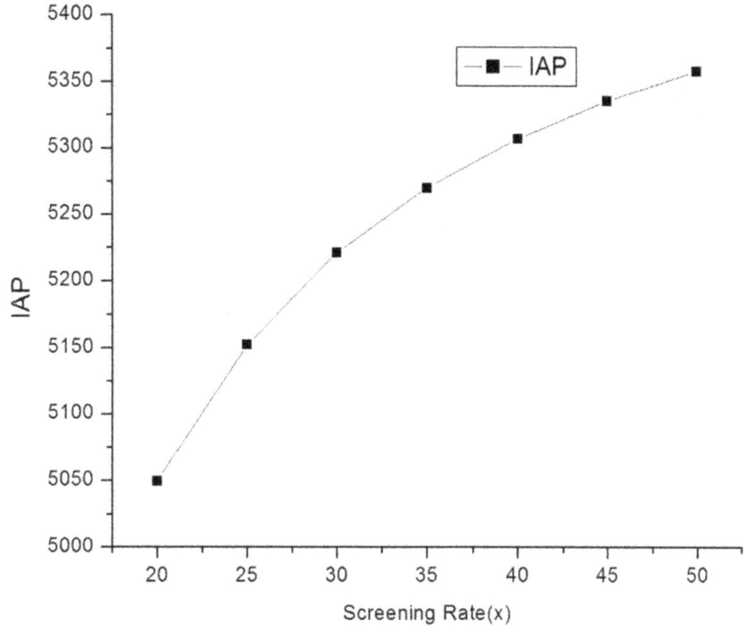

Figure 13 Graphical representation of change the optimum value of IAP with *x*.

responsible for performing the raw materials into finished product and make them available to satisfy customers' demand in time. In comparing with the existing literature on the supply chain, the following are the main contributions in the proposed model:

Inspection cost is incurred during the production run time, and the manufacturer continuously inspects as well as separates the perfect quality items, less perfect quality items, repairable items which are transformed into perfect quality items after some rework, and reject items. Reworked cost is considered by the manufacturer to repair a certain percentage of imperfect quality items. The demand rate of the customers for perfect quality items and less perfect quality items are respectively assumed as stock dependent and selling price dependent. Here retailer have two showrooms PW_1 and PW_2 of finite capacities at a busy market place, and the market demands of perfect and less perfect quality items are respectively met through the showrooms PW_1 and PW_2. Retailer has a secondary warehouse SW of infinite capacity, away from busy market place, to store the excess amount perfect quality items from where the items are continuously transferred to the showroom. It is considered that the holding cost per unit per unit time at SW is less than the holding cost at PW_1 per unit per unit time. The repairing costs of corrective and preventive maintenance should also be considered, as these costs increase the unit production cost. Inventory and production decisions are made at the supplier, manufacturer, and retailer levels. Actually in this paper, the coordination between production and inventory decisions has been established across the supply chain so that the integrated average profit of the chain is maximum.

Appendix 1

$$
\begin{aligned}
T &= t_4 + \frac{1}{\beta_1}\log\left[\left(1-\frac{D_r}{\alpha_1}\right)\left(1-e^{-\beta_1 t_3}\right)\right] \\
&= t_2 + \frac{S-W_1}{\alpha_1+\beta_1 W_1} + \frac{1}{\beta_1}\log\left[\left(1-\frac{D_r}{\alpha_1}\right)\left(1-e^{-\beta_1 t_3}\right)\right] \\
&= \frac{P}{D_r}\left[\frac{1}{\alpha}(1-\beta)\left(1-e^{-\alpha t_1}\right)+\beta t_1 + \frac{S-W_1}{\alpha_1+\beta_1 W_1} + \frac{1}{\beta_1}\log\left[\left(1-\frac{D_r}{\alpha_1}\right)\left(1-e^{-\beta_1 t_3}\right)\right]\right] \\
&= \frac{P}{D_r}t_1 + \frac{S-W_1}{\alpha_1+\beta_1 W_1} + \frac{1}{\beta_1}\log\left[\left(1-\frac{D_r}{\alpha_1}\right)\left(1-e^{-\beta_1 t_3}\right)\right] \\
&= \frac{1-\theta}{D_r}R + t_0. \qquad \text{[using, (6), (12a), (12b) and } 1-e^{-\alpha t_1}\approx \alpha t_1]
\end{aligned}
$$

$$
\begin{aligned}
\text{APS}(R,P) &= \frac{1}{T}\left[w_s(R-R\theta)+w_s'R\theta-(A_s+c_sR)-s_cR-h_s\left[\frac{(R-R\theta)t_1}{2}+\frac{R^2\theta}{x}\right]-I_{cs}(T-t_1)\right] \\
&= \frac{1}{T}\left[-(A_s+I_{cs}t_0)+\left[w_s(1-\theta)+w_s'-c_s-s_c-\frac{I_{cs}(1-\theta)}{D_r}\right]R+I_{cs}(1-\theta)\frac{R}{P}\right. \\
&\quad \left. -h_s(1-\theta)^2\frac{R^2}{2P}-\frac{h_s\theta}{x}R^2\right] \\
&= \frac{1}{T}\left[-Z_{0s}+Z_{1s}R+Z_{2s}\frac{R}{P}-Z_{3s}\frac{R^2}{2P}-Z_{4s}R^2\right].
\end{aligned}
$$

where $Z_{0s}=(A_s+I_{cs}t_0)$ $Z_{1s}=\left[w_s(1-\theta)+w_s'-c_s-s_c-\frac{I_{cs}(1-\theta)}{D_r}\right]$
$Z_{2s}=I_{cs}(1-\theta)$, $Z_{3s}=h_s(1-\theta)^2$, $Z_{4s}=\frac{h_s\theta}{x}$.

$$\begin{aligned}
\mathrm{APM}(R,P) = &\frac{1}{T}\Bigg[\left(s_{\mathrm{m}}D_{\mathrm{r}}t_2+s'_{\mathrm{m}}D'_{\mathrm{r}}t'_2\right)-\{w_{\mathrm{s}}+C(P)+I_{\mathrm{sm}}\}Pt_1-r_{\mathrm{cm}}\left\{-\frac{P\beta}{\alpha}\left(1-e^{-\alpha t_1}\right)+P\beta t_1\right\}\\
&-A_{\mathrm{m}}-h_{\mathrm{m}}\left\{\frac{P}{\alpha}(1-\beta)t_1-\frac{P}{\alpha^2}(1-\beta)\left(1-e^{-\alpha t_1}\right)+P\beta\frac{t_1^2}{2}-D_{\mathrm{r}}\frac{t_2^2}{2}\right\}\\
&-I_{\mathrm{cm}}(T-t_2)-h'_{\mathrm{m}}\Bigg\{\frac{P}{\alpha^2}\gamma(1-\beta)\left(1-e^{-\alpha t_1}\right)-\frac{\gamma}{\alpha}(1-\beta)Pt_1\\
&+(\gamma(1-\beta)P-D'_{\mathrm{r}})\frac{t_1^2}{2}+\frac{D'_{\mathrm{r}}}{2}\left(t_1-t'_2\right)^2\Bigg\}\Bigg]\\
=&\frac{1}{T}\Bigg[\left(s_{\mathrm{m}}+\frac{I_{\mathrm{cm}}}{D_{\mathrm{r}}}\right)\left\{\frac{P}{\alpha}(1-\beta)\left(1-e^{-\alpha t_1}\right)+P\beta t_1\right\}-\{w_{\mathrm{s}}+C(P)+I_{\mathrm{sm}}\}Pt_1-A_{\mathrm{m}}\\
&+s'_{\mathrm{m}}\left\{\frac{-P\gamma}{\alpha}(1-\beta)\left(1-e^{-\alpha t_1}\right)+P\gamma(1-\beta)t_1\right\}-r_{\mathrm{cm}}\left[-\frac{P\beta}{\alpha}\left(1-e^{-\alpha t_1}\right)+P\beta t_1\right]\\
&-I_{\mathrm{cm}}T-h_{\mathrm{m}}\left\{\frac{P}{\alpha}(1-\beta)t_1-\frac{P}{\alpha^2}(1-\beta)(1-e^{-\alpha t_1})+P\beta\frac{t_1^2}{2}-D_{\mathrm{r}}\frac{t_2^2}{2}\right\}\\
&-h'_{\mathrm{m}}\Bigg\{\frac{P}{\alpha^2}\gamma(1-\beta)\left(1-e^{-\alpha t_1}\right)-\frac{\gamma}{\alpha}(1-\beta)Pt_1+(\gamma(1-\beta)P-D'_{\mathrm{r}})\frac{t_1^2}{2}\\
&+\frac{D'_{\mathrm{r}}}{2}(t_1-t'_2)^2\Bigg\}\Bigg]
\end{aligned}$$

$$\begin{aligned}
=&\frac{1}{T}\Bigg[\left(s_{\mathrm{m}}+\frac{I_{\mathrm{cm}}}{D_{\mathrm{r}}}-w_{\mathrm{s}}-I_{\mathrm{sm}}-R_1\right)Pt_1-I_{\mathrm{cm}}T-A_{\mathrm{m}}-Gt_1-HP^2t_1\\
&+\left[s'_{\mathrm{m}}(1-\beta)\alpha\gamma-\left(s_{\mathrm{m}}+\frac{I_{\mathrm{cm}}}{D_{\mathrm{r}}}\right)(1-\beta)\alpha-h_{\mathrm{m}}-r_{\mathrm{cm}\alpha\beta}\right]\frac{Pt_1{}^2}{2}\\
&+\frac{h_{\mathrm{m}}}{2D_{\mathrm{r}}}P^2t_1{}^2+h'_{\mathrm{m}}D'_{\mathrm{r}}t_1{}^2\Bigg],\qquad [\text{using } 1-e^{-\alpha t_1}\approx\alpha t_1]\\
=&\frac{1}{T}\Bigg[-I_{\mathrm{cm}}t_0-G(1-\theta)\frac{R}{P}+(s_{\mathrm{m}}+\frac{I_{\mathrm{cm}}}{D_{\mathrm{r}}}-w_{\mathrm{s}}-I_{\mathrm{sm}}-R_1-\frac{I_{\mathrm{cm}}}{D_{\mathrm{r}}})(1-\theta)R\\
&-H(1-\theta)PR^2+\left\{s'_{\mathrm{m}}(1-\beta)\alpha\gamma-(s_{\mathrm{m}}+\frac{I_{\mathrm{cm}}}{D_{\mathrm{r}}})(1-\beta)\alpha-h_{\mathrm{m}}-r_{\mathrm{cm}\alpha\beta}\right\}(1-\theta)^2\frac{R^2}{2P}\\
&+\frac{h_{\mathrm{m}}}{2D_{\mathrm{r}}}(1-\theta)^2R^2+h'_{\mathrm{m}}D'_{\mathrm{r}}(1-\theta)^2\frac{R^2}{P^2}\Bigg],\qquad [\text{using (3)}]\\
=&\frac{1}{T}\left[-Z_{0m}-Z_{1m}\frac{R}{P}+Z_{2m}R+Z_{3m}\frac{R^2}{P^2}+Z_{4m}\frac{R^2}{P}+Z_{5m}R^2-Z_{6m}PR^2\right].
\end{aligned}$$

where $Z_{0m}=I_{\mathrm{cm}}t_0-A_{\mathrm{m}}\ Z_{1m}=G(1-\theta)$

$$\begin{aligned}
Z_{2m}&=(s_{\mathrm{m}}-w_{\mathrm{s}}-I_{\mathrm{sm}}-R_1)(1-\theta),\ Z_{3m}=h'_{\mathrm{m}}D'_{\mathrm{r}}(1-\theta)^2\\
Z_{4m}&=\left[s'_{\mathrm{m}}(1-\beta)\alpha\gamma-\left(s_{\mathrm{m}}+\frac{I_{\mathrm{cm}}}{D_{\mathrm{r}}}\right)(1-\beta)\alpha-h_{\mathrm{m}}-r_{\mathrm{cm}\alpha\beta}\right]\frac{(1-\theta)^2}{2}\\
Z_{5m}&=\frac{h_{\mathrm{m}}}{2D_{\mathrm{r}}}(1-\theta)^2,\ Z_{6m}=H(1-\theta).
\end{aligned}$$

$$
\begin{aligned}
\mathrm{APR}(R,P) = {} & \frac{1}{T}\left[(s_{\mathrm{r}}-c_{\mathrm{tp}})\left[(\alpha_1+\beta_1 W_1)(t_4-t_3)+D_{\mathrm{r}}t_3-\frac{D_{\mathrm{r}}}{\beta_1}+\frac{D_{\mathrm{r}}-\alpha_1}{\beta_1}e^{-\beta_1 t_3}+\frac{\alpha_1}{\beta_1}e^{-\beta_1(t_4-T)}\right]\right. \\
& +\left(s'_{\mathrm{r}}-c'_{\mathrm{tp}}\right)D'_{\mathrm{r}}t'_2-\frac{h_{\mathrm{rs}}}{2}\left[(D_{\mathrm{r}}-(\alpha_1+\beta_1 W_1))(t_2-t_3)^2+(\alpha_1+\beta_1 W_1)(t_2-t_4)^2\right] \\
& -h_{\mathrm{r}}\left\{\frac{D_{\mathrm{r}}-\alpha_1}{\beta_1}\left(t_3+\frac{e^{-\beta_1 t_3}}{\beta_1}-\frac{1}{\beta_1}\right)+W_1(t_4-t_3)-\frac{\alpha_1}{\beta_1}\left\{(T-t_4)+\frac{1}{\beta_1}-\frac{e^{-\beta_1(t_4-T)}}{\beta_1}\right\}\right\} \\
& -\frac{h'_{\mathrm{r}}}{2}\left\{\left(D'_{\mathrm{r}}-D'_{\mathrm{c}}\right){t'_2}^2+D'_{\mathrm{c}}(t'_2-T')^2\right\}-A_{\mathrm{r}}-A'_{\mathrm{r}} \\
& -s_{\mathrm{m}}\left\{\frac{P}{\alpha}(1-\beta)t_1-\frac{P}{\alpha^2}(1-\beta)\left(1-e^{-\alpha t_1}\right)+(P\beta)\frac{t_1^2}{2}-D'_{\mathrm{r}}\frac{t_2^2}{2}\right\} \\
& \left.-s'_{\mathrm{m}}\left\{\frac{P\gamma}{\alpha^2}(1-\beta)\left(1-e^{-\alpha t_1}\right)-\frac{P\gamma}{\alpha}(1-\beta)t_1+\frac{D'_{\mathrm{r}}}{2}\left(t_1-t'_2\right)^2+\left(P\gamma(1-\beta)-D'_{\mathrm{r}}\right)\frac{t_1^2}{2}\right\}\right] \\
= {} & \frac{1}{T}\left[Z_{0r}+Z_{1r}R+Z_{2r}\frac{R^2}{2P}+Z_{3r}R^2+Z_{4r}\frac{R^3}{P}+Z_{5r}\frac{R^4}{P^2}\right]. \\
& \left[\text{using (3), (7), (15) and } 1-e^{-\alpha t_1}\approx \alpha t_1\right]
\end{aligned}
$$

where $Z_{0r}=-h_{\mathrm{rs}}\left[\frac{(s-W_1)^2}{2(D_{\mathrm{r}}-(\alpha_1+\beta_1 W_1))}+\frac{(s-W_1)^2}{2(\alpha_1+\beta_1 W_1)}\right]-A_{\mathrm{r}}-A'_{\mathrm{r}}$
$-h_{\mathrm{r}}\left[-\frac{\alpha_1}{\beta_1^2}\log\left(1+\frac{\beta_1 W_1}{\alpha_1}\right)-\frac{1}{\beta_1}\left(\frac{D_{\mathrm{r}}-\alpha_1}{\beta_1}-\frac{D_{\mathrm{r}}W_1}{\alpha_1+\beta_1 W_1}\right)\log\left(1-\frac{W_1\beta_1}{D_{\mathrm{r}}-\alpha_1}\right)\right]$
$Z_{1r}=(s_{\mathrm{r}}-c_{\mathrm{tp}}-\frac{h_{\mathrm{r}}W_1}{\alpha_1+\beta_1 W_1})(1-\theta)$
$Z_{2r}=\left[\left(s'_{\mathrm{r}}-c'_{\mathrm{tp}}\right)\gamma\alpha(1-\beta)+\left(s_{\mathrm{r}}-c_{\mathrm{tp}}-\frac{h_{\mathrm{r}}W_1}{\alpha_1+\beta_1 W_1}\right)\alpha(1-\beta)-\frac{W_{\mathrm{m}}}{2}\right](1-\theta)^2$
$Z_{3r}=\frac{s_{\mathrm{m}}}{2D_{\mathrm{r}}}(1-\theta)^2,\ Z_{4r}=\left[\frac{s'_{\mathrm{m}}\gamma\alpha}{2}(1-\beta)-\frac{s_{\mathrm{m}}\gamma}{2D_{\mathrm{r}}}(1-\beta)\right](1-\theta)^3$
$Z_{5r}=\left[s_{\mathrm{m}}\frac{\gamma^2}{4D_{\mathrm{r}}}(1-\beta)^2-s'_{\mathrm{m}}\frac{\gamma^2\alpha^2}{8D_{\mathrm{r}}}(1-\beta)^2-\frac{h'_{\mathrm{r}}}{8D'_{\mathrm{r}}}\left\{(D'_{\mathrm{r}}-D'_{\mathrm{c}})+\frac{(D'_{\mathrm{r}}-D'_{\mathrm{c}})^2}{D'^2_{\mathrm{c}}}\right\}\gamma^2\alpha^2(1-\beta)^2\right].$

$$
\begin{aligned}
\mathrm{IAP}(R,P) = {} & [\mathrm{APS}+\mathrm{APM}+\mathrm{APR}] \\
= {} & \frac{1}{T}\left[(Z_{0r}-Z_{0m}-Z_{0s})+(Z_{2s}-Z_{1m})\frac{R}{P}+(Z_{1s}+Z_{1r}+Z_{2m})R-(Z_{3s}+2Z_{4m}+Z_{2r})\frac{R^2}{2P}\right. \\
& \left.+Z_{3m}\frac{R^2}{P^2}-Z_{6m}PR^2-(Z_{4s}-Z_{3r}-Z_{5m})R^2+Z_{4r}\frac{R^3}{P}+Z_{5r}\frac{R^4}{P^2}\right] \\
= {} & \frac{1}{T}\left[Z_0+Z_1\frac{R}{P}+Z_2R-Z_3\frac{R^2}{2P}+Z_4\frac{R^2}{P^2}-Z_5PR^2-Z_6R^2+Z_7\frac{R^3}{P}+Z_8\frac{R^4}{P^2}\right].
\end{aligned}
$$

where $Z_0=(Z_{0r}-Z_{0m}-Z_{0s}), Z_1=(Z_{2s}-Z_{1m}),\ Z_2=(Z_{1s}+Z_{1r}+Z_{2m})\ Z_3=(Z_{3s}+2Z_{4m}+Z_{2r}),\ Z_4=Z_{3m},\ Z_5=Z_{6m},\ Z_7=Z_{4R},\ Z_8=Z_{5R},\ Z_6=(Z_{4s}-Z_{3r}-Z_{5m}).$

$$
\begin{aligned}
\frac{\partial}{\partial R}(\mathrm{APM}) = {} & \frac{1}{T}\left[-Z_{1m}\frac{1}{P}+Z_{2m}+2Z_{3m}\frac{R}{P^2}+2Z_{4m}\frac{R}{P}+2Z_{5m}R-2PRZ_{6m}\right] \\
& -\frac{(1-\theta)}{D_{\mathrm{r}}T^2}\left[-Z_{0m}-Z_{1m}\frac{R}{P}+Z_{2m}R+Z_{3m}\frac{R^2}{P^2}+Z_{4m}\frac{R^2}{P}+Z_{5m}R^2-Z_{6m}PR^2\right] \\
= {} & \frac{(1-\theta)}{D_{\mathrm{r}}T}Z_{0m}+\left[\frac{(1-\theta)R}{D_{\mathrm{r}}TP}-\frac{1}{P}\right]Z_{1m}+\left[1-\frac{(1-\theta)R}{D_{\mathrm{r}}T}\right]Z_{2m}+\left[\frac{2R}{P^2}-\frac{(1-\theta)R^2}{D_{\mathrm{r}}TP^2}\right]Z_{3m} \\
& +\left[\frac{2R}{P}-\frac{(1-\theta)R^2}{D_{\mathrm{r}}TP}\right]Z_{4m}+\left[2R-\frac{(1-\theta)R^2}{D_{\mathrm{r}}T}\right]Z_{5m}+\left[\frac{(1-\theta)R^2P}{D_{\mathrm{r}}T}-2PR\right]Z_{6m}.
\end{aligned}
$$

$$
\begin{aligned}
\frac{\partial}{\partial P}(\text{APM}) &= \frac{1}{T}\left[Z_{1m}\frac{R}{P^2} - 2Z_{3m}\frac{R^2}{P^3} - Z_{4m}\frac{R^2}{P^2} - Z_{6m}R^2\right].\\
\frac{\partial^2}{\partial P^2}(\text{APM}) &= 2\left[-Z_{1m}\frac{R}{P^3} + 3Z_{3m}\frac{R^2}{P^4} + Z_{4m}\frac{R^2}{P^3}\right]\\
\frac{\partial^2}{\partial R\partial P}(\text{APM}) &= \frac{1}{T}\left[Z_{1m}\frac{1}{P^2} - 4Z_{3m}\frac{R}{P^3} - 2Z_{4m}\frac{R}{P^2} - 2Z_{6m}R\right]\\
\frac{\partial^2}{\partial R^2}(\text{APM}) &= \frac{2}{T}\left[Z_{3m}\frac{1}{P^2} + Z_{4m}\frac{1}{P} + Z_{5m} - Z_{6m}P\right]\\
&\quad -\frac{2(1-\theta)}{D_{\text{r}}T^2}\left[-Z_{1m}\frac{1}{P} + Z_{2m} + 2Z_{3m}\frac{R}{P^2} + 2Z_{4m}\frac{R}{P} + 2Z_{5m}R - Z_{6m}2PR\right]\\
&\quad +\frac{2(1-\theta)^2}{D_r^2T^3}\left[-Z_{0m} - Z_{1m}\frac{R}{P} + Z_{2m}R + Z_{3m}\frac{R^2}{P^2} + Z_{4m}\frac{R^2}{P} + Z_{5m}R^2 - Z_{6m}PR^2\right]\\
&= \frac{1}{T}\left[-\frac{(1-\theta)^2}{D_{\text{r}}{}^2T^2}Z_{0m} + \left\{\frac{(1-\theta)}{D_{\text{r}}PT} - \frac{(1-\theta)^2R}{D_{\text{r}}{}^2T^2P}\right\}Z_{1m} + \left\{\frac{(1-\theta)^2R}{D_{\text{r}}{}^2T^2} - \frac{(1-\theta)}{D_{\text{r}}T}\right\}Z_{2m}\right.\\
&\quad + \left\{\frac{(1-\theta)^2R^2}{D_{\text{r}}{}^2T^2P^2} + \frac{1}{P^2} - \frac{2(1-\theta)R}{D_{\text{r}}TP^2}\right\}Z_{3m} + \left\{\frac{(1-\theta)^2}{D_{\text{r}}{}^2T^2P} + \frac{1}{P} - \frac{2(1-\theta)R}{D_{\text{r}}TP}\right\}Z_{4m}\\
&\quad \left. + \left\{1 - \frac{2(1-\theta)R}{D_{\text{r}}PT} + \frac{(1-\theta)^2R^2}{D_{\text{r}}{}^2T^2}\right\}Z_{5m} - \left\{P - \frac{2(1-\theta)PR}{D_{\text{r}}T} + \frac{(1-\theta)^2R^2P}{D_{\text{r}}{}^2T^2}\right\}Z_{6m}\right].
\end{aligned}
$$

$$
\begin{aligned}
\frac{\partial}{\partial R}(\text{IAP}) &= \frac{1}{T}\left[Z_1\frac{1}{P} + Z_2 - Z_3\frac{R}{P} + 2Z_4\frac{R}{P^2} - 2Z_5PR - 2Z_6R + 3Z_7\frac{R^2}{P} + 4Z_8\frac{R^3}{P^2}\right]\\
&\quad -\frac{(1-\theta)}{D_{\text{r}}T^2}\left[Z_0 + Z_1\frac{R}{P} + Z_2R - Z_3\frac{R^2}{2P} + Z_4\frac{R^2}{P^2} - Z_5PR^2 - Z_6R^2 + Z_7\frac{R^3}{P} + Z_8\frac{R^4}{P^2}\right]\\
\frac{\partial^2}{\partial R^2}(\text{IAP}) &= \frac{1}{T}\left[-Z_3\frac{1}{P} + 2Z_4\frac{1}{P^2} - 2Z_5P - 2Z_6 + 6Z_7\frac{R}{P} + 12Z_8\frac{R^2}{P^2}\right]\\
&\quad -\frac{2(1-\theta)}{D_{\text{r}}T^2}\left[Z_1\frac{1}{P} + Z_2 - Z_3\frac{R}{P} + 2Z_4\frac{R}{P^2} - 2Z_5PR - 2Z_6R + 3Z_7\frac{R^2}{P} + 4Z_8\frac{R^3}{P^2}\right]\\
&\quad +\frac{2(1-\theta)^2}{D_{\text{r}}{}^2T^3}\left[Z_0 + Z_1\frac{R}{P} + Z_2R - Z_3\frac{R^2}{2P} + Z_4\frac{R^2}{P^2} - Z_5PR^2 - Z_6R^2 + Z_7\frac{R^3}{P} + Z_8\frac{R^4}{P^2}\right]\\
&= \frac{1}{T}\left[\frac{2(1-\theta)^2}{D_{\text{r}}{}^2T^2}Z_0 + \left\{\frac{2(1-\theta)^2R}{D_{\text{r}}{}^2T^2P} - \frac{2(1-\theta)}{D_{\text{r}T}P}\right\}Z_1 + \left\{\frac{2(1-\theta)^2R}{D_{\text{r}}{}^2T^2} - \frac{2(1-\theta)}{D_{\text{r}T}}\right\}Z_2\right.\\
&\quad - \left\{1 + \frac{2(1-\theta)^2R^2}{D_{\text{r}}{}^2T^2P} - \frac{2(1-\theta)R}{D_{\text{r}T}P}\right\}Z_3 + \left\{\frac{2}{P^2} + \frac{2(1-\theta)^2R^2}{D_{\text{r}}{}^2T^2P^2} - \frac{4(1-\theta)R}{D_{\text{r}T}P}\right\}Z_4\\
&\quad - \left\{2P + \frac{2(1-\theta)^2PR^2}{D_{\text{r}}{}^2T^2} - \frac{4(1-\theta)PR}{D_{\text{r}T}}\right\}Z_5 - \left\{2 + \frac{2(1-\theta)^2R^2}{D_{\text{r}}{}^2T^2} - \frac{4(1-\theta)R}{D_{\text{r}T}}\right\}Z_6\\
&\quad \left. + \left\{\frac{6R}{P} + \frac{2(1-\theta)^2R^3}{D_{\text{r}}{}^2T^2P} - \frac{6(1-\theta)R^2}{D_{\text{r}T}P}\right\}Z_7 + \left\{\frac{12R^2}{P^2} + \frac{2(1-\theta)^2R^4}{D_{\text{r}}{}^2T^2P^2} - \frac{8(1-\theta)R^3}{D_{\text{r}T}P^2}\right\}Z_8\right].\\
\frac{\partial}{\partial P}(\text{IAP}) &= \frac{1}{T}\left[-Z_1\frac{R}{P^2} + Z_3\frac{R^2}{2P^2} - 2Z_4\frac{R^2}{P^3} - Z_5R^2 - Z_7\frac{R^3}{P^2} - 2Z_8\frac{R^4}{P^3}\right].\\
\frac{\partial^2}{\partial P^2}(\text{IAP}) &= \frac{1}{T}\left[2Z_1\frac{R}{P^3} - Z_3\frac{R^2}{P^3} + 6Z_4\frac{R^2}{P^4} + 2Z_7\frac{R^3}{P^3} + 6Z_8\frac{R^4}{P^4}\right].\\
\frac{\partial^2}{\partial R\partial P}(\text{IAP}) &= \frac{1}{T}\left[-Z_1\frac{1}{P^2} + Z_3\frac{R}{P^2} - 4Z_4\frac{R}{P^3} - 2Z_5R - 3Z_7\frac{R^2}{P^2} - 8Z_8\frac{R^3}{P^3}\right]\\
&\quad -\frac{(1-\theta)}{D_{\text{r}}T^2}\left[-Z_1\frac{R}{P^2} + Z_3\frac{R^2}{2P^2} - 2Z_4\frac{R^2}{P^3} - Z_5R^2 - Z_7\frac{R^3}{P^2} - 2Z_8\frac{R^4}{P^3}\right]\\
&= \frac{1}{T}\left[\left\{\frac{(1-\theta)R}{D_{\text{r}}TP^2} - \frac{1}{P^2}\right\}Z_1 + \left\{\frac{R}{P^2} - \frac{(1-\theta)R^2}{2D_{\text{r}}TP^2}\right\}Z_3 + \left\{\frac{2(1-\theta)R^2}{D_{\text{r}}TP^3} - \frac{4R}{P^3}\right\}Z_4\right.\\
&\quad \left. + \left\{\frac{(1-\theta)R^2}{D_{\text{r}}T} - 2R\right\}Z_5 + \left\{\frac{(1-\theta)R^2}{D_{\text{r}}TP^2} - \frac{3R^2}{P^2}\right\}Z_7 + \left\{\frac{2(1-\theta)R^2}{D_{\text{r}}TP^3} - \frac{8R^3}{P^3}\right\}Z_8\right].
\end{aligned}
$$

Appendix 2

$$\frac{d}{dP}(\text{APM}) = \frac{1}{T}\left[Z_{1m}\frac{R}{P^2} - 2Z_{3m}\frac{R^2}{P^3} - Z_{4m}\frac{R^2}{P^2} - Z_{6m}R^2\right].$$

$$\frac{d^2}{dP^2}(\text{APM}) = \frac{1}{T}\left[-Z_{1m}\frac{R}{P^3} + 3Z_{3m}\frac{R^2}{P^4} + Z_{4m}\frac{R^2}{P^3}\right].$$

$$\frac{d}{dP}(\text{IAP}) = \frac{1}{T}\left[-Z_1\frac{R}{P^2} + Z_3\frac{R^2}{2P^2} - 2Z_4\frac{R^2}{P^3} - Z_5R^2 - Z_7\frac{R^3}{P^2} - 2Z_8\frac{R^4}{P^3}\right].$$

$$\frac{d^2}{dP^2}(\text{IAP}) = \frac{1}{T}\left[2Z_1\frac{R}{P^3} - Z_3\frac{R^2}{P^3} + 6Z_4\frac{R^2}{P^4} + 2Z_7\frac{R^3}{P^3} + 6Z_8\frac{R^4}{P^4}\right].$$

Acknowledgements
The authors thanks to the reviewers and members of the editorial board for their valuable comments and constructive suggestions to improve the content of this research work.

Author details
[1]Department of Applied Mathematics with Oceanology and Computer Programming, Vidyasagar University, Midnapore, West Bengal 721102, India. [2]Department of Mathematics, Mahishadal Raj College, Mahishadal, West Bengal, 721628, India.

References

1. Weng, ZK: Channel coordination and quantity discounts. Manag. Sci. **41**, 1509–1522 (1999)
2. Munson, LC, Rosenblatt, JM: Coordinating a three level supply chain with quantity discounts. IIE Trans. **33**, 371–384 (2001)
3. Yang, CP, Wee, MH: An arborescent inventory model in a supply chain system. Prod. Plann. Control. **12**, 728–735 (2001)
4. Khouja, M: Optimizing inventory decisions in a multistage multi customer supply chain. Trans. Res. Part E. **39**, 193–208 (2003)
5. Yao, Y, Evers, PT, Dresner, ME: Supply chain integration in vendor-managed inventory. Decis. Support Syst. **43**, 663–674 (2007)
6. Chaharsooghi, SK, Heydari, J, Zegordi, SH: A reinforcement learning model for supply chain ordering management: an application to the beer game. Decis. Support Syst. **45**, 949–959 (2008)
7. Wang, WT, Wee, HM, Tsao, HSJ: Revisiting the note on supply chain integration in vendor-managed inventory. Decis. Support Syst. **48**, 419–420 (2010)
8. Salameh, MK, Jaber, MY: Economic production quantity model for items with imperfect quality. Int. J. Prod. Econ. **64**, 59–64 (2000)
9. Goyal, SK, Cardenas-Barron, LE: Note on: economic production quantity model for items with imperfect quality—a practical approach. Int. J. Prod. Econ. **77**, 85–87 (2002)
10. Yu, JCP, Wee, HM, Chen, JM: Optimal ordering policy for a deteriorating item with imperfect quality and partial back ordering. J. Chin. Inst. Ind. Eng. **22**, 509–520 (2005)
11. Liu, JJ, Yang, P: Optimal lot-sizing in an imperfect production system with homogeneous reworkable jobs. Eur. J. Oper. Res. **91**, 517–527 (1996)
12. Panda, D, Maiti, M: Multi-item inventory models with price dependent demand under flexibility and reliability consideration and imprecise space constraint: a geometric programming approach. Math. Comput. Model. **49**, 1733–1749 (2009)
13. Ma, W-N, Gong, D-C, Lin, GC: An optimal production cycle time for imperfect production process with scrap. Math. Comput. Model. **52**, 724–737 (2010)
14. Sana, SS: An economic production lot size model in an imperfect production system. Eur. J. Oper. Res. **201**, 158–170 (2010)
15. Sana, SS: A production-inventory model of imperfect quality products in a three-layer supply chain. Decis. Support Syst. **50**, 539–547 (2011)
16. Hartely, RV: Operations Research - a Managerial Emphasis, Good Year Publishing Company, California, pp. 315–317 (1976)
17. Sarma, KVS: A deterministic inventory model with two levels of storage and an optimum release rule. Opsearch **20**, 175–180 (1983)
18. Dave, U: On the EOQ models with two levels of storage. Opsearch. **25**, 190–196 (1988)
19. Goswami, A, Chaudhuri, KS: An economic order quantity model for items with two level of storage for a linear trend in demand. J. Oper. Res. Soc. **43**, 157–167 (1992)
20. Pakkala, TPM, Achary, KK: Discrete time inventory model for deteriorating items with two warehouses. Opsearch **29**, 90–103 (1992)
21. Bhunia, AK, Maiti, M: A two warehouses inventory model for deteriorating items with a linear trend in demand and shortages. Journal of Operational Research society. **49**, 287–292 (2007)

22. Benkherouf, L: A deterministic order level inventory model for deteriorating items with two storage facilities. Int. J. Prod. Econ. **48**, 167–175 (1997)
23. Zhou, YW: An optimal EOQ model for deteriorating items with two warehouses and time varying demand. Mathematica Applicata. **10**, 19–23 (1998)
24. Kar, SK, Bhunia, AK, Maiti, M: Deterministic inventory model with levels of storage, a linear trend in demand and a fixed time horizon. Comp. Oper. Res. **28**, 1315–1331 (2001)
25. Chung, K, Huang, T: The optimal retailer's ordering policies for deteriorating items with limited storage capacity under trade credit financing. Int. J. Prod. Econ. **106**, 127–146 (2007)
26. Dey, JK, Mondal, SK, Maiti, M: Two storage inventory problem with dynamic demand and interval valued lead-time over finite time horizon under inflation and time-value of money. Eur. J. Oper. Res. **185**, 170–194 (2008)
27. Liang, Y, Zhou, F: A two-warehouse inventory model for deteriorating items under conditionally permissible delay in payment. Appl. Math. Model. **35**, 2221–2231 (2011)
28. Hariga, M: Inventory models for multi-warehouse systems under fixed and flexible space leasing contracts. Comput. Ind. Eng. **61**, 744–751 (2011)
29. Dubois, D, Prade, H: Rough fuzzy sets and fuzzy rough sets. Int. J. Gen. Syst. **17**, 91–208 (1990)
30. Morsi, NN, Yakout, MM: Axiomatics for fuzzy rough sets. Fuzzy Set. Syst. **100**, 327–342 (1998)
31. Radzikowska, AM, Kerre, EE: A comparative study of rough sets. Fuzzy Set. Syst. **126**, 137–155 (2002)
32. Xu, J, Zhou, X: Fuzzy-Like Multiple Objective Decision Making. Springer, Berlin (2009)
33. Liu, G, Sai, Y: Invertible approximation operators of generalized rough sets and fuzzy rough sets. Inf. Sci. **180**, 2221–2229 (2010)
34. Chen, D, Kwong, S, He, Q, Wang, H: Geometrical interpretation and applications of membership functions with fuzzy rough set. Fuzzy Set and Syst. **193**, 122–135 (2012)

7

Subadditivity of chance measure

Yongchao Hou

Correspondence: hyc@chu.edu.cn
Department of Mathematical Sciences, Chaohu University, Anhui 238000, China

Abstract

Chance theory is a mathematical methodology for dealing with indeterminacy phenomena involving uncertainty and randomness. In this paper, some properties of chance space are investigated. Based on this, the subadditivity theorem, null-additivity theorem, and asymptotic theorem of chance measure are proved.

Keywords: Uncertainty theory; Chance theory; Chance measure; Subadditivity

Introduction

Uncertainty theory founded by Liu [1] in 2007 is a branch of axiomatic mathematics based on normality, duality, subadditivity, and product axioms. After that, many researchers widely studied the uncertainty theory and made significative progress. Liu [1] presented the concept of uncertain variable and uncertainty distribution. Then, a sufficient and necessary condition of uncertainty distribution was proved by Peng and Iwamura [2] in 2010. In addition, a measure inversion theorem was proposed by Liu [3] from which the uncertain measures of some events can be calculated via the uncertainty distribution. After proposing the concept of independence [4], Liu [3] presented the operational law of uncertain variables. In order to sort uncertain variables, Liu [3] proposed the concept of expected value of uncertain variable. A useful formula was presented by Liu and Ha [5] to calculate the expected values of monotone functions of uncertain variables. Based on the expected value, Liu [1] presented the concepts of variance, moments, and distance of uncertain variables. In order to characterize the uncertainty of uncertain variables, Liu [4] proposed the concept of entropy in 2009. Dai and Chen [6] verified the positive linearity of entropy and presented some formulas for calculating the entropy of monotone function of uncertain variables. Chen and Dai [7] discussed the maximum entropy principle for selecting the uncertainty distribution that has maximum entropy and satisfies the prescribed constraints. In order to make an extension of entropy, Chen et al. [8] proposed a concept of cross-entropy for comparing an uncertainty distribution against a reference uncertainty distribution. Liu [9] introduced a paradox of stochastic finance theory based on uncertainty theory and uncertain differential equation. In addition, an uncertain integral was proposed by Chen and Ralescu [10] presented with respect to the general Liu process.

In 2013, Liu [11] proposed chance theory by giving the concepts of uncertain random variable and chance measure in order to describe the situation that uncertainty and randomness appear in a system. Some related concepts of uncertain random variables such as chance distribution, expected value, and variance were also presented by Liu [11].

As an important contribution to chance theory, Liu [12] presented an operational law of uncertain random variables. After that, uncertain random variables were discussed widely. Yao and Gao [13] provided a law of large numbers for uncertain random variables. Gao and Yao [14] gave some concepts and theorems of uncertain random process. In addition, Yao and Gao [13] proposed an uncertain random process as a generalization of both stochastic process and uncertain process. As applications of chance theory, Liu [12] proposed uncertain random programming. Uncertain random risk analysis was presented by Liu and Ralescu [15]. Besides, chance theory was applied into many fields, and many achievements were obtained, such as uncertain random reliability analysis [16], uncertain random logic [17], uncertain random graph [18], and uncertain random network [18].

In this paper, some properties of chance space are investigated. Based on this, the subadditivity theorem, null-additivity theorem, and asymptotic theorem of chance measure are proposed.

Preliminary

As a branch of axiomatic mathematics, uncertainty theory aims to deal with human uncertainty. In this section, we will provide a brief introduction to uncertain variables and uncertain random variables, which will be used throughout this paper.

Uncertain variables

Definition 1. (Liu [1]) Let Γ be a non-empty set and $\mathcal{L}$ be a σ-algebra on Γ. Each element in $\mathcal{L}$ is called an event. A set function $\mathcal{M}$ from $\mathcal{L}$ to $[0,1]$ is called an uncertain measure if it satisfies the following axioms:

Axiom 1. (Normality Axiom) $\mathcal{M}\{\Gamma\} = 1$ for the universal set Γ.

Axiom 2. (Duality Axiom) $\mathcal{M}\{\Lambda\} + \mathcal{M}\{\Lambda^c\} = 1$ for any event Λ.

Axiom 3. (Subadditivity Axiom) For every countable sequence of events $\Lambda_1, \Lambda_2, \cdots$, we have

$$\mathcal{M}\left\{\bigcup_{i=1}^{\infty} \Lambda_i\right\} \leq \sum_{i=1}^{\infty} \mathcal{M}\{\Lambda_i\}.$$

The triplet $(\Gamma, \mathcal{L}, \mathcal{M})$ *is called an uncertainty space.* In 2009, Liu [4] defined product uncertain measure via the fourth axiom of uncertainty theory.

Axiom 4. (Product Axiom) Let $(\Gamma_k, \mathcal{L}_k, \mathcal{M}_k)$ be uncertainty spaces for $k = 1, 2, \cdots$ Then, the product uncertain measure $\mathcal{M}$ is an uncertain measure satisfying

$$\mathcal{M}\left\{\prod_{k=1}^{\infty} \Lambda_k\right\} = \bigwedge_{k=1}^{\infty} \mathcal{M}_k\{\Lambda_k\}$$

where Λ_k are arbitrarily chosen events from $\mathcal{L}_k$ for $k = 1, 2, \cdots$, respectively.

An uncertain variable is a real-valued function on an uncertainty space, which is defined as follows.

Definition 2. (Liu [1]) Let $(\Gamma, \mathcal{L}, \mathcal{M})$ be an uncertainty space. An uncertain variable is a measurable function from an uncertainty space $(\Gamma, \mathcal{L}, \mathcal{M})$ to the set of real numbers, i.e., for any Borel set B of real numbers, the set $\xi^{-1}(B) = \{\gamma \in \Gamma \mid \xi(\gamma) \in B\}$ is an event.

In order to describe uncertain variables, a concept of uncertainty distribution was introduced by Liu [1].

Definition 3. (Liu [1]) The uncertainty distribution Φ of an uncertain variable ξ is defined by

$$\Phi(x) = \mathcal{M}\{\xi \leq x\}$$

for any real number x.

Definition 4. (Liu [4]) The uncertain variables $\xi_1, \xi_2, \cdots, \xi_n$ are said to be independent if

$$\mathcal{M}\left\{\bigcap_{i=1}^{n}(\xi_i \in B_i)\right\} = \bigwedge_{i=1}^{n} \mathcal{M}\{\xi_i \in B_i\}$$

for any Borel sets $B_1, B_2, \cdots, B_n$ of real numbers.

Theorem 1. *(Liu [1]) Assume that $\xi_1, \xi_2, \cdots, \xi_n$ are independent uncertain variables with regular uncertainty distributions $\Phi_1, \Phi_2, \cdots, \Phi_n$, respectively. If $f(x_1, x_2, \cdots, x_n)$ is strictly increasing with respect to $x_1, x_2, \cdots, x_m$ and strictly decreasing with respect to $x_{m+1}, x_{m+2}, \cdots, x_n$, then $\xi = f(\xi_1, \xi_2, \cdots, \xi_n)$ is an uncertain variable with inverse uncertainty distribution*

$$\Psi^{-1}(\alpha) = f(\Phi_1^{-1}(\alpha), \cdots, \Phi_m^{-1}(\alpha), \Phi_{m+1}^{-1}(1-\alpha), \cdots, \Phi_n^{-1}(1-\alpha)).$$

To represent the average value of an uncertain variable in the sense of uncertain measure, the expected value is defined as follows.

Definition 5. (Liu [1]) Let ξ be an uncertain variable. Then, the expected value of ξ is defined by

$$E[\xi] = \int_0^{+\infty} \mathcal{M}\{\xi \geq r\}\mathrm{d}r - \int_{-\infty}^{0} \mathcal{M}\{\xi \leq r\}\mathrm{d}r$$

provided that at least one of the two integrals is finite.

Definition 6. (Liu [1]) Let ξ be an uncertain variable with uncertainty distribution Φ. If the expected value exists, then

$$E[\xi] = \int_0^{+\infty} (1 - \Phi(x))\mathrm{d}x - \int_{-\infty}^{0} \Phi(x)\mathrm{d}x. \tag{1}$$

For calculating the expected value by inverse uncertainty distribution, Liu and Ha [5] proved the following theorem.

Theorem 2. *(Liu and Ha [5]) Assume that* $\xi_1, \xi_2, \cdots, \xi_n$ *are independent uncertain variables with regular uncertainty distributions* $\Phi_1, \Phi_2, \cdots, \Phi_n$*, respectively. If* $f(x_1, x_2, \cdots, x_n)$ *is strictly increasing with respect to* $x_1, x_2, \cdots, x_m$ *and strictly decreasing with respect to* $x_{m+1}, x_{m+2}, \cdots, x_n$*, then the uncertain variable* $\xi = f(\xi_1, \xi_2, \cdots, \xi_n)$ *has an expected value*

$$E[\xi] = \int_0^1 f(\Phi_1^{-1}(\alpha), \cdots, \Phi_m^{-1}(\alpha), \Phi_{m+1}^{-1}(1-\alpha), \cdots, \Phi_n^{-1}(1-\alpha))\mathrm{d}\alpha. \tag{2}$$

Uncertain random variables

In 2013, Liu [11] first proposed chance theory, which is a mathematical methodology for modeling complex systems with both uncertainty and randomness, including chance measure, uncertain random variable, chance distribution, operational law, expected value, and so on. The chance space is referred to the product $(\Gamma, \mathcal{L}, \mathcal{M}) \times (\Omega, \mathcal{A}, \Pr)$, in which $(\Gamma, \mathcal{L}, \mathcal{M})$ is an uncertainty space and $(\Omega, \mathcal{A}, \Pr)$ is a probability space.

Definition 7. (Liu [11]) Let $(\Gamma, \mathcal{L}, \mathcal{M}) \times (\Omega, \mathcal{A}, \Pr)$ be a chance space, and let $\Theta \in \mathcal{L} \times \mathcal{A}$ be an event. Then, the chance measure of Θ is defined as

$$\mathrm{Ch}\{\Theta\} = \int_0^1 \Pr\{\omega \in \Omega \mid \mathcal{M}\{\gamma \in \Gamma | (\gamma, \omega) \in \Theta\} \geq r\}\mathrm{d}r.$$

Notation: For a real number r, the set $\Theta_r = \{\omega \in \Omega \mid \mathcal{M}\{\gamma \in \Gamma | (\gamma, \omega) \in \Theta\} \geq r\}$ is a subset of Ω but not necessarily an event in $\mathcal{A}$. In this case, $\Pr\{\Theta_r\}$ is assigned by

$$\Pr\{\Theta_r\} = \begin{cases} \inf\limits_{A \in \mathcal{A}, A \supset \Theta_r} \Pr\{A\}, & \text{if } \inf\limits_{A \in \mathcal{A}, A \supset \Theta_r} \Pr\{A\} < 0.5 \\ \sup\limits_{A \in \mathcal{A}, A \subset \Theta_r} \Pr\{A\}, & \text{if } \sup\limits_{A \in \mathcal{A}, A \subset \Theta_r} \Pr\{A\} > 0.5 \\ 0.5, & \text{otherwise} \end{cases} \tag{3}$$

Liu [11] proved that a chance measure satisfies normality, duality, and monotonicity properties, that is

(a) $\mathrm{Ch}\{\Gamma \times \Omega\} = 1, \mathrm{Ch}\{\emptyset\} = 0$;
(b) $\mathrm{Ch}\{\Theta\} + \mathrm{Ch}\{\Theta^c\} = 1$ for any event Θ;
(c) $\mathrm{Ch}\{\Theta_1\} \leq \mathrm{Ch}\{\Theta_2\}$ for any event $\Theta_1 \subset \Theta_2$.

First, we give an equivalent definition of $\Pr\{\cdot\}$ in (3).

Lemma 1. *Let* $(\Gamma, \mathcal{A}, \mathcal{M}) \times (\Omega, \mathcal{A}, \Pr)$ *be a chance space, and let* $\Theta \in \mathcal{L} \times \mathcal{A}$ *be an event. Denote that* $\Theta_B = \{\omega \in \Omega \mid \mathcal{M}\{\gamma \in \Gamma | (\gamma, \omega) \in \Theta\} \in B\}$ *for any Borel set B. Then, we have*

$$\Pr\{\Theta_B\} = \inf_{A \in \mathcal{A}, A \supset \Theta_B} \Pr\{A\} \wedge \left(\sup_{A \in \mathcal{A}, A \subset \Theta_B} \Pr\{A\} \vee 0.5 \right) \tag{4}$$

$$\Pr\{\Theta_B\} = \sup_{A \in \mathcal{A}, A \subset \Theta_B} \Pr\{A\} \vee \left(\inf_{A \in \mathcal{A}, A \supset \Theta_B} \Pr\{A\} \wedge 0.5 \right) \tag{5}$$

Proof. The argument breaks down into three cases.

Case 1: $\inf\limits_{A\in\mathcal{A},A\supset\Theta_B} \Pr\{A\} < 0.5$. In this case, note that $\left(\sup\limits_{A\in\mathcal{A},A\subset\Theta_B} \Pr\{A\} \vee 0.5\right) \geq 0.5$. Then, we have

$$\inf_{A\in\mathcal{A},A\supset\Theta_B} \Pr\{A\} \wedge \left(\sup_{A\in\mathcal{A},A\subset\Theta_B} \Pr\{A\} \vee 0.5\right) = \inf_{A\in\mathcal{A},A\supset\Theta_B} \Pr\{A\}.$$

Case 2: $\sup\limits_{A\in\mathcal{A},A\subset\Theta_B} \Pr\{A\} > 0.5$. Then, we have

$$\begin{aligned}
&\inf_{A\in\mathcal{A},A\supset\Theta_B} \Pr\{A\} \wedge \left(\sup_{A\in\mathcal{A},A\subset\Theta_B} \Pr\{A\} \vee 0.5\right)\\
&= \inf_{A\in\mathcal{A},A\supset\Theta_B} \Pr\{A\} \wedge \sup_{A\in\mathcal{A},A\subset\Theta_B} \Pr\{A\}\\
&= \sup_{A\in\mathcal{A},A\subset\Theta_B} \Pr\{A\}.
\end{aligned}$$

Case 3: Otherwise. It means $\inf\limits_{A\in\mathcal{A},A\supset\Theta_B} \Pr\{A\} \geq 0.5$ and $\sup\limits_{A\in\mathcal{A},A\subset\Theta_B} \Pr\{A\} \leq 0.5$. Then, we have

$$\begin{aligned}
&\inf_{A\in\mathcal{A},A\supset\Theta_B} \Pr\{A\} \wedge \left(\sup_{A\in\mathcal{A},A\subset\Theta_B} \Pr\{A\} \vee 0.5\right)\\
&= \inf_{A\in\mathcal{A},A\supset\Theta_B} \Pr\{A\} \wedge 0.5 = 0.5.
\end{aligned}$$

The equality (4) is proved. Note that

$$\begin{aligned}
&\inf_{A\in\mathcal{A},A\supset\Theta_B} \Pr\{A\} \wedge \left(\sup_{A\in\mathcal{A},A\subset\Theta_B} \Pr\{A\} \vee 0.5\right)\\
&= \left(\inf_{A\in\mathcal{A},A\supset\Theta_B} \Pr\{A\} \wedge \sup_{A\in\mathcal{A},A\subset\Theta_B} \Pr\{A\}\right) \vee \left(\inf_{A\in\mathcal{A},A\supset\Theta_B} \Pr\{A\} \wedge 0.5\right)\\
&= \sup_{A\in\mathcal{A},A\subset\Theta_B} \Pr\{A\} \vee \left(\inf_{A\in\mathcal{A},A\supset\Theta_B} \Pr\{A\} \wedge 0.5\right).
\end{aligned}$$

Hence, the equality (5) holds. □

Lemma 2. *Let* $(\Gamma, \mathcal{L}, \mathcal{M}) \times (\Omega, \mathcal{A}, \Pr)$ *be a chance space, and let* $\Theta \in \mathcal{L} \times \mathcal{A}$ *be an event. Denote that* $\Theta_B = \{\omega \in \Omega \mid \mathcal{M}\{\gamma \in \Gamma | (\gamma, \omega) \in \Theta\} \in B\}$ *for any Borel set B. Then, we have*

$$\Pr\{\Theta_B\} + \Pr\{\Theta_{B^c}\} = 1$$

Proof. According to the equivalent definition of $\Pr\{\cdot\}$ in Lemma 1, we have

$$\begin{aligned}
\Pr\{\Theta_{B^c}\} &= \inf_{A\in\mathcal{A},A\supset\Theta_{B^c}} \Pr\{A\} \wedge \left(\sup_{A\in\mathcal{A},A\subset\Theta_{B^c}} \Pr\{A\} \vee 0.5 \right) \\
&= \inf_{A\in\mathcal{A},A^c\subset\Theta_B} \Pr\{A\} \wedge \left(\sup_{A\in\mathcal{A},A^c\subset\Theta_B} \Pr\{A\} \vee 0.5 \right) \\
&= \inf_{A\in\mathcal{A},A\subset\Theta_B} \Pr\{A^c\} \wedge \left(\sup_{A\in\mathcal{A},A\subset\Theta_B} \Pr\{A^c\} \vee 0.5 \right) \\
&= \inf_{A\in\mathcal{A},A\subset\Theta_B} (1-\Pr\{A\}) \wedge \left(\sup_{A\in\mathcal{A},A\subset\Theta_B} (1-\Pr\{A\}) \vee 0.5 \right) \\
&= \left(1 - \sup_{A\in\mathcal{A},A\subset\Theta_B} \Pr\{A\}\right) \wedge \left(\left(1 - \inf_{A\in\mathcal{A},A\subset\Theta_B} \Pr\{A\}\right) \vee 0.5 \right) \\
&= \left(1 - \sup_{A\in\mathcal{A},A\subset\Theta_B} \Pr\{A\}\right) \wedge \left(1 - \inf_{A\in\mathcal{A},A\subset\Theta_B} \Pr\{A\} \wedge 0.5 \right) \\
&= 1 - \sup_{A\in\mathcal{A},A\subset\Theta_B} \Pr\{A\} \vee \left(\inf_{A\in\mathcal{A},A\supset\Theta_B} \Pr\{A\} \wedge 0.5 \right) \\
&= 1 - \Pr\{\Theta_B\}
\end{aligned}$$

The lemma is proved. □

Lemma 3. *Let $(\Gamma, \mathcal{L}, \mathcal{M}) \times (\Omega, \mathcal{A}, \Pr)$ be a chance space, and let $\Theta_1, \Theta_2 \in \mathcal{L} \times \mathcal{A}$ be two events satisfying $\Theta_1 \subset \Theta_2$. Then, we have*

$$\Pr\{\Theta_1\} \le \Pr\{\Theta_2\}. \tag{6}$$

Proof. $\Theta_1 \subset \Theta_2$, we have

$$\inf_{A\in\mathcal{A},A\supset\Theta_1} \Pr\{A\} \le \inf_{A\in\mathcal{A},A\supset\Theta_2} \Pr\{A\},$$

$$\sup_{A\in\mathcal{A},A\subset\Theta_1} \Pr\{A\} \le \sup_{A\in\mathcal{A},A\subset\Theta_2} \Pr\{A\}.$$

According to Lemma 1, we have

$$\begin{aligned}
\Pr\{\Theta_1\} &= \inf_{A\in\mathcal{A},A\supset\Theta_1} \Pr\{A\} \wedge \left(\sup_{A\in\mathcal{A},A\subset\Theta_1} \Pr\{A\} \vee 0.5 \right) \\
&\le \inf_{A\in\mathcal{A},A\supset\Theta_2} \Pr\{A\} \wedge \left(\sup_{A\in\mathcal{A},A\subset\Theta_2} \Pr\{A\} \vee 0.5 \right) = \Pr\{\Theta_2\}.
\end{aligned}$$

The lemma is proved. □

Theorem 3. *(Subadditivity Theorem) The chance measure is subadditive. That is, for any countable sequence of events $\Theta_1, \Theta_2, \cdots$, we have*

$$\mathrm{Ch}\left\{\bigcup_{i=1}^{\infty} \Theta_i\right\} \le \sum_{i=1}^{\infty} \mathrm{Ch}\{\Theta_i\}.$$

Proof. For each ω, it follows from the subadditivity of uncertain measure that

$$\mathcal{M}\left\{\gamma \in \Gamma | (\gamma, \omega) \in \bigcup_{i=1}^{\infty} \Theta_i\right\} \leq \sum_{i=1}^{\infty} \mathcal{M}\{\gamma \in \Gamma | (\gamma, \omega) \in \Theta_i\}.$$

Thus, for any real number r, we have

$$\left\{\omega \in \Omega | \mathcal{M}\left\{\gamma \in \Gamma | (\gamma, \omega) \in \bigcup_{i=1}^{\infty} \Theta_i\right\} \geq r\right\} \subset \left\{\omega \in \Omega | \sum_{i=1}^{\infty} \mathcal{M}\{\gamma \in \Gamma | (\gamma, \omega) \in \Theta_i\} \geq r\right\}$$

According to Lemma 3, we have

$$\Pr\left\{\omega \in \Omega | \mathcal{M}\left\{\gamma \in \Gamma | (\gamma, \omega) \in \bigcup_{i=1}^{\infty} \Theta_i\right\} \geq r\right\} \leq \Pr\left\{\omega \in \Omega | \sum_{i=1}^{\infty} \mathcal{M}\{\gamma \in \Gamma | (\gamma, \omega) \in \Theta_i\} \geq r\right\}$$

By the definition of chance measure, we get

$$\begin{aligned}
\mathrm{Ch}\left\{\bigcup_{i=1}^{\infty} \Theta_i\right\} &= \int_0^1 \Pr\left\{\omega \in \Omega | \mathcal{M}\left\{\gamma \in \Gamma | (\gamma, \omega) \in \bigcup_{i=1}^{\infty} \Theta_i\right\} \geq r\right\} \mathrm{d}r \\
&\leq \int_0^1 \Pr\left\{\omega \in \Omega | \sum_{i=1}^{\infty} \mathcal{M}\{\gamma \in \Gamma | (\gamma, \omega) \in \Theta_i\} \geq r\right\} \mathrm{d}r \\
&\leq \int_0^{+\infty} \Pr\left\{\omega \in \Omega | \sum_{i=1}^{\infty} \mathcal{M}\{\gamma \in \Gamma | (\gamma, \omega) \in \Theta_i\} \geq r\right\} \mathrm{d}r \\
&= \sum_{i=1}^{\infty} \int_0^{+\infty} \Pr\{\omega \in \Omega | \mathcal{M}\{\gamma \in \Gamma | (\gamma, \omega) \in \Theta_i\} \geq r\} \mathrm{d}r \\
&= \sum_{i=1}^{\infty} \int_0^1 \Pr\{\omega \in \Omega | \mathcal{M}\{\gamma \in \Gamma | (\gamma, \omega) \in \Theta_i\} \geq r\} \mathrm{d}r \\
&= \sum_{i=1}^{\infty} \mathrm{Ch}\{\Theta_i\}.
\end{aligned}$$

That is, the chance measure is subadditive. □

Null-additivity is a direct deduction from the above theorem. In fact, a more general theorem can be proved as follows.

Theorem 4. *Let* $(\Gamma, \mathcal{L}, \mathcal{M}) \times (\Omega, \mathcal{A}, \Pr)$ *be a chance space and* $\Theta_1, \Theta_2, \cdots$ *be a sequence of events with* $\mathrm{Ch}\{\Theta_i\} \to 0$ *as* $i \to \infty$*. Then, for any event* Θ*, we have*

$$\lim_{i\to\infty} \mathrm{Ch}\{\Theta \cup \Theta_i\} = \lim_{i\to\infty} \mathrm{Ch}\{\Theta \setminus \Theta_i\} = \mathrm{Ch}\{\Theta\}.$$

Proof. By using the monotonicity and subadditivity of chance measure, we have

$$\mathrm{Ch}\{\Theta\} \leq \mathrm{Ch}\{\Theta \cup \Theta_i\} \leq \mathrm{Ch}\{\Theta\} + \mathrm{Ch}\{\Theta_i\} \tag{7}$$

for each i. For $\mathrm{Ch}\{\Theta_i\} \to 0$ as $i \to \infty$, we get $\mathrm{Ch}\{\Theta \cup \Theta_i\} \to \mathrm{Ch}\{\Theta\}$. Note that $\Theta \setminus \Theta_i \subset \Theta \subset ((\Theta \setminus \Theta_i) \cup \Theta_i)$. We have

$$\mathrm{Ch}\{\Theta \setminus \Theta_i\} \leq \mathrm{Ch}\{\Theta\} \leq \mathrm{Ch}\{\Theta \setminus \Theta_i\} + \mathrm{Ch}\{\Theta_i\}. \tag{8}$$

Hence, $\lim_{i\to\infty} \mathrm{Ch}\{\Theta \setminus \Theta_i\} = \mathrm{Ch}\{\Theta\}$. □

Remark. From the above theorem, we know that the chance measure is null-additive. That means $\text{Ch}\{\Theta_1 \cup \Theta_2\} = \text{Ch}\{\Theta_1\} + \text{Ch}\{\Theta_2\}$ if either $\text{Ch}\{\Theta_1\} = 0$ or $\text{Ch}\{\Theta_2\} = 0$. □

Theorem 5. *(Asymptotic Theorem) Let* $(\Gamma, \mathcal{L}, \mathcal{M}) \times (\Omega, \mathcal{A}, \text{Pr})$ *be a chance space. For any events* $\Theta_1, \Theta_2, \cdots$, *we have*

$$\lim_{i\to\infty} \text{Ch}\{\Theta_i\} > 0, \quad \textit{if } \Theta_i \uparrow \Gamma \times \Omega, \tag{9}$$

$$\lim_{i\to\infty} \text{Ch}\{\Theta_i\} < 1, \quad \textit{if } \Theta_i \downarrow \emptyset. \tag{10}$$

Proof. Assume $\Theta_i \uparrow \Gamma \times \Omega$. Since $\Gamma \times \Omega = \cup_i \Theta_i$, it follows from the subadditivity of chance measure that

$$1 = \text{Ch}\{\Gamma \times \Omega\} \le \sum_{i=1}^{\infty} \text{Ch}\{\Theta_i\}.$$

Note that $\text{Ch}\{\Theta_i\}$ is increasing with respect to i. We get $\lim_{i\to\infty} \text{Ch}\{\Theta_i\} > 0$. If $\Theta_i \downarrow \emptyset$, then $\Theta_i^c \uparrow \Gamma \times \Omega$. By using inequality (9) and the duality of chance measure, we have

$$\lim_{i\to\infty} \text{Ch}\{\Theta_i\} = 1 - \lim_{i\to\infty} \text{Ch}\{\Theta_i^c\} < 1.$$

The theorem is proved. □

Competing interests
This paper proposed several properties of chance space. Besides, the subadditivity theorem, null-additivity theorem, and asymptotic theorem of chance measure were proved.

Acknowledgements
This work was supported by the National Natural Science Foundation of China Grant No.61273044 and University Science Research Project of Anhui Province No. KJ2011B105.

References
1. Liu, B: Uncertainty Theory, 2nd Edition. Springer, Berlin (2007)
2. Peng, Z, Iwamura, K: A sufficient and necessary condition of uncertainty distribution, J. Interdisciplin. Math. **13**(3), 277–285 (2010)
3. Liu, B: Uncertainty Theory: A Branch of Mathematics for Modeling Human Uncertainty. Springer, Berlin (2010)
4. Liu, B: Some research problems in uncertainty theory. J. Uncertain Syst. **3**(1), 3–10 (2009)
5. Liu, Y, Ha, M: Expected value of function of uncertain variables. J. Uncertain. Syst. **4**(3), 181–186 (2010)
6. Dai, W, Chen, X: Entropy of function of uncertain variables. Math. Comput. Modell. **55**(3–4), 754–760 (2012)
7. Chen, X, Dai, W: Maximum entropy principle for uncertain variables. Int. J. Fuzzy. Syst. **13**(3), 232–236 (2011)
8. Chen, X, Kar, S, Ralescu, D: Cross-entropy measure of uncertain variables. Inf. Sci. **201**, 53–60 (2012)
9. Liu, B: Toward uncertain finance theory. J. Uncertain. Anal. Appl. **1**(1) (2013). doi:10.1186/2195-5468-1-1
10. Chen, X, Ralescu, D: Liu process and uncertain calculus. J. Uncertain. Anal. Appl. **1**(3) (2013). doi:10.1186/2195-5468-1-3
11. Liu, Y: Uncertain random variables: a mixture of uncertainty and randomness. Soft Comp. **17**(4), 625–634 (2013)
12. Liu, Y: Uncertain random programming with applications. Fuzzy Optim. Decis. Ma. **12**(2), 153–169 (2013)
13. Yao, K, Gao, J: Law of large numbers for uncertain random variables. http://orsc.edu.cn/online/120401.pdf (2012). Accessed 1 April 2012
14. Gao, J, Yao, K: Some concepts and theorems of uncertain random process. Int. J. Intell. Syst. (2014, in press)
15. Liu, Y, Ralescu, D: Risk index in uncertain random risk analysis. Int. J. Uncertain. Fuzz. (2014, in press)
16. Wen, M, Kang, R: Reliability analysis in uncertain random system. http://orsc.edu.cn/online/120419.pdf (2012). Accessed 19 April 2012
17. Liu, Y: Uncertain random logic and uncertain random entailment. Technical Report (2013)
18. Liu, B: Uncertain random graph and uncertain random network. J. Uncertain Syst. **8**(1), 3–12 (2014)

8

Solving multi-choice multi-objective transportation problem: a utility function approach

Gurupada Maity and Sankar Kumar Roy*

*Correspondence:
sankroy2006@gmail.com
Department of Applied Mathematics with Oceanology and Computer Programming, Vidyasagar University, Midnapore, West Bengal 721102, India

Abstract

This paper explores the study of multi-choice multi-objective transportation problem (MCMTP) under the environment of utility function approach. MCMTP is converted to multi-objective transportation problems (MOTP) by transforming the multi-choice parameters like cost, demand, and supply to real-valued parameters. A general transformation procedure using binary variables is illustrated to reduce MCMTP into MOTP. Most of the MOTP are solved by goal programming (GP) approach. Using GP, the solution of MOTP may not be satisfied all the time by the decision maker (DM) when the proposed problem contains interval-valued aspiration level. To overcome this difficulty, here we propose the approaches of revised multi-choice goal programming (RMCGP) and utility function into the MOTP and then compared the solution between them. Finally, numerical examples are presented to show the feasibility and usefulness of our paper.

Keywords: Transportation problem; Multi-choice programming; Multi-objective decision making; Goal programming; Multi-choice programming; Utility function

Introduction

The transportation problem is the central nerve system to keep the balance in economical world from ancient day until today. In earlier days, transportation problem developed with the assumption that the supply, demand, and cost parameters are exactly known. But in real-life applications, all the parameters of the transportation problem are not generally defined precisely. Keeping this point of view, in this paper, we have incorporated with multi-choice multi-objective transportation problem (MCMTP) considering the parameters of transportation problem as multi-choice type.

Instead of single choice, if there may be several choices involved associated with the transportation parameters like cost, supply, or demand, then the decision maker is confused to select the proper choice for these parameters. In this circumstances, the study of transportation problem creates a new direction which is called multi-choice multi-objective transportation problem. Chang [1] proposed a multi-choice goal programming approach to solve the mathematical programming. Again in the subsequent year, Chang [2] proposed another multi-choice goal programming approach in revised form. Though the multi-choice concept discussed in both the papers of Chang [1,2] is totally related to the goals of objective functions, recently, Mahapatra et al. [3] and Roy et al. [4] discussed the multi-choice stochastic transportation problem involving extreme value distribution and exponential distribution in which the multi-choice concept involved only in the cost parameters.

In this paper, we have designed a general transformation technique to reduce the multi-choice parameters like cost, time, and demand to single-valued parameters. Using this technique, the MCMTP problem can be reduced to MOTP.

Goal programming (GP), an analytical approach, is devised to address the decision-making problem where targets have been assigned to all objective functions. The objective functions are conflicting and commensurable to each other, and the DM is interested to minimize the non-achievement of the corresponding goals. In other words, the DM derived an optimal solution with this strategy of GP which is satisfactory. However, using GP, the solution procedure for MOTP has some limitation. The main limitation behind GP is that the priority of goals for the DM is not easily considered. It seems far from reality. In the recent past, the notion of utility function is introduced by several researchers such as [5], Yu et al. [6], and Podinovski [7]. Recently, multi-choice goal programming (MCGP) has been proposed by Chang [1,2] to solve multi-objective decision-making problems with aspiration level. However, to the best of our knowledge, no works have been done on utility function to solve MOTP with the DM's preferences. The main motivation of this study is to investigate the better solution of MOTP by using utility function approach and then compare the solution to other methods such as GP and RMCGP.

Charnes et al. [8] introduced the concept of GP further developed by several researchers such as Charnes and Cooper [9], Lee [10], Ignizio [11], Tamiz et al. [12], and Romero [13]. In long back, the main concept of GP was to minimize the deviation between the achievement goals and the achievement levels. The mathematical model of multi-objective decision making (MODM) can be considered in the following form:

GP

$$\min \sum_{i=1}^{K} w_i |Z^i(X) - g_i|$$
$$\text{subject to} \quad x \in F \ \ (F \text{ is the feasible set})$$

where w_i are the weights attached to the deviation of the achievement function. $Z^i(X)$ is the ith objective function of the ith goal, and g_i is the aspiration level of the ith goal. $|Z^i(X) - g_i|$ represents the deviation of the ith goal. Later on, a modification on GP is provided and denoted as weighted goal programming (WGP) which can be displayed in the following form:

WGP

$$\min \sum_{i=1}^{K} w_i (d_i^+ + d_i^-)$$
$$\text{subject to} \quad Z^i(X) - d_i^+ + d_i^- = g_i$$
$$d_i^+ \geq 0, d_i^- \geq 0, \ \ i \in \{1, 2, \ldots, K\}$$
$$x \in F \ \ (F \text{ is the feasible set})$$

where d_i^+ and d_i^- are over and under achievements of the ith goal, respectively.

However, the conflicts of resources and the incompleteness of available information make it almost impossible for DMs to set the specific aspiration levels and choose the better decision. To overcome this situation, MCGP approach has been presented by Chang [1] with a new direction to solve MODM problem. In the next year, Chang [2] proposed the revised form of MCGP defined as RMCGP to solve MODM. The mathematical model of MODM using RMCGP is defined as follows:

RMCGP

$$\min \sum_{i=1}^{K} w_i(d_i^+ + d_i^-) + \alpha_i(e_i^+ + e_i^-)$$

$$\text{subject to } \; Z^i(X) - d_i^+ + d_i^- = y_i, \quad i = 1, 2, \ldots, K$$

$$y_i - e_i^+ + e_i^- = g_{i,\max}, \text{ or } g_{i,\min} \quad i = 1, 2, \ldots, K$$

$$g_{i,\min} \leq y_i \leq_{i,\max}, \quad i = 1, 2, \ldots, K$$

$$d_i^+, d_i^-, e_i^+, e_i^- \geq 0, \quad i = 1, 2, \ldots, K$$

$$x \in F \;\; (F \text{ is the feasible set})$$

where y_i is the continuous variable associated with ith goal which restricted between the upper ($g_{i,\max}$) and lower ($g_{i,\min}$) bounds, e_i^+ and e_i^- are positive and negative deviations attached to the ith goal of $|y_i - g_{i,\max}|$, and α_i is the weight attached to the sum of the deviations of $|y_i - g_{i,\max}|$; other variables are defined as in WGP.

The main motivation of this paper is to investigate the better solution of MCMTP by using utility function approach and then compare the solutions to other methods such as GP and RMCGP.

The remainder of the paper is organized as follows: In Section "Mathematical model", mathematical models are formulated for MOTP and MCMTP and solution procedures have been discussed with utility function approach. In the next section "Numerical examples", we demonstrate the usefulness of the proposed model with realistic examples. Finally, conclusion is presented regarding our consideration.

Mathematical model

The mathematical model of multi-objective transportation problem (MOTP) can be considered as follows:

Model 1

$$\min : Z^t = \sum_{i=1}^{m} \sum_{j=1}^{n} C_{ij}^t x_{ij}, \quad t = 1, 2, \ldots, K \tag{1}$$

$$\text{subject to } \sum_{j=1}^{n} x_{ij} \leq a_i, \quad i = 1, 2, \ldots, m \tag{2}$$

$$\sum_{i=1}^{m} x_{ij} \geq b_j, \quad j = 1, 2, \cdots, n \tag{3}$$

$$\text{and } \; x_{ij} \geq 0, \quad \forall \; i \text{ and } j \tag{4}$$

Here C_{ij}^t, a_i, b_j are the cost, supply, and demand parameters of tth objective function in MOTP, respectively, and $\sum_{i=1}^{m} a_i \geq \sum_{j=1}^{n} b_j$ is the feasibility condition. According to the nature of the problem, the decision maker has the right to choose the goals of the objective functions. Assuming that these goals are g_t $(t = 1, 2, \ldots, K)$ of K objective functions, the goals are defined as interval valued as $g_t = [g_{t,\min}, g_{t,\max}]$, $(t = 1, 2, \ldots, K)$.

In many real-life situations, the multiple choices in the transportation parameters like cost, demand, and source create complexities for the DM to make a decision. Multi-choice costs may occur due to several routes for transporting the goods. Due to weather condition or different seasons, the demands or the supply become multi-choices in nature. In the atmosphere of multi-choice transportation parameters, the mathematical model of MCMTP is defined as follows:

Model M1

$$\min : Z^t = \sum_{i=1}^{m}\sum_{j=1}^{n}\left(\tilde{C}_{ij}^{t1} \text{ or } \tilde{C}_{ij}^{t2} \text{ or } \cdots \text{ or } \tilde{C}_{ij}^{tr}\right) x_{ij}, \quad t = 1, 2, \ldots, K \tag{5}$$

$$\text{subject to} \quad \sum_{j=1}^{n} x_{ij} \le \tilde{a}_i^1 \text{ or } \tilde{a}_i^2 \text{ or } \cdots \text{ or } \tilde{a}_i^p, \quad i = 1, 2, \ldots, m \tag{6}$$

$$\sum_{i=1}^{m} x_{ij} \ge \tilde{b}_j^1 \text{ or } \tilde{b}_j^2 \text{ or } \cdots \text{ or } \tilde{b}_j^q, \quad j = 1, 2, \ldots, n \tag{7}$$

$$x_{ij} \ge 0, \ \forall \ i \text{ and } j. \tag{8}$$

Here $\left(\tilde{C}_{ij}^{t1} \text{ or } \tilde{C}_{ij}^{t2} \text{ or } \cdots \text{ or } \tilde{C}_{ij}^{tr}\right)$, $(\tilde{a}_i^1 \text{ or } \tilde{a}_i^2 \text{ or } \cdots \text{ or } \tilde{a}_i^p)$, and $\left(\tilde{b}_j^1 \text{ or } \tilde{b}_j^2 \text{ or } \cdots \text{ or } \tilde{b}_j^q\right)$ are the multi-choice cost, supply, and demand parameters for the tth objective function, respectively. In a objective function, respectively. In a transportation problem, the total demand should be less or equal to the total capacity of supply to get a feasible solution. In the present case for multi-choice of supply and demands, the information of total capacity of supply in the origins and demands in the destinations is not precisely calculated. So we have selected here the maximum possible supply in the origins and consequently the minimum demand in the destinations and then formulated the feasibility condition as $\sum_{i=1}^{m} \max\left\{\tilde{a}_i^1, \tilde{a}_i^2, \ldots, \tilde{a}_i^p\right\} \ge \sum_{j=1}^{n} \min\left\{\tilde{b}_j^1, \tilde{b}_j^2, \ldots, \tilde{b}_j^q\right\}$. This feasibility condition is the best possible wide range of feasible region of the MCMTP. However, the feasibility condition can be remodeled as per as decision maker's choice.

Transformation technique for multi-choice parameters like cost, supply, and demand to the equivalent form

When there are multiple choice of parameters such as cost, supply, and demand, we should select a single choice satisfying supply and demand restrictions. The selection of choices should be done in such a way that the whole problem is optimized. Introduction of binary variables is an important concept to select a choice in the problem.

If we have to choose one among t number of possibilities, then we use p number of binary variables where $2^{p-1} < t \le 2^p$. Let $t = {}^pC_0 + {}^pC_1 + {}^pC_2 + \cdots + {}^pC_d + k$, for some d satisfying $1 \le d \le p, 0 \le k < {}^pC_{d+1}$. Let us take p binary variables $z_j^1, z_j^2, \ldots, z_j^p$ to deduce a formula which will select one among the t values $c_j^1, c_j^2, \ldots, c_j^t$. Let us form a function with p binary variables, $f_0(z) = \left(z_j^1 z_j^2 \ldots z_j^p\right) c_j^1$ where $z = \left(z_j^1, z_j^2, \ldots, z_j^p\right)$. When each $z_j^i = 1$ for $i = 1, 2, \ldots p$, $f_0(z) = c_j^1$. Thus, $f_0(z) = c_j^1$, when $z_j^1 + z_j^2 + \cdots + z_j^p = p$. Again, let us assume a function

$$f_1(z) = \left(1 - z_j^1\right) z_j^2 \ldots z_j^p c_j^2 + \left(1 - z_j^2\right) z_j^1 z_j^3 \ldots z_j^p c_j^3 + \cdots + \left(1 - z_j^p\right) z_j^1 \ldots z_j^{p-1} c_j^{1+{}^pC_1}$$

When $z_j^1 + z_j^2 + \cdots + z_j^p = p - 1$, $f_1(z)$ gives output one of the following c_j^ts: $c_j^2, c_j^3, \ldots, c_j^{1+{}^pC_1}$. Similarly, we consider

$$f_2(z) = \left(1 - z_j^1\right)\left(1 - z_j^2\right) z_j^3 \ldots z_j^p c_j^{1+{}^pC_1+1} + \left(1 - z_j^1\right)\left(1 - z_j^3\right) z_j^2 z_j^4 \ldots z_j^p c_j^{1+{}^pC_1+2} + \cdots + \left(1 - z_j^1\right)\left(1 - z_j^p\right) z_j^2 \ldots z_j^{p-1} c_j^{1+{}^pC_1+(p-2)} + \left(1 - z_j^2\right)\left(1 - z_j^3\right) z_j^1 z_j^4 \ldots z_j^p c_j^{1+{}^pC_1+(p-2)+1} +$$
$$\vdots$$
$$+\left(1 - z_j^{p-1}\right)\left(1 - z_j^p\right) z_j^1 \ldots z_j^{p-2} c_j^{1+{}^pC_1+{}^pC_2}$$

When $z_j^1 + z_j^2 + \cdots + z_j^p = p - 2$, the above function $f_2(z)$ gives one among the following c_j^ts: $c_j^{1+{}^pC_1+1}, c_j^{1+{}^pC_1+2}, \ldots, c_j^{1+{}^pC_1+{}^pC_2}$.

Proceeding in the same manner, we find

$$f_d(z) = \left(1 - z_j^1\right)\left(1 - z_j^2\right) \cdots \left(1 - z_j^d\right) z_j^{d+1} \cdots z_j^p c_j^{1+{}^pC_1+{}^pC_2+\cdots+{}^pC_{d-1}+1}$$
$$+\left(1 - z_j^1\right)\left(1 - z_j^2\right) \cdots \left(1 - z_j^{d-1}\right)\left(1 - z_j^{d+1}\right) z_j^d z_j^{d+2} \cdots z_j^p c_j^{1+{}^pC_1+{}^pC_2+\cdots+{}^pC_{d-1}+2} +$$
$$\vdots$$
$$+\left(1 - z_j^{p-d+1}\right)\left(1 - z_j^{p-d+2}\right)\left(1 - z_j^p\right) z_j^1 \ldots z_j^{p-d} c_j^{1+{}^pC_1+{}^pC_2+\cdots+{}^pC_d}$$

When $z_j^1 + z_j^2 + \cdots + z_j^p = p - d$, the above function gives one among the following c_j^ts: $c_j^{1+{}^pC_1+{}^pC_2+\cdots{}^pC_{d-1}+1}, c_j^{1+{}^pC_1+{}^pC_2+\cdots{}^pC_{d-1}+2}, \ldots, c_j^{1+{}^pC_1+{}^pC_2+\cdots{}^pC_{d-1}+{}^pC_d}$.

When $k = 0$, the function $f(z) = f_0(z) + f_1(z) + \cdots + f_d(z)$ gives one of the value c_j^t for all z satisfying $p - d \le z_j^1 + z_j^2 + \cdots + z_j^p \le p$.

If $k \neq 0$, then $k <^p C_{d+1}$ and we formulate the function

$$f_{d+1}(z) = \left(1 - z_j^{i_1}\right)\left(1 - z_j^{i_2}\right) \ldots \left(1 - z_j^{i_d}\right)\left(1 - z_j^{i_{d+1}}\right) z_j^{d+2} \ldots z_j^n c_j^{t-k+1}$$
$$+\left(1 - z_j^{i_1}\right)\left(1 - z_j^{i_2}\right) \ldots \left(1 - z_j^{i_d}\right)\left(1 - z_j^{i_{d+2}}\right) z_j^{d+1} z_j^{d+3} \ldots z_j^p c_j^{t-k+2}$$
$$+ \cdots + \left(\text{terms up to } c_j^t\right)$$

When $z_j^1 + z_j^2 + \cdots + z_j^p = p - (d + 1)$, $f_{d+1}(z)$ can give one output among ${}^pC_{d+1}$ number of choices. Here we have used ${}^pC_{d+1} - k$ restrictions to restrict its possible outputs in k numbers. Let the kth term occur at $i_1 = i'_1, i_2 = i'_2, \ldots, i_{d+1} = i'_{d+1}$, then the restrictions are

$$p - (d + 1) \le z_j^{i_1} + z_j^{i_2} + \cdots + z_j^{i_p} \le p$$
$$z_j^{i_1} + z_j^{i_2} + \cdots + z_j^{i_{d+1}} \ge 1, \text{ for all } i_1 = i'_1, i_2 = i'_2, \ldots i_d = i'_d, i_p \ge i_{d+1} > i'_{d+1};$$
$$z_j^{i_1} + z_j^{i_2} + \cdots + z_j^{i_{d+1}} \ge 1, \text{ for all } i_1 = i'_1, i_2 = i'_2, \ldots, i_{d-1} = i'_{d-1}, i_{p-1} \ge i_d > i'_d;$$
$$\vdots$$
$$z_j^{i_1} + z_j^{i_2} + \cdots + z_j^{i_{d+1}} \ge 1, \text{ for all } i_{p-d-1} \ge i_1 > i'_1.$$

Thus, $f(z) = f_0(z) + f_1(z) + \cdots + f_d(z) + f_{d+1}(z)$ gives the generalized selection function of the multi-choice c_j^ts.

Without loss of any generality in treating the value of $c_j^t = 1$ and using the product and summation notation, we have formulated the following formulae to select the crisp value of multi-choice parameters:

$$\prod_{i=1}^{p} z_j^i + \sum_{i_1=1}^{p}\left[\left(1 - z_j^{i_1}\right) \prod_{i=1, i \neq i_1}^{p} z_j^i\right] + \sum_{i_2=2 i_2 > i_1}^{p} \sum_{i_1=1}^{p}\left[\left(1 - z_j^{i_1}\right)\left(1 - z_j^{i_2}\right) \prod_{i=1 (i \neq i_1, i_2)}^{p} z_j^i\right]$$
$$+ \cdots + \sum_{\substack{i_d=d \\ i_d > i_{(d-1)}}}^{p} \sum_{\substack{i_{d-1}=d-1 \\ i_{d-1} > i_{d-2}}}^{p} \cdots \sum_{i_1=1}^{p}\left[\left(1 - z_j^{i_1}\right)\left(1 - z_j^{i_2}\right) \ldots \left(1 - z_j^{i_d}\right) \prod_{i=1, i \neq (i_1, \ldots, i_d)}^{p} z_j^i\right]$$

where $p - d \le z_j^{i_1} + z_j^{i_2} + \cdots + z_j^{i_p} \le p$ for all $i_1 < i_2 < \cdots < i_p$.

When $k \neq 0$, we add first k terms with the above function from the following formula:

$$\left(1 - z_j^{i_1}\right)\left(1 - z_j^{i_2}\right)\ldots\left(1 - z_j^{i_d}\right)\left(1 - z_j^{i_{d+1}}\right) \prod_{i=1, i \neq (i_1,..,i_d,i_{d+1})}^{p} z_j^i$$

$$+\left(1 - z_j^{i_1}\right)\left(1 - z_j^{i_2}\right)\ldots\left(1 - z_j^{i_d}\right)\left(1 - z_j^{i_{d+2}}\right) \prod_{i=1, i \neq (i_1,..,i_d,i_{d+2})}^{p} z_j^i$$

$$+\cdots\cdots+\left(1 - z_j^{i_1}\right)\left(1 - z_j^{i_2}\right)\ldots\left(1 - z_j^{i_d}\right)\left(1 - z_j^{i_p}\right) \prod_{i=1, i \neq (i_1,..,i_d,i_p)}^{p} z_j^i$$

$$+\left(1 - z_j^{i_1}\right)\left(1 - z_j^{i_2}\right)\ldots\left(1 - z_j^{i_{d+1}}\right)\left(1 - z_j^{i_{d+2}}\right) \prod_{i=1, i \neq (i_1,..,i_{d+1},i_{d+2})}^{p} z_j^i$$

$$+\left(1 - z_j^{i_1}\right)\left(1 - z_j^{i_2}\right)\ldots\left(1 - z_j^{i_{d+1}}\right)\left(1 - z_j^{i_{d+3}}\right) \prod_{i=1, i \neq (i_1,..,i_{d+1})}^{p} z_j^i +$$

$$\vdots$$

$$+\left(1 - z_j^{i_{p-(d+1)}}\right)\left(1 - z_j^{i_{p-(d-1)}}\right)\ldots\left(1 - z_j^{i_{p-1}}\right)\left(1 - z_j^{i_p}\right) \prod_{\substack{i=1 \\ i \neq (i_{p-d-1}, i_{p-d+1}, \ldots, i_p)}}^{p} z_j^i$$

Assuming that $i_1 < i_2 < \cdots\cdot < i_p$ and let kth term occurred at $i'_1, i'_2, \ldots, i'_{d+1}$, then the restrictions are

$$p - (d+1) \le z_j^{i_1} + z_j^{i_2} + \cdots + z_j^{i_p} \le p$$
$$z_j^{i_1} + z_j^{i_2} + \cdots + z_j^{i_{d+1}} \ge 1, \text{ for all } i_1 = i'_1, i_2 = i'_2, \ldots i_d = i'_d, i_p \ge i_{d+1} > i'_{d+1};$$
$$z_j^{i_1} + z_j^{i_2} + \cdots + z_j^{i_{d+1}} \ge 1, \text{ for all } i_1 = i'_1, i_2 = i'_2, \ldots, i_{d-1} = i'_{d-1}, i_{p-1} \ge i_d > i'_d;$$
$$\vdots$$
$$z_j^{i_1} + z_j^{i_2} + \cdots + z_j^{i_{d+1}} \ge 1, \text{ for all } i_{p-d-1} \ge i_1 > i'_1.$$

$$\text{Let } \tilde{C}_{ij}^t = \sum_{g=1}^{T} (\text{term})^g \, C_{ij}^{tg} \; i = 1, 2, \ldots, m; j = 1, 2, \ldots, n \tag{9}$$

where $(\text{term})^g$ (for $g = 1, 2, \ldots, T$) are the T number of terms in the functions of the binary variables mentioned in above. Similarly,

$$\tilde{a}_i = \sum_{g=1}^{P} (\text{term})^g \, a_i^g \; i = 1, 2, \ldots, m \tag{10}$$

$$\text{and } \tilde{b}_j = \sum_{g=1}^{Q} (\text{term})^g \, b_j^g \; j = 1, 2, \ldots, n \tag{11}$$

where $(\text{term})^g$ (for $g = 1, 2, \ldots, P$) is the P number of terms in the functions of the binary variables mentioned above to reduce the P number of choices a_i^g to single choice a'_i, and $(\text{term})^g$ (for $g = 1, 2, \ldots, Q$) is the Q number of terms in functions of binary variables mentioned above to reduce the Q number of choices b_j^g to single choice b'_j.

Reduction of MCMTP to MOTP

The MCMTP as given in the *Model M1* transformed to a MOTP by transforming the multi-choice parameters in the objective functions (5) and the multi-choice supplies and demands in constraints (6) and (7) to single-valued ones, using the technique described in subsection "Transformation technique for multi-choice parameters like cost, supply, and demand to the equivalent form". Thus, the equivalent MOTP of Model M1 is given in the following model:

Model M2

$$\min: Z^t = \sum_{i=1}^{m}\sum_{j=1}^{n} C_{ij}^{\prime t} x_{ij}, \ t = 1, 2, \ldots, K \tag{12}$$

$$\text{subject to} \quad \sum_{j=1}^{n} x_{ij} \leq a_i', \ i = 1, 2, \ldots, m \tag{13}$$

$$\sum_{i=1}^{n} x_{ij} \geq b_j', \ j = 1, 2, \ldots, n \tag{14}$$

$$\text{and} \ x_{ij} \geq 0, \quad \forall \ i \text{ and } j \tag{15}$$

Here $C_{ij}^{\prime t}, a_i', b_j'$ are the reduced cost, supply, and demand parameters of tth objective function in MOTP, respectively, and $\sum_{i=1}^{m} \max\left\{\tilde{a}_i^1, \tilde{a}_i^2, \ldots, \tilde{a}_i^p\right\} \geq \sum_{j=1}^{n} \min\left\{\tilde{b}_j^1, \tilde{b}_j^2, \ldots, \tilde{b}_j^q\right\}$ is the feasibility condition.

The transportation problem *Model M2* is same as the problem described in Model 1. We can solve the Model M2 as the procedure described using the different techniques to solve Model 1.

Solution procedure

The approaches such as goal programming and revised multi-choice goal programming are used to solve the MOTP, which are defined as follows:

A. Goal programming approach

Let us briefly discuss the goal programming approach for solving MOTP (see *Model 1A*). If d_t^+ and d_t^- be positive and negative deviations corresponding to the tth goal of the objective function, then the mathematical model is defined as follows:

Model 1A

$$\min \sum_{t=1}^{K} w_t \left(d_t^+ + d_t^-\right) \tag{16}$$

$$\text{subject to} \quad Z^t(X) - d_t^+ + d_t^- = y_t, \ t = 1, 2, \ldots, K \tag{17}$$

$$g_{t,\min} \leq y_t \leq g_{t,\max}, \ t = 1, 2, \ldots, K \tag{18}$$

$$d_t^+, d_t^- \geq 0, \ t = 1, 2, \ldots, K \tag{19}$$

$$\text{and} \ (2) \text{ to } (4)$$

B. Revised multi-choice goal programming approach

In the similar way, the RMCGP is introduced to solve the MOTP. Let us assume that the multiple goals are considered to the objective functions and this can be achieved by considering the following model (see *Model 1B*) as

Model 1B

$$\min \sum_{t=1}^{K} w_t \left(d_t^+ + d_t^-\right) + \alpha_t \left(e_t^+ + e_t^-\right) \tag{20}$$

$$\text{subject to} \quad Z^t (X) - d_t^+ + d_t^- = y_t, \quad t = 1, 2, \ldots, K \tag{21}$$

$$y_t - e_t^+ + e_t^- = g_{t,\max}, \text{ or } g_{t,\min} \quad t = 1, 2, \ldots, K \tag{22}$$

$$g_{t,\min} \le y_t \le g_{t,\max}, \quad t = 1, 2, \ldots, K \tag{23}$$

$$d_t^+, d_t^-, e_t^+, e_t^- \ge 0, \quad t = 1, 2, \ldots, K \tag{24}$$

$$\text{and} \quad (2) \text{ to } (4)$$

where tth aspiration level is defined as y_t which is the continuous variable that lies between the upper $\left(g_{t,\max}\right)$ and lower $\left(g_{t,\min}\right)$ bounds. Again, e_t^+ and e_t^- are positive and negative deviations attached to tth goal of $|y_t - g_{t,\max}|$, and α_t is the weight attached to the sum of the deviations of $|y_t - g_{t,\max}|$.

Utility function approach to solve MOTP

Here, the concept of utility function has been addressed to solve MOTP. A short introduction is presented here and then we discuss the methodology for solving MOTP using utility function.

Utility function

In this paper, introduction of utility is taken to be correlative to 'Desire' or 'Want'. It has been already argued that desire cannot be measured directly, but only indirectly, by the outward phenomena in which the context is presented.

Definition 1. *The utility function describes a function $U : X \longrightarrow \Re$ which assigns a real number to every outcome in such a way that it captures DM's preferences over the desired goals of the objectives, where X is the set of feasible points and $\Re$ is the set of real numbers.*

The purpose of this study is to derive the achievement function of MOTP under the light of utility function for the DM according to the priority of goals. In our proposed approach, the DM wants to maximize his/her expected utility. For the sake of simplicity, two popular utility functions (linear and S-shaped) are considered as follows.

Linear utility function $u_i\left(y_i\right)$ for decision-making (management) problems can be found in Lai and Hwang [14] and S-shaped utility function (for the same purpose) has been proposed by Chang [15]. The utility function is generally considered in three cases as follows:

Case 1: Left linear utility function (LLUF)

$$u_i\left(y_i\right) = \begin{cases} 1, & \text{if } y_i \le g_{i,\min} \\ \frac{g_{i,\max} - y_i}{g_{i,\max} - g_{i,\min}}, & \text{if } g_{i,\min} \le y_i \le g_{i,\max}, \quad i = 1, 2, \ldots, K \\ 0, & \text{if if } y_i \ge g_{i,\max} \end{cases}$$

Case 2: Right linear utility function (RLUF)

$$u_i\left(y_i\right) = \begin{cases} 1, & \text{if } y_i \ge g_{i,\max} \\ \frac{y_i - g_{i,\min}}{g_{i,\max} - g_{i,\min}}, & \text{if } g_{i,\min} \le y_i \le g_{i,\max}, \quad i = 1, 2, \ldots, K \\ 0, & \text{if if } y_i \le g_{i,\min} \end{cases}$$

Case 3: S-shaped utility function

$$u_i(y_i) = \begin{cases} 0, & \text{if } y_i \le g_{i2} \\ \frac{y_i - g_{i2}}{g_{i8} - g_{i2}}, & \text{if } g_{i2} \le y_i \le g_{i4} \\ \frac{y_i - g_{i3}}{g_{i6} - g_{i3}}, & \text{if } g_{i4} \le y_i \le g_{i5} \\ \frac{y_i - g_{i1}}{g_{i7} - g_{i1}}, & \text{if } g_{i5} \le y_i \le g_{i7} \end{cases}, i = 1, 2, \ldots, K$$

where $g_{i,\min}$ and $g_{i,\max}$ are lower and upper bounds corresponding to the ith goal respectively. The graphs of above utility functions are drawn in the following figures (see Figures 1,2 and 3).

Model formulation for case 1

The DM would like to increase the utility value $u_t(y_t)$ as much as possible in the case of LLUF (Figure 1). In order to achieve this goal, the value of y_t should be as close to the target value $g_{t,\min}$ as possible. The MOTP from Model 1A can be reformulated using the proposed LLUF as follows:

Model 2A

$$\min \sum_{t=1}^{K} w_t\left(d_t^+ + d_t^-\right) + \beta_t f_t^- \tag{25}$$

$$\text{subject to} \quad Z^t(X) - d_t^+ + d_t^- = y_t, \quad t = 1, 2, \ldots, K \tag{26}$$

$$g_{t,\min} \le y_t \le g_{t,\max}, \quad t = 1, 2, \ldots, K \tag{27}$$

$$u_t \le \frac{g_{t,\max} - y_t}{g_{t,\max} - g_{t,\min}}, \quad t = 1, 2, \ldots, K \tag{28}$$

$$u_t + f_t^- = 1, \quad t = 1, 2, \ldots, K \tag{29}$$

$$u_t, f_t^- \ge 0, \quad t = 1, 2, \ldots, K \tag{30}$$

and (2) to (4)

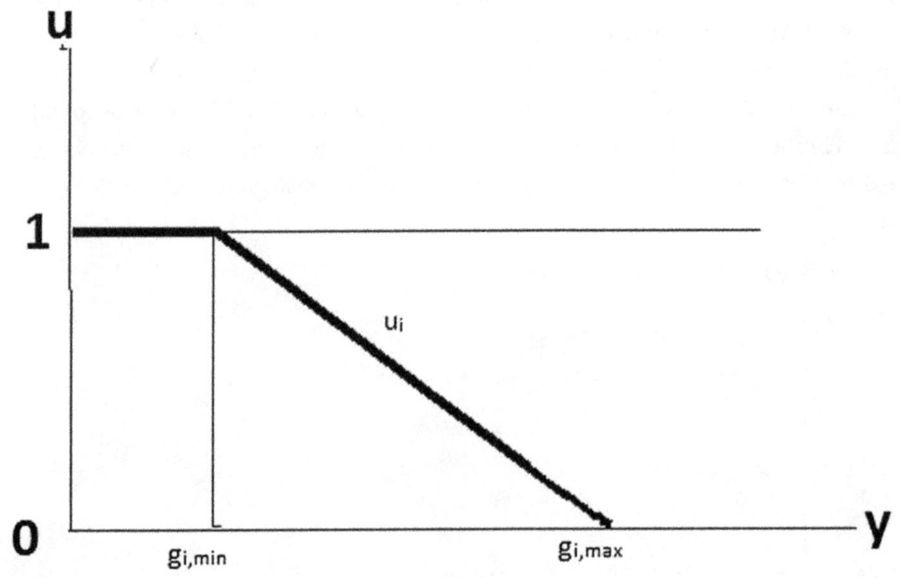

Figure 1 LLUF.

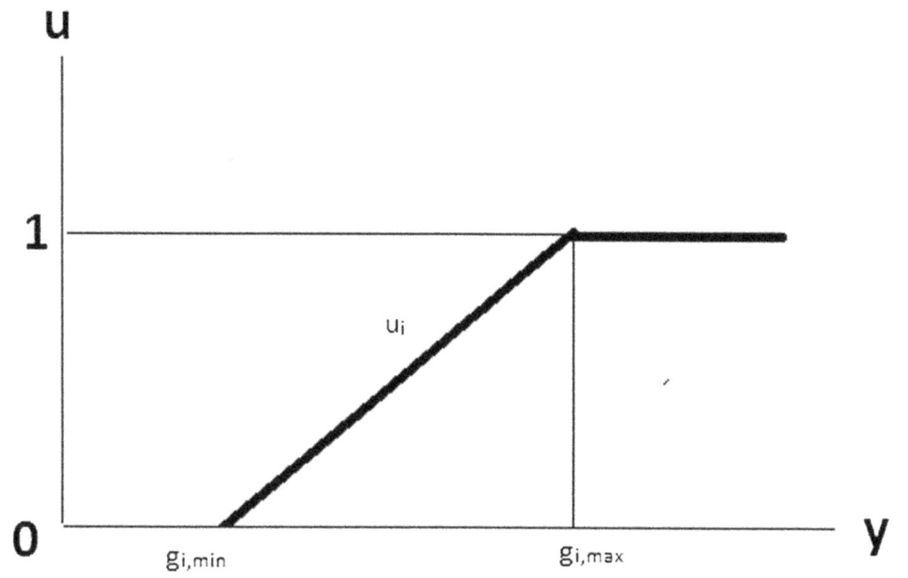

Figure 2 RLUF.

where β_t is the weight attached to deviation f_t^-. The role of weight β_t can be seen as the preferential component for the utility value u_t.

Proposition 1. *Achievement of optimal utility in the LLUF* (Figure 1) *is equivalent to the optimal solution of Model 2A.*

Proof. When u_t approaches to the highest value 1, then the deviation $f_t^- \to 0$ of the utility function (from Equation 29), because f_t^- should be minimized in the objective function to obtain the optimal solution of Model 2A. This represents y_t approach to $g_{t,\min}$ (from Equation 28), and $Z_t(X)$ is also closer to $g_{t,\min}$ (from Equation 26) because d_t^+ and

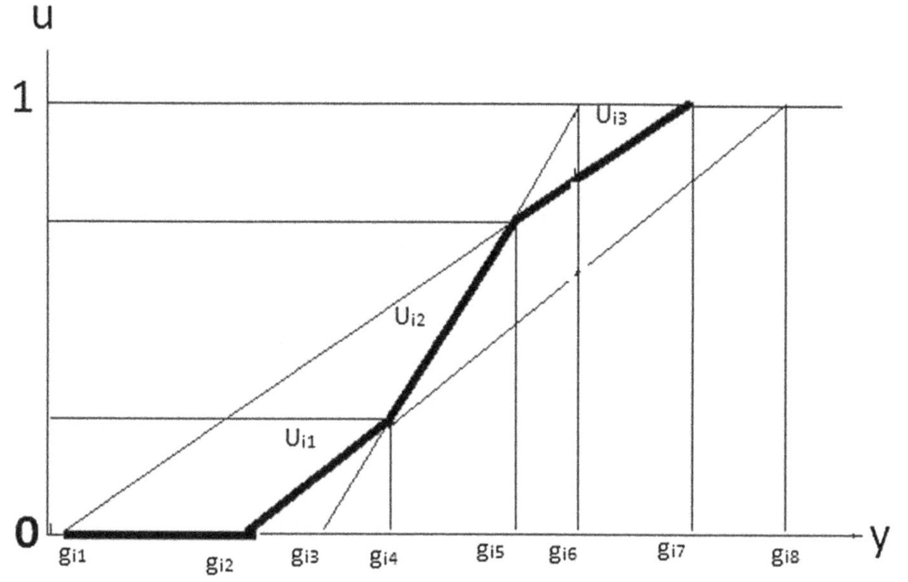

Figure 3 S-shaped utility function.

d_t^- should also be minimized in the objective function. It is obvious that the behavior of Model 2A and the level of utility are achieved. This completes the proof. □

Model formulation for case 2

The DM would like to increase the utility value $u_t\,(y_t)$ as much as possible in the case of RLUF (Figure 2). In order to achieve this goal, the value of y_t should be as close to the target value $g_{t,\max}$ as possible. The MOTP from Model 1A can be reformulated using the proposed RLUF as follows:

$$\textit{Model 2B} \qquad \min \sum_{t=1}^{K} w_t\left(d_t^+ + d_t^-\right) + \beta_t f_t^-$$

$$\begin{aligned}
\text{subject to} \quad Z^t(X) - d_t^+ + d_t^- &= y_t, \quad t = 1,2,\ldots,K \\
g_{t,\min} \le y_t \le g_{t,\max}&, \quad t = 1,2,\ldots,K \\
u_t &\le \frac{y_t - g_{t,\min}}{g_{t,\max} - g_{t,\min}} \\
u_t + f_t^- &= 1 \\
u_t, f_t^- &\ge 0 \\
\text{and (2) to (4)}&
\end{aligned}$$

where β_t is the weight attached to the deviation f_t^-. The role of weight β_t can be seen as a preferential component for the utility value u_t.

Proposition 2. *Achievement of optimal utility in the RLUF* (Figure 2) *is equivalent to the optimal solution of Model 2B.*

Proof. Similar way can be followed as we have done in Proposition 1. □

The advantages of the use of LLUF and RLUF in the decision-making problems are as follows:

(1) The DM can easily formulate their MOTP by taking into account their preference mappings with utility functions in real situation.
(2) The two linear utility models represented as linear form which can be easily solved using software.

Due to variation of deviation variables d_t^+, d_t^-, and f^- in different ranges, biasness may occur towards the objective functions with larger magnitude. Normalization technique may help to remove this biasness. Several normalization approaches such as percentage, Euclidean, summation, and zero-one notarizations (Tamiz et al. [12]; Kettani et al. [16]) are available to execute this. According to the normalization technique proposed by Tamiz et al. [12], Model 2A can be redesigned as follows:

$$\textit{Model 2AN} \qquad \min \sum_{t=1}^{K}\left[\frac{w_t\left(d_t^+ + d_t^-\right) + \beta_t f_t^-}{\phi_t}\right]$$

$$\text{subject to} \quad (26) \text{ to } (30) \text{ and } (2) \text{ to } (4)$$

where ϕ_t is the normalization constant for tth goal.

In order to solve this problem, utility normalization concept is introduced as follows: Let $d_t^+, d_t^- \in [0, \bar{u}_t]$ and $f_t^- \in [0,1]$ where $\bar{u}_t$ is the upper bound of d_t^+ and d_t^-. The normalized weights w_t and β_t can be easily obtained as $w_t = \frac{1}{1+\bar{u}_t}$ and $\beta_t = \frac{\bar{u}_t}{1+\bar{u}_t}$. This technique of normalization ensures that deviation variables d_t^+, d_t^-, and f_t^- approximated the same magnitude. Similarly, the same methodology can be applied to the Model 2B.

The utility value for S-shaped utility function can be expressed as a sum of linear utility functions (RLUF or LLUF) by introducing binary variables [17]. But Chang [15] proposed in his paper that the utility value for S-shaped utility function can be considered without using the binary variables and this is shown in the following model (see *Model 2C*):

Model 2C

$$\min \sum_{t=1}^{K} w_t \left[p_{t1} + p_{t2} + p_{t3}\right] + \beta_t f_t^-$$

$$\text{subject to } \; Z^t(X) - d_t^+ + d_t^- = y_t, \quad t = 1,2,\ldots,K$$

$$g_{t,\min} \le y_t \le g_{t,\max}, \quad t = 1,2,\ldots,K,$$

$$u_t = [\,u_t(g_{t4}) - u_t(g_{t2})\,]\frac{p_{t1} - p_{t2}}{g_{t4} - g_{t2}} + [\,u_t(g_{t5}) - u_t(g_{t4})\,]\frac{p_{t2} - p_{t3}}{g_{t5} - g_{t4}}$$

$$+[\,u_t(g_{t7}) - u_t(g_{t5})\,]\frac{p_{t3}}{g_{t7} - g_{t5}}, t = 1,2,\ldots,K$$

$$y_t - p_{t1} + n_{t1} = g_{t2}, t = 1,2,\ldots,K$$

$$y_t - p_{t2} + n_{t2} = g_{t4}, t = 1,2,\ldots,K$$

$$y_t - p_{t3} + n_{t3} = g_{t5}, t = 1,2,\ldots,K$$

$$u_t + f_t^- = 1, t = 1,2,\ldots,K$$

$$u_t, p_{tl}, n_{tl} \ge 0, t = 1,2,\ldots,K, l = 1,2,3$$

$$\text{and (2) to (4)}$$

MCMTP which occurred in many real-life situations can be reduced to MOTP and then the problem can be reduced to the models such as 2A, 2B, and 2C, with interval goals under the consideration of utility functions related to these goals. Solving the formulated problem, the DM obtained the satisfactory solution.

Numerical examples

Here we have presented two numerical examples; the first one explores the applicability of MOTP and the second one represents the applicability of MCMTP.

Case 1

Let us consider the following MOTP (see *Model 3*) with three objectives:

Model 3

Goal 1: $Z^1 = 7x_{11} + 8x_{12} + 7.5x_{13} + 8x_{21} + 7.2x_{22} + 8.4x_{23} + 9x_{31} + 8x_{32} + 7.7x_{33}$
with goal as [170, 220], more is better, follows RLUF.

Goal 2: $Z^2 = 50x_{11} + 65x_{12} + 62x_{13} + 60x_{21} + 55x_{22} + 58x_{23} + 65x_{31} + 60x_{32} + 58x_{33}$
with goal as [1,550; 1,800], less is better, follows LLUF.

Goal 3: $Z^3 = 10x_{11} + 8x_{12} + 9x_{13} + 8.5x_{21} + 9.5x_{22} + 8.5x_{23} + 9.5x_{31} + 8.8x_{32} + 9x_{33}$
with goal as [200, 290], more is better, follows S-shaped utility function as given in Figure 4.

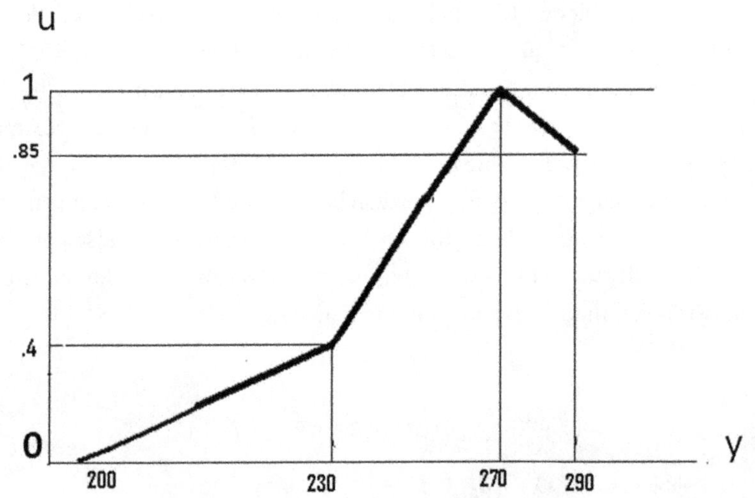

Figure 4 S-shaped utility function for the goal 3 of the proposed example.

$$\text{subject to } x_{11} + x_{12} + x_{13} \leq 10 \tag{31}$$

$$x_{21} + x_{22} + x_{23} \leq 9 \tag{32}$$

$$x_{31} + x_{32} + x_{33} \leq 11 \tag{33}$$

$$x_{11} + x_{21} + x_{31} \geq 9 \tag{34}$$

$$x_{12} + x_{22} + x_{32} \geq 8 \tag{35}$$

$$x_{13} + x_{23} + x_{33} \geq 10 \tag{36}$$

$$x_{ij} \geq 0, \quad \forall\ i, j = 1, 2, 3 \tag{37}$$

To achieve the goals in the proposed problem (see Model 3), we may formulate the following models.

In the proposed problem, the deviations of goals 1, 2, 3 are 50, 250, 90, respectively. By considering the weights $w_1 = \frac{1}{50}, w_2 = \frac{1}{250}, w_3 = \frac{1}{90}$ for the Model 1A, the above Model 3 reduces to the following model (see *Model 3A*) as

Model 3A

$$\min\ \frac{1}{50}\left(d_1^+ + d_1^-\right) + \frac{1}{250}\left(d_2^+ + d_2^-\right) + \frac{1}{90}\left(d_3^+ + d_3^-\right)$$

$$\begin{aligned}
\text{subject to } Z^1 &= 7x_{11} + 8x_{12} + 7.5x_{13} + 8x_{21} + 7.2x_{22} + 8.4x_{23} \\
&\quad + 9x_{31} + 8x_{32} + 7.7x_{33} - d_1^+ + d_1^- = y_1 \\
&170 \leq y_1 \leq 220 \\
Z^2 &= 50x_{11} + 65x_{12} + 62x_{13} + 60x_{21} + 55x_{22} + 58x_{23} \\
&\quad + 65x_{31} + 60x_{32} + 58x_{33} - d_2^+ + d_2^- = y_2 \\
&1550 \leq y_2 \leq 1800 \\
Z^3 &= 10x_{11} + 8x_{12} + 9x_{13} + 8.5x_{21} + 9.5x_{22} + 8.5x_{23} \\
&\quad + 9.5x_{31} + 8.8x_{32} + 9x_{33} - d_3^+ + d_3^- = y_3 \\
&200 \leq y_3 \leq 290 \\
&d_t^+, d_t^- \geq 0, \ t = 1, 2, 3 \\
&\text{and (31) to (37)}
\end{aligned}$$

Again, considering the same weights w_t as used in Model 3A for all $t = 1, 2, 3$ and setting $\alpha_t = w_t$ for $t = 1, 2, 3$ for deviation of goals and using Model 1B, Model 3 reduces to the following model (see *Model 3B*) as

Model 3B

$$
\begin{aligned}
\min\ & \frac{1}{50}\left(d_1^+ + d_1^-\right) + \frac{1}{250}\left(d_2^+ + d_2^-\right) + \frac{1}{90}\left(d_3^+ + d_3^-\right) \\
& + \frac{1}{50}\left(e_1^+ + e_1^-\right) + \frac{1}{250}\left(e_2^+ + e_2^-\right) + \frac{1}{90}\left(e_3^+ + e_3^-\right) \\
\text{subject to}\ & 7x_{11} + 8x_{12} + 7.5x_{13} + 8x_{21} + 7.2x_{22} + 8.4x_{23} \\
& + 9x_{31} + 8x_{32} + 7.7x_{33} - d_1^+ + d_1^- = y_1 \\
& y_1 - e_1^+ + e_1^- = 220 \\
& 170 \leq y_1 \leq 220 \\
& 50x_{11} + 65x_{12} + 62x_{13} + 60x_{21} + 55x_{22} + 58x_{23} \\
& + 65x_{31} + 60x_{32} + 58x_{33} - d_2^+ + d_2^- = y_2 \\
& y_2 - e_2^+ + e_2^- = 1550 \\
& 1550 \leq y_2 \leq 1800 \\
& 10x_{11} + 8x_{12} + 9x_{13} + 8.5x_{21} + 9.5x_{22} + 8.5x_{23} \\
& + 9.5x_{31} + 8.8x_{32} + 9x_{33} - d_3^+ + d_3^- = y_3 \\
& y_3 - e_3^+ + e_3^- = 290 \\
& 200 \leq y_3 \leq 290 \\
& d_t^+, d_t^-, e_t^+, e_t^- \geq 0, \quad t = 1, 2, 3 \\
& \text{and (31) to (37)}
\end{aligned}
$$

Using the concept of utility function described in section "Transformation technique for multi-choice parameters like cost, supply, and demand to the equivalent form", Model 3 can be reformulated as follows.

The consideration of utility function depends on the DM. Here, we assume that goals 1, 2, and 3 follow the utility functions LLUF (Figure 1), RLUF (Figure 2), and S-shaped utility function as given in Figure 4, respectively. In the given example, the upper bound of variations $d_1^+, d_1^-, d_2^+, d_2^-, d_3^+, d_3^-$ are 50, 50, 250, 250, 90, 90, respectively, and the upper bounds of f_1^-, f_2^-, f_3^- are 1. We find the weights as described in Section "Mathematical model" as follows: $w_1 = \frac{1}{50}, w_2 = \frac{1}{250}, w_3 = \frac{1}{90}, \beta_1 = \frac{50}{51}, \beta_2 = \frac{250}{251}, \beta_3 = \frac{90}{91}$.

With these supplied data, Model 3 can be reformulated as follows (see *Model 3C*):

Model 3C

$$
\begin{aligned}
\min\ & w_1\left(d_1^+ + d_1^-\right) + \beta_1 f_1^- + w_2\left(d_2^+ + d_2^-\right) + \beta_2 f_2^- + w_3\left(d_{31}^+ + d_{32}^+ + d_{32}^+\right) + \beta_3 f_3^- \\
\text{subject to}\ & 7x_{11} + 8x_{12} + 7.5x_{13} + 8x_{21} + 7.2x_{22} + 8.4x_{23} \\
& + 9x_{31} + 8x_{32} + 7.7x_{33} - d_1^+ + d_1^- = y_1 \\
& u_1 \leq \frac{220 - y_1}{50} \\
& f_1^- + u_1 = 1 \\
& 170 \leq y_1 \leq 220
\end{aligned}
$$

$$50x_{11}+65x_{12}+62x_{13}+60x_{21}+55x_{22}+58x_{23}$$
$$+65x_{31}+60x_{32}+58x_{33}-d_2^{+}+d_2^{-}=y_2$$
$$u_2 \le \frac{y_2-1550}{250}$$
$$f_2^{-}+u_2=1$$
$$1550 \le y_2 \le 1800$$
$$10x_{11}+8x_{12}+9x_{13}+8.5x_{21}+9.5x_{22}+8.5x_{23}$$
$$+9.5x_{31}+8.8x_{32}+9x_{33}-d_3^{+}+d_3^{-}=y_3$$
$$u_3=(.4-0)\left(d_{31}-d_{32}\right)/30+(1-.4)\left(d_{32}-d_{33}\right)/40+(.85-1)\,d_{33}/20$$
$$y_3-d_{31}+dn_{31}=230, y_3-d_{32}+dn_{32}=270, y_3-d_{33}+dn_{33}=290$$
$$d_{31}dn_{31}=0, d_{32}dn_{32}=0, d_{33}dn_{33}=0,\ \ f_3^{-}+u_3=1$$
$$u_t \ge 0, f_t^{-} \ge 0\ \forall\ t=1,2,3$$
and (31) to (37)

Results and discussion for problem given in case 1

Using LINGO software, we solved Models 3A, 3B, and 3C and reported the solution as follows:

The optimal solution of Model 3A is reported as
$x_{11}=0, x_{12}=9, x_{13}=1, x_{21}=0, x_{22}=0, x_{23}=9, x_{31}=6, x_{32}=0, x_{33}=0$; $Z^1=209.1, Z^2=1559, Z^3=214.5$.

The optimal solution of Model 3B is as follows:
$x_{11}=10, x_{12}=0, x_{13}=0, x_{21}=0, x_{22}=9, x_{23}=0, x_{31}=0, x_{32}=0, x_{33}=11$; $Z^1=219.50, Z^2=1633, Z^3=284.5$.

The optimal solution of Model 3C is also as follows:
$x_{11}=10, x_{12}=0, x_{13}=0, x_{21}=0, x_{22}=5, x_{23}=4, x_{31}=0, x_{32}=3, x_{33}=7$; $Z^1=217.5, Z^2=1593, Z^3=270.9$.

Here, the solution obtained in Model 3B is better compared with the solution of Model 3A, but the DM is not satisfied because in the proposed problem, satisfying the goal is not only the important notion but is also a utility factor to the DM which is important for the decision-making (management) problem. When the utility value is more important rather than the benefit, then the solutions obtained in Model 3A or in Model 3B are not satisfactory to the DM to make an appropriate decision. The marketing survey indicates that the higher utility value of goal 3 will increase the number of customers to the network service provider company. The solution obtained in Model 3C demonstrated the high utility value of goal 3, whenever the other two models failed to give satisfactory results (Table 1). In this context, we may suggest that the utility function approach provided better result compared with other results obtained in classical techniques like GP and RMCGP.

Table 1 Comparison of achieved goals obtained from the different methods

Method	Achievement of goal 1 (%)	Achievement of goal 2 (%)	Achievement of goal 3 (%)
GP	75	95	40
RMCGP	99	60	90
Utility approach	98	80	99

Case 2

Let us consider the following MCMTP (see Model 3) with two objectives:

Model 4

Goal 1: $z_1 = \{5 \text{ or } 7\}\, x_{11} + 8x_{12} + \{7 \text{ or } 6 \text{ or } 10\}\, x_{13} + \{6 \text{ or } 8\}\, x_{21} + 8x_{22} + 10x_{23}$
with goal as [150,200], more is better, but follows S-shape utility function (Figure 5).

Goal 2: $z_2 = 15x_{11} + \{18 \text{ or } 16\}\, x_{12} + 17x_{13} + 16x_{21} + \{18 \text{ or } 20\}\, x_{22} + 20x_{23}$
with goal as [400,500], less is better, follows LLUF:

$$\text{subject to} \quad x_{11} + x_{12} + x_{13} \le \{11 \text{ or } 13 \text{ or } 12 \text{ or } 16\} \tag{38}$$

$$x_{21} + x_{22} + x_{23} \le \{14 \text{ or } 13\} \tag{39}$$

$$x_{11} + x_{21} \ge \{8 \text{ or } 7\} \tag{40}$$

$$x_{12} + x_{22} \ge \{7 \text{ or } 8 \text{ or } 6\} \tag{41}$$

$$x_{13} + x_{23} \ge 9 \tag{42}$$

$$x_{ij} \ge 0 \forall\ i = 1, 2 \text{ and } j = 1, 2, 3. \tag{43}$$

Model 4 is equivalent to the following model (see *Model 5*).

Model 5

$$\text{Goal 1: } Z_1 = \left\{5z_{11}^{11} + 7\left(1 - z_{11}^{11}\right)\right\} x_{11} + 8x_{12} + \left\{7z_{11}^{12}z_{11}^{13} + 6z_{11}^{12}\left(1 - z_{11}^{13}\right)\right.$$
$$\left. + 10z_{11}^{13}\left(1 - z_{11}^{12}\right)\right\} x_{13} + \left\{6z_{11}^{21} + 8\left(1 - z_{11}^{21}\right)\right\} x_{21} + 8x_{22} + 10x_{23}$$

with goal as [150,200], more is better, but follows S-shape utility function (Figure 5).

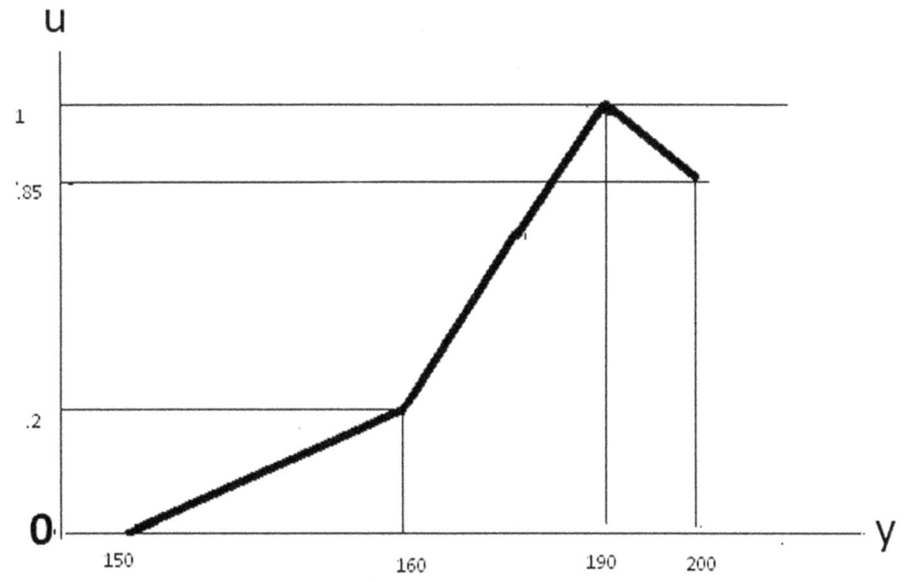

Figure 5 S-shaped utility function for goal 2 in case 2.

Goal 2: $Z_2 = 15x_{11} + \left\{18z_{12}^{11} + 16\left(1 - z_{12}^{11}\right)\right\}x_{12} + 17x_{13} + 16x_{21} + \left\{18z_{12}^{12} + 20\left(1 - z_{12}^{12}\right)\right\}x_{22} + 20x_{23}$

with goal as [400,500], less is better, follows LLUF:

$$\text{subject to} \quad x_{11} + x_{12} + x_{13} \leq \left\{11z_1^{11}z_1^{12} + 13z_1^{11}\left(1 - z_1^{12}\right) + 12z_1^{12}\left(1 - z_1^{11}\right) + 16\left(1 - z_1^{11}\right)\left(1 - z_1^{12}\right)\right\} \quad (44)$$

$$x_{21} + x_{22} + x_{23} \leq \left\{14z_2^{11} + 16\left(1 - z_2^{11}\right)\right\} \quad (45)$$

$$x_{11} + x_{21} \geq \left\{8z^{11} + 7\left(1 - z^{11}\right)\right\} \quad (46)$$

$$x_{12} + x_{22} \geq \left\{7z^{21}z^{22} + 8z^{21}\left(1 - z^{22}\right) + 6z^{22}\left(1 - z^{21}\right)\right\} \quad (47)$$

$$x_{13} + x_{23} \geq 9 \quad (48)$$

$$x_{ij} \geq 0 \forall \; i = 1, 2 \text{ and } j = 1, 2, 3. \quad (49)$$

$$z_{11}^{12} + z_{11}^{13} \geq 1 \quad (50)$$

$$z^{21} + z^{22} \geq 1 \quad (51)$$

$$z_{11}^{11}, z_{11}^{12}, z_{11}^{13}, z_{12}^{11}, z_{12}^{12} = 0 \text{ or } 1 \quad (52)$$

$$z_1^{11}, z_1^{12}, z_2^{11} = 0 \text{ or } 1 \quad (53)$$

$$z^{11}, z^{21}, z^{22} = 0 \text{ or } 1 \quad (54)$$

In the given problem in Model 5, the deviations of goal 1, goal 2 are 50, 100 respectively. By considering the weights $w_1 = \frac{1}{50}, w_2 = \frac{1}{100}$ for the Model 1A, Model 5 reduces to *Model 5A* as

Model 5A

$$\min \; \frac{1}{50}\left(d_1^+ + d_1^-\right) + \frac{1}{100}\left(d_2^+ + d_2^-\right)$$

$$\text{subject to} \quad \left\{5z_{11}^{11} + 7\left(1 - z_{11}^{11}\right)\right\}x_{11} + 8x_{12} + \left\{7z_{11}^{12}z_{11}^{13} + 6z_{11}^{12}\left(1 - z_{11}^{13}\right) + 10z_{11}^{13}\left(1 - z_{11}^{12}\right)\right\}x_{13} + \left\{6z_{11}^{21} + 8\left(1 - z_{11}^{21}\right)\right\}x_{21} + 8x_{22} + 10x_{23} - d_1^+ + d_1^- = y_1,$$

$$150 \leq y_1 \leq 200,$$

$$15x_{11} + \left\{18z_{12}^{11} + 16\left(1 - z_{12}^{11}\right)\right\}x_{12} + 17x_{13} + 16x_{21} + \left\{18z_{12}^{12} + 20\left(1 - z_{12}^{12}\right)\right\}x_{22} + 20x_{23} - d_2^+ + d_2^- = y_2,$$

$$400 \leq y_2 \leq 500,$$

$$d_t^+, d_t^- \geq 0, \; t = 1, 2$$

and (44) to (54)

Again, considering the same weights w_t as used in Model 3A for all $t = 1, 2$ and the weights $\alpha_t = w_t$ for $t = 1, 2$ for deviation of goals and using the Model 1B, Model 5 reduces to *Model 5B* as

Model Ex. 5B

$$\begin{aligned}
\min\ & \frac{1}{50}\left(d_1^+ + d_1^-\right) + \frac{1}{100}\left(d_2^+ + d_2^-\right) + \\
& + \frac{1}{50}\left(e_1^+ + e_1^-\right) + \frac{1}{100}\left(e_2^+ + e_2^-\right) \\
\text{subject to}\ & \left\{5z_{11}^{11} + 7\left(1 - z_{11}^{11}\right)\right\} x_{11} + 8x_{12} + \left\{7z_{11}^{12}z_{11}^{13} + 6z_{11}^{12}\left(1 - z_{11}^{13}\right)\right. \\
& \left. + 10z_{11}^{13}\left(1 - z_{11}^{12}\right)\right\} x_{13} + \left\{6z_{11}^{21} + 8\left(1 - z_{11}^{21}\right)\right\} x_{21} + 8x_{22} + 10x_{23} - d_1^+ + d_1^- = y_1, \\
& y_1 - e_1^+ + e_1^- = 200, \\
& 150 \le y_1 \le 200, \\
& 15x_{11} + \left\{18z_{12}^{11} + 16\left(1 - z_{12}^{11}\right)\right\} x_{12} + 17x_{13} + 16x_{21} \\
& + \left\{18z_{12}^{12} + 20\left(1 - z_{12}^{12}\right)\right\} x_{22} + 20x_{23} - d_2^+ + d_2^- = y_2, \\
& y_2 - e_2^+ + e_2^- = 400, \\
& 400 \le y_2 \le 500, \\
& d_t^+, d_t^-, e_t^+, e_t^- \ge 0,\ t = 1, 2 \\
& \text{and (44) to (54)}
\end{aligned}$$

Let us solve the proposed problem (see Model 3) using the concept of utility function. The consideration of utility function depends on the DM. Here we assume that goal 1 and goal 2 follow the S-shaped utility function given in Figure 5 and the utility functions LLUF (Figure 1) , respectively. In the given example, the upper bound of variations $d_1^+, d_1^-; d_2^+, d_2^-$ are 50, 100 respectively, and the upper bounds of f_1^-, f_2^- are 1. We find the weights as suggested in Section "Mathematical model" as follows: $w_1 = \frac{1}{50}$, $\beta_1 = \frac{50}{51}$, $w_2 = \frac{1}{100}$, $\beta_2 = \frac{100}{101}$.

With these supplied data , Model 5 can be formulated as follows:

Model 5C

$$\begin{aligned}
\min\ & w_1\left(d_{11}^+ + d_{12}^+ + d_{13}^+\right) + \beta_1 f_1^- + w_2\left(d_2^+ + d_2^-\right) + \beta_2 f_2^- \\
\text{subject to}\ & \left\{5z_{11}^{11} + 7\left(1 - z_{11}^{11}\right)\right\} x_{11} + 8x_{12} + \left\{7z_{11}^{12}z_{11}^{13} + 6z_{11}^{12}\left(1 - z_{11}^{13}\right)\right. \\
& \left. + 10z_{11}^{13}\left(1 - z_{11}^{12}\right)\right\} x_{13} + \left\{6z_{11}^{21} + 8\left(1 - z_{11}^{21}\right)\right\} x_{21} + 8x_{22} + 10x_{23} - d_1^+ + d_1^- = y_1, \\
& u_1 = (.2 - 0)\left(d_{11} - d_{12}\right)/10 + (1 - .2)\left(d_{12} - d_{13}\right)/30 + (.85 - 1)\, d_{13}/10 \\
& f_3 - d_{11} + dn_{11} = 160, f_3 - d_{12} + dn_{12} = 190, f_3 - d_{13} + dn_{13} = 200 \\
& d_{11}dn_{11} = 0, d_{12}dn_{12} = 0, d_{13}dn_{13} = 0, \\
& f_1 + u_1 = 1, \\
& 15x_{11} + \left\{18z_{12}^{11} + 16\left(1 - z_{12}^{11}\right)\right\} x_{12} + 17x_{13} + 16x_{21} \\
& + \left\{18z_{12}^{12} + 20\left(1 - z_{12}^{12}\right)\right\} x_{22} + 20x_{23} - d_2^+ + d_2^- = y_2, \\
& u_2 \le \frac{200 - y_2}{100}, \\
& f_2 + u_2 = 1, \\
& 400 \le y_2 \le 500, \\
& u_t \ge 0, f_t \ge 0\ \forall\ t = 1, 2 \\
& \text{and (44) to (54)}
\end{aligned}$$

Results and discussion for problem given in case 2

Solving the model presented in Model 5A, the optimal solution of the Model 5A is reported as $x_{11} = 7, x_{12} = 5, x_{13} = 0, x_{21} = 0, x_{22} = 1, x_{23} = 10$, and the values of the objective functions are $Z^1 = 197, Z^2 = 405$.

The selection of the choices corresponding to the optimal solution is as follows:

$c_{11}^1 = 7, c_{12}^1 = 8, c_{13}^1 = 10, c_{21}^1 = 8, c_{22}^1 = 8, c_{23}^1 = 10$
$c_{11}^2 = 15, c_{12}^2 = 16, c_{13}^2 = 17, c_{21}^2 = 16, c_{22}^2 = 18, c_{23}^2 = 20$
$a_1 = 16, a_2 = 14, b_1 = 7, b_2 = 6, b_3 = 9$

Solving the model presented in Model 5B, we have listed the following solution:

$x_{11} = 7, x_{12} = 6, x_{13} = 0, x_{21} = 0, x_{22} = 0, x_{23} = 10$, and the values of the objective functions are $Z^1 = 197, Z^2 = 401$.

The selection of the choices corresponding to the optimal solution is as follows:

$c_{11}^1 = 7, c_{12}^1 = 8, c_{13}^1 = 10, c_{21}^1 = 6, c_{22}^1 = 8, c_{23}^1 = 10$
$c_{11}^2 = 15, c_{12}^2 = 16, c_{13}^2 = 17, c_{21}^2 = 16, c_{22}^2 = 18, c_{23}^2 = 20$
$a_1 = 16, a_2 = 14, b_1 = 7, b_2 = 8, b_3 = 9$

Solving the model presented in Model 5C, we obtained the solution listed below:

$x_{11} = 6, x_{12} = 3, x_{13} = 0, x_{21} = 0, x_{22} = 3, x_{23} = 10$, and the values of the objective functions are $Z^1 = 190, Z^2 = 404$.

The selection of the choices corresponding the optimal solution is as follows:

$c_{11}^1 = 7, c_{12}^1 = 8, c_{13}^1 = 10, c_{21}^1 = 6, c_{22}^1 = 8, c_{23}^1 = 10$
$c_{11}^2 = 15, c_{12}^2 = 18, c_{13}^2 = 17, c_{21}^2 = 16, c_{22}^2 = 20, c_{23}^2 = 20$
$a_1 = 12, a_2 = 14, b_1 = 7, b_2 = 6, b_3 = 9$

Table 2 helps us to conclude that the solution of the MCMTP obtained in Model 5B is better compared with the solution of Model 5A, but the DM is not satisfied because in the proposed problem satisfying the goal is not only the important notion, but is also a utility factor to the DM which is important for the decision-making (management) problem. When the utility value is more important rather than the benefit, then the solutions obtained in Model 5A or in Model 5B are not satisfactory to the DM to make the appropriate decision. The solution obtained in the Model 5C demonstrated the high utility value of goal 2, whenever the other two models failed to give satisfactory results. In this context, we may suggest that the utility function approach provided better result compared with other results.

Conclusions

In this paper, we have considered MCMTP where the cost, demand, and supply coefficients are multi-choice type. Another important notion of this study is to give an impression of goal preferences of the DM. The approach of utility function is the most useful skill for representing the DM's preferences. The concept of utility in this paper

Table 2 Comparison of achieved goals obtained from different methods

Method	Achievement of goal 1 (%)	Achievement of goal 2 (%)
GP	85	96
RMCGP	85	98
Utility approach	100	95

proposes a new approach for extending the utilization of real-life MCMTP. The MCMTP gives a new direction to handle the real-life transportation problems when the transportation parameters are multi-choice in nature. The numerical examples presented in this paper explored the applicability and suitability for solving MOTP and MCMTP and also for representing the DM's preferences. In addition, the proposed method can be used as a decision-making aid for multi-choice multi-objective decision-making problem that occurred in the real-life purposes, like economical, agricultural, industrial management, and military. In particular, in case of incomplete information, the DM can use the proposed method to set the goals according to their own utility functions, while the proposed method can easily find the better solution than the previous methods (GP, RMCGP) used to solve MCMTP which is shown by the proposed examples in this paper.

References

1. Chang, C-T: Multi-choice goal programming. Omega. **35**, 389–396 (2007)
2. Chang, C-T: Revised multi-choice goal programming. Appl. Math. Model. **32**, 2587–2595 (2008)
3. Mahapatra, DR, Roy, SK, Biswal, MP: Multi-choice stochastic transportation problem involving extreme value distribution. Appl. Math. Modell. **37**, 2230–2240 (2013)
4. Roy, SK, Mahapatra, DR, Biswal, MP: Multi-choice stochastic transportation problem with exponential distribution. J. Uncert. Syst. **6**(3), 200–213 (2013)
5. Al-nowaihi, A, Bradley, I, Dhami, S: The utility function under prospect theory. Econ. Lett. **99**, 337–339 (2008)
6. Yu, BW-T, Pang, WK, Troutt, MD, Hou, SH: Objective comparisons of the optimal portfolios corresponding to different utility functions. Eur. J. Oper. Res. **199**, 604–610 (2009)
7. Podinovski, VV: Set choice problems with incomplete information about the preferences of the decision maker. Eur. J. Oper. Res. **207**, 371–379 (2010)
8. Charnes, A, Cooper, WW, Ferguson, RO: Optimal estimation of executive compensation by linear programming. Manage. Sci. **1**, 138–151 (1955)
9. Charnes, A, Cooper, WW: Management Model and Industrial Application of Linear Programming, Vol. 1. Wiley, New York (1961)
10. Lee, SM: Goal Programming for Decision Analysis. Auerbach, Philadelphia (1972)
11. Ignizio, JP: Introduction to Linear Goal Programming. Sage, Beverly Hills (1985)
12. Tamiz, M, Jones, D, Romero, C: Goal programming for decision making: an overview of the current state-of-the-art. Eur. J. Oper. Res. **111**, 567–581 (1998)
13. Romero, C: Extended lexicographic goal programming: a unifying approach. Omega. **29**, 63–71 (2001)
14. Lai, Y-J, Hwang, C-L: Fuzzy Multiple Objective Decision Making: Methods and Applications. Springer, Berlin (1994)
15. Chang, C-T: An approximation approach for representing S-shaped membership functions. IEEE Trans. Fuzzy Syst. **18**, 412–424 (2010)
16. Kettani, O, Aouni, B, Martel, JM: The double role of the weight factor in the goal programming model. Comput. Oper. Res. **31**, 1833–1845 (2004)
17. Chang, C-T: Mixed binary interval goal programming. J. Oper. Res. Soc. **57**, 469–473 (2006)

9

New parametric generalized exponential fuzzy divergence measure

Vijay Prakash Tomar and Anshu Ohlan*

* Correspondence:
anshu.gahlawat@yahoo.com
Department of Mathematics,
Deenbandhu Chhotu Ram
University of Science and
Technology, Murthal 131039,
Haryana, India

Abstract

The present paper introduces a parametric generalized exponential measure of fuzzy divergence of order α with the proof of its validity. A particular case of proposed fuzzy divergence measure is studied. Some properties of the new divergence measure between different fuzzy sets are proved. We establish a relation between exponential fuzzy entropy of order α and our fuzzy divergence measure. Further, a numerical example is given for the comparative study of the new divergence measure with some of existing measures. Finally, application of the measure to strategic decision-making is discussed and a comparative study of the method of strategic decision-making with the existing methods is presented. It is noted that the new measure of fuzzy divergence and the method of strategic decision-making comprise greater simplicity, consistency and flexibility in applications due to the presence of the parameter.

Keywords: Fuzzy set; Fuzziness; Fuzzy entropy; Fuzzy directed divergence; Decision making

Introduction

Entropy is one of the key measures of information first used by Shannon [1]. Entropy as a measure of fuzziness was introduced by Zadeh [2]. Fuzzy entropy is an important concept for measuring fuzzy information. A measure of the fuzzy entropy of a fuzzy set is a measure of the fuzziness of the set. Kapur [3] argues that fuzzy entropy measures uncertainty due to fuzziness of information, while probabilistic entropy measures uncertainty due to the information being available in terms of a probability distribution only. The concept of fuzzy sets proposed by Zadeh [2] has proven useful in the context of pattern recognition, image processing, speech recognition, bioinformatics, fuzzy aircraft control, feature selection, decision-making, etc.

During the last six decades, entropy, as a very important notion for measuring fuzziness degree or uncertain information in fuzzy set theory, has received a great attention. Fuzzy sets gained a vital attention from researchers for their application in various fields. For example, De Luca and Termini [4] introduced the measure of fuzzy entropy corresponding to Shannon [1] entropy. Later on, Bhandari and Pal [5] defined the exponential fuzzy entropy corresponding to Pal and Pal [6] exponential entropy. Verma and Sharma [7] generalized the Pal and Pal [6] exponential fuzzy entropy of order $\alpha > 0$. Kullback and Leibler [8] obtained the measure of directed divergence between two probability distributions. Bhandari and Pal [5] presented some axioms to describe the measure of directed divergence between fuzzy sets, which is proposed corresponding to Kullback and Leibler [8] measure of directed divergence.

Thereafter, many other researchers have studied the fuzzy divergence measures in different ways and provide their application in different areas. For example, Fan and Xie [9] introduced the divergence measure based on exponential operation and corresponding to the fuzzy exponential entropy of Bhandari and Pal [5] and studied its relation with divergence measure introduced in [5]. Ghosh et al. [10] gave application of fuzzy divergence measure in the area of automated leukocyte recognition. In 2002, Montes et al. [11] studied the special classes of divergence measures and used the link between fuzzy and probabilistic uncertainty. Prakash et al. [12] proposed two fuzzy divergence measures corresponding to Ferreri [13] probabilistic measure of directed divergence. Corresponding to Renyi [14] generalized measure of directed divergence and Sharma and Mittal [15] generalized measure of directed divergence, Bajaj and Hooda [16] proposed the measure of fuzzy-directed divergence respectively. Bhatia and Singh [17] proposed four fuzzy divergence measures; one of them was corresponding to Taneja [18] arithmetic-geometric divergence measure.

Inspired by the abovementioned work, we introduce a generalized methodology for measuring the degree of difference between two fuzzy sets. We present a new parametric generalized exponential measure of fuzzy divergence and study the essential properties of this measure in order to check its authenticity.

The paper consists of eight sections. The introductory section is followed by a discussion on some well-known concepts and the notation related to fuzzy set theory. In the 'New parametric generalized exponential measure of fuzzy divergence' section, we introduce a parametric generalized fuzzy exponential measure of divergence corresponding to generalized fuzzy entropy given by Verma and Sharma [7]. In the 'Properties of generalized exponential fuzzy divergence measure' section, we first provide some interesting properties of the proposed measure of fuzzy divergence and then a relation between generalized exponential fuzzy entropy and the proposed fuzzy divergence measure is established. In the 'A comparative study' section, the comparison of the proposed divergence with some of existing generalized measures of fuzzy divergence is presented with the help of table and graph. The application of the proposed parametric generalized exponential measure of fuzzy divergence to strategic decision-making is illustrated with the help of a numerical example in the 'Application of parametric generalized exponential measure of fuzzy divergence in strategic decision-making' section. The 'Application of fuzzy TOPSIS and MOORA methods for strategic decision-making: a comparative analysis' section presents the application of the proposed measure of fuzzy divergence in the existing methods of strategic decision-making and a comparative analysis between the proposed method of strategic decision-making and the existing methods. Finally, the paper is concluded in the 'Concluding remarks' section.

Preliminaries

This section is devoted to introduce some well-known concepts and the notations related to fuzzy divergence measures. First of all, we will focus on the theory of fuzzy sets. Then, we will recall the axiomatic definition of a divergence measure for fuzzy sets.

Fuzzy sets

Fuzzy sets are used to solve a lot of real-world problems. Fuzziness, a feature of uncertainty, results from the lack of sharp difference of being or not being a member of the set, i.e., the boundaries of the set under consideration are not sharply defined.

A fuzzy set A defined on a universe of discourse X is given as Zadeh [2]:

$$A = \{\langle x, \mu_A(x)\rangle / x \in X\}$$

where $\mu_A : X \rightarrow [0, 1]$ is the membership function of A. The membership value $\mu_A(x)$ describes the degree of the belongingness of $x \in X$ in A. When $\mu_A(x)$ is valued in $\{0, 1\}$, it is the characteristic function of a crisp (i.e., non-fuzzy) set. Zadeh [19] gave some notions related to fuzzy sets, some of them which we shall need in our discussion are as follows:

(1) Compliment: $\bar{A}$ = Compliment of $A \Leftrightarrow \mu_{\bar{A}}(x) = 1 - \mu_A(x)$ for all $x \in X$.
(2) Union: $A \cup B$ = Union of A and $B \Leftrightarrow \mu_{A \cup B}(x) = \max\{\mu_A(x), \mu_B(x)\}$ for all $x \in X$.
(3) Intersection: $A \cap B$ = Intersection of A and $B \Leftrightarrow \mu_{A \cap B}(x) = \min\{\mu_A(x), \mu_B(x)\}$ for all $x \in X$

Fuzzy divergence measure

In fuzzy context, several measures have been proposed in order to measure the degree of difference between two fuzzy sets. A general study of the axiomatic definition of a divergence measure for fuzzy sets was presented in Bouchon-Meunier et al. [20] and as a particular case was widely studied in Montes et al. [11].

Bhandari et al. [21] introduced the measure of fuzzy directed divergence corresponding to Kullback and Leibler [8] measure of directed divergence, as

$$I(A:B) = \sum_{i=1}^{n}\left[\mu_A(x_i)\log\frac{\mu_A(x_i)}{\mu_B(x_i)} + (1-\mu_A(x_i))\log\frac{1-\mu_A(x_i)}{1-\mu_B(x_i)}\right] \tag{1}$$

and also provides the essential conditions for a measure of divergence.

The measure of fuzzy divergence between two fuzzy sets gives the difference between two fuzzy sets, and this measure of distance/difference between two fuzzy sets is called the fuzzy divergence measure.

Fan and Xie [9] proposed the fuzzy information of discrimination of A against B corresponding to the exponential fuzzy entropy of Pal and Pal [6] and is defined by

$$I(A,B) = \sum_{i=1}^{n}\left[1-(1-\mu_A(x_i))e^{\mu_A(x_i)-\mu_B(x_i)} - \mu_A(x_i)e^{(\mu_B(x_i)-\mu_A(x_i))}\right] \tag{2}$$

Finally, we may mention some other generalized measures of fuzzy divergence with which we compare our study.

Kapur [3] presented a fuzzy divergence measure corresponding to Havada-Charvat [22] measure of directed divergence which is given by

$$I_\alpha(A,B) = \frac{1}{\alpha-1}\sum_{i=1}^{n}\left[\mu_A^\alpha\mu_B^{1-\alpha} + (1-\mu_A)^\alpha(1-\mu_B)^{1-\alpha} - 1\right],\ \alpha \neq 1,\ \alpha > 0. \tag{3}$$

Prakash et al. [12] proposed a fuzzy divergence measure corresponding to Ferreri's [13] probabilistic measure of directed divergence given by

$$I_a(A:B) = \sum_{i=1}^{n}\left[\mu_A(x_i)\log\frac{\mu_A(x_i)}{\mu_B(x_i)} + (1-\mu_A(x_i))\log\frac{1-\mu_A(x_i)}{1-\mu_B(x_i)}\right]$$
$$-\frac{1}{a}\sum_{i=1}^{n}\left[(1+a\mu_A(x_i))\log\frac{1+a\mu_A(x_i)}{1+a\mu_B(x_i)} + \{1+a(1-\mu_A(x_i))\}\log\frac{1+a(1-\mu_A(x_i))}{1+a(1-\mu_B(x_i))}\right] \quad (4)$$

Corresponding to Renyi [14] generalized measure of directed divergence Bajaj and Hooda [16] provided the generalized fuzzy divergence measure which is given by

$$D_\alpha(A,B) = \frac{1}{\alpha-1}\sum_{i=1}^{n}\log\left[\mu_A^\alpha\mu_B^{1-\alpha} + (1-\mu_A)^\alpha(1-\mu_B)^{1-\alpha}\right],\ \alpha\neq1,\ \alpha>0. \quad (5)$$

New parametric generalized exponential measure of fuzzy divergence

We now propose a new parametric generalized exponential measure of divergence between fuzzy sets A and B corresponding to generalized exponential fuzzy entropy of order $\alpha > 0$ given by Verma and Sharma [7] as

$$I_{E_\alpha}(A,B) = \sum_{i=1}^{n}\left[1-(1-\mu_A(x_i))e^{((1-\mu_B(x_i))^\alpha-(1-\mu_A(x_i))^\alpha)} - \mu_A(x_i)e^{\left(\mu_B^\alpha(x_i)-\mu_A^\alpha(x_i)\right)}\right] \quad (6)$$

Theorem 1 $I_{E_\alpha}(A,B)$ is a valid measure of fuzzy directed divergence.

Proof: It is clear from (6) that

(i) $I_{E_\alpha}(A,B)\geq0$

(ii) $I_{E_\alpha}(A,B) = 0$ if $\mu_A(x_i) = \mu_B(x_i)$

(iii) $I_{E_\alpha}(A,B)\neq I_{E_\alpha}(B,A)$

But $J_{E_\alpha}(A,B) = I_{E_\alpha}(A,B) + I_{E_\alpha}(B,A)$ is symmetric.

(iv) We now check the convexity of $I_{E_\alpha}(A,B)$:

$$\frac{\partial I_{E_\alpha}}{\partial\mu_A(x_i)} = (1-\alpha(1-\mu_A(x_i))^\alpha)e^{((1-\mu_B(x_i))^\alpha-(1-\mu_A(x_i))^\alpha)} + (\alpha-1)e^{\left(\mu_B^\alpha(x_i)-\mu_A^\alpha(x_i)\right)}$$

$$\frac{\partial^2 I_{E_\alpha}}{\partial\mu_A^2(x_i)} = \alpha(\alpha+1)\left[(1-\mu_A(x_i))^{\alpha-1}e^{(1-\mu_B(x_i))^\alpha-(1-\mu_A(x_i))^\alpha} + \mu_A^{\alpha-1}(x_i)e^{\mu_B^\alpha(x_i)-\mu_A^\alpha(x_i)}\right]$$
$$-\alpha^2\left[(1-\mu_A(x_i))^{2\alpha-1}e^{(1-\mu_B(x_i))^\alpha-(1-\mu_A(x_i))^\alpha} + \mu_A^{2\alpha-1}(x_i)e^{\mu_B^\alpha(x_i)-\mu_A^\alpha(x_i)}\right] > 0 \text{ for } \alpha>0.$$

Similarly, $\dfrac{\partial^2 I_{E_\alpha}}{\partial\mu_B^2(x_i)} > 0$ for $\alpha > 0$.

Thus, $I_{E_\alpha}(A,B)$ is a convex function of fuzzy sets A and B and therefore $I_{E_\alpha}(A,B)$ is a valid measure of fuzzy-directed divergence.

Particular case: For $\alpha = 1$, $I_{E_\alpha}(A,B)$ reduces to $I(A,B)$ given in (2).

Properties of generalized exponential fuzzy divergence measure

The generalized exponential fuzzy divergence measure $I_{E_\alpha}(A,B)$ defined above has the following properties:

Theorem 2 (a) $I_{E_\alpha}(A\cup B,A) + I_{E_\alpha}(A\cap B,A) = I_{E_\alpha}(B,A)$

(b) $I_{E_\alpha}(A\cup B,C) + I_{E_\alpha}(A\cap B,C) = I_{E_\alpha}(A,C) + I_{E_\alpha}(B,C)$

(c) $I_{E_\alpha}(\overline{A\cup B},\overline{A\cap B}) = I_{E_\alpha}(\bar{A}\cap\bar{B},\bar{A}\cup\bar{B},)$

Proof: 2(a) Let us consider the sets

$$X_1 = \{x/x \in X, \mu_A(x_i) \geq \mu_B(x_i)\} \tag{7}$$

$$\text{and } X_2 = \{x/x \in X, \mu_A(x_i) < \mu_B(x_i)\} \tag{8}$$

Using the notions explained above in the 'Preliminaries' section:

$$\begin{aligned}
&\text{In set } X_1, A \cup B = \text{Union of } A \text{ and } B \Leftrightarrow \mu_{A \cup B}(x) = \max\{\mu_A(x), \mu_B(x)\} = \mu_A(x)\\
&A \cap B = \text{Intersection of } A \text{ and } B \Leftrightarrow \mu_{A \cap B}(x) = \min\{\mu_A(x), \mu_B(x)\} = \mu_B(x)\\
&\text{In set } X_2, A \cup B = \text{Union of } A \text{ and } B \Leftrightarrow \mu_{A \cup B}(x) = \max\{\mu_A(x), \mu_B(x)\} = \mu_B(x)\\
&A \cap B = \text{Intersection of } A \text{ and } B \Leftrightarrow \mu_{A \cap B}(x) = \min\{\mu_A(x), \mu_B(x)\} = \mu_A(x)
\end{aligned}$$

We have

$$\begin{aligned}
I_{E_\alpha}(A \cup B, A) + I_{E_\alpha}(A \cap B, A) &= \sum_{i=1}^{n}\Big[1-(1-\mu_{A\cup B}(x_i))e^{((1-\mu_A(x_i))^\alpha-(1-\mu_{A\cup B}(x_i))^\alpha)}\\
&\qquad -\mu_{A\cup B}(x_i)e^{\left(\mu_A^\alpha(x_i)-\mu_{A\cup B}^\alpha(x_i)\right)}\Big]\\
&\quad +\sum_{i=1}^{n}\Big[1-(1-\mu_{A\cap B}(x_i))e^{((1-\mu_A(x_i))^\alpha-(1-\mu_{A\cap B}(x_i))^\alpha)}\\
&\qquad -\mu_{A\cap B}(x_i)e^{\left(\mu_A^\alpha(x_i)-\mu_{A\cap B}^\alpha(x_i)\right)}\Big]\\
&= \Bigg\{\sum_{X_1}\Big[1-(1-\mu_A(x_i))e^{((1-\mu_A(x_i))^\alpha-(1-\mu_A(x_i))^\alpha)}\\
&\qquad -\mu_A(x_i)e^{\left(\mu_A^\alpha(x_i)-\mu_A^\alpha(x_i)\right)}\Big]\\
&\quad +\sum_{X_2}\Big[1-(1-\mu_B(x_i))e^{((1-\mu_A(x_i))^\alpha-(1-\mu_B(x_i))^\alpha)}\\
&\qquad -\mu_B(x_i)e^{\left(\mu_A^\alpha(x_i)-\mu_B^\alpha(x_i)\right)}\Big]\Bigg\}\\
&\quad +\Bigg\{\sum_{X_1}\Big[1-(1-\mu_B(x_i))e^{((1-\mu_A(x_i))^\alpha-(1-\mu_B(x_i))^\alpha)}\\
&\qquad -\mu_B(x_i)e^{\left(\mu_A^\alpha(x_i)-\mu_B^\alpha(x_i)\right)}\Big]\\
&\quad +\sum_{X_2}\Big[1-(1-\mu_A(x_i))e^{((1-\mu_A(x_i))^\alpha-(1-\mu_A(x_i))^\alpha)}\\
&\qquad -\mu_A(x_i)e^{\left(\mu_A^\alpha(x_i)-\mu_A^\alpha(x_i)\right)}\Big]\Bigg\}\\
&= \sum_{X_1}\Big[1-(1-\mu_B(x_i))e^{((1-\mu_A(x_i))^\alpha-(1-\mu_B(x_i))^\alpha)}\\
&\qquad -\mu_B(x_i)e^{\left(\mu_A^\alpha(x_i)-\mu_B^\alpha(x_i)\right)}\Big]\\
&\quad +\sum_{X_2}\Big[1-(1-\mu_B(x_i))e^{((1-\mu_A(x_i))^\alpha-(1-\mu_B(x_i))^\alpha)}\\
&\qquad -\mu_B(x_i)e^{\left(\mu_A^\alpha(x_i)-\mu_B^\alpha(x_i)\right)}\Big]\\
&= \sum_{i=1}^{n}\Big[1-(1-\mu_B(x_i))e^{((1-\mu_A(x_i))^\alpha-(1-\mu_B(x_i))^\alpha)}\\
&\qquad -\mu_B(x_i)e^{\left(\mu_A^\alpha(x_i)-\mu_B^\alpha(x_i)\right)}\Big] = I_{E_\alpha}(B, A)
\end{aligned}$$

Thus, $I_{E_\alpha}(A \cup B, A) + I_{E_\alpha}(A \cap B, A) = I_{E_\alpha}(B, A)$

Hence, the proof of 2(a) holds.

$$
\begin{aligned}
&2(\mathrm{b})\, I_{E_\alpha}(A\cup B, C) + I_{E_\alpha}(A\cap B, C)\\
&= \sum_{i=1}^{n}\left[1-(1-\mu_{A\cup B}(x_i))e^{((1-\mu_C(x_i))^\alpha-(1-\mu_{A\cup B}(x_i))^\alpha)}-\mu_{A\cup B}(x_i)e^{\left(\mu_C^\alpha(x_i)-\mu_{A\cup B}^\alpha(x_i)\right)}\right]\\
&\quad+\sum_{i=1}^{n}\left[1-(1-\mu_{A\cap B}(x_i))e^{((1-\mu_C(x_i))^\alpha-(1-\mu_{A\cap B}(x_i))^\alpha)}-\mu_{A\cap B}(x_i)e^{\left(\mu_C^\alpha(x_i)-\mu_{A\cap B}^\alpha(x_i)\right)}\right]\\
&= \sum_{X_1}\left[1-(1-\mu_A(x_i))e^{((1-\mu_C(x_i))^\alpha-(1-\mu_A(x_i))^\alpha)}-\mu_A(x_i)e^{\left(\mu_C^\alpha(x_i)-\mu_A^\alpha(x_i)\right)}\right]\\
&\quad+\sum_{X_2}\left[1-(1-\mu_B(x_i))e^{((1-\mu_C(x_i))^\alpha-(1-\mu_B(x_i))^\alpha)}-\mu_B(x_i)e^{\left(\mu_C^\alpha(x_i)-\mu_B^\alpha(x_i)\right)}\right]\\
&\quad+\sum_{X_1}\left[1-(1-\mu_B(x_i))e^{((1-\mu_C(x_i))^\alpha-(1-\mu_B(x_i))^\alpha)}-\mu_B(x_i)e^{\left(\mu_C^\alpha(x_i)-\mu_B^\alpha(x_i)\right)}\right]\\
&\quad+\sum_{X_2}\left[1-(1-\mu_A(x_i))e^{((1-\mu_C(x_i))^\alpha-(1-\mu_A(x_i))^\alpha)}-\mu_A(x_i)e^{\left(\mu_C^\alpha(x_i)-\mu_A^\alpha(x_i)\right)}\right]\\
&= \sum_{i=1}^{n}\left[1-(1-\mu_A(x_i))e^{((1-\mu_C(x_i))^\alpha-(1-\mu_A(x_i))^\alpha)}-\mu_A(x_i)e^{\left(\mu_C^\alpha(x_i)-\mu_A^\alpha(x_i)\right)}\right]\\
&\quad+\sum_{i=1}^{n}\left[1-(1-\mu_B(x_i))e^{((1-\mu_C(x_i))^\alpha-(1-\mu_B(x_i))^\alpha)}-\mu_B(x_i)e^{\left(\mu_C^\alpha(x_i)-\mu_B^\alpha(x_i)\right)}\right]\\
&= I_{E_\alpha}(A, C) + I_{E_\alpha}(B, C)
\end{aligned}
$$

Hence, the proof of 2(b) holds.

$$
\begin{aligned}
2(\mathrm{c})\, I_{E_\alpha}\left(\overline{A\cup B}, \overline{A\cap B}\right) &= \sum_{i=1}^{n}\Big[1-(1-\mu_{\overline{A\cup B}}(x_i))e^{\left((1-\mu_{\overline{A\cap B}}(x_i))^\alpha-(1-\mu_{\overline{A\cup B}}(x_i))^\alpha\right)}\\
&\qquad -\mu_{\overline{A\cup B}}(x_i)e^{\left(\mu^\alpha_{\overline{A\cap B}}(x_i)-\mu^\alpha_{\overline{A\cup B}}(x_i)\right)}\Big]
\end{aligned}
$$

$$
\begin{aligned}
&= \sum_{i=1}^{n}\left[1-\mu_{A\cup B}(x_i)e^{\left(\mu_{A\cap B}^\alpha(x_i)-\mu_{A\cup B}^\alpha(x_i)\right)}-(1-\mu_{A\cup B}(x_i))e^{((1-\mu_{A\cap B}(x_i))^\alpha-(1-\mu_{A\cup B}(x_i))^\alpha)}\right]\\
&= \sum_{X_1}\left[1-\mu_A(x_i)e^{\left(\mu_B^\alpha(x_i)-\mu_A^\alpha(x_i)\right)}-(1-\mu_A(x_i))e^{((1-\mu_B(x_i))^\alpha-(1-\mu_A(x_i))^\alpha)}\right]\\
&\quad+\sum_{X_2}\left[1-\mu_B(x_i)e^{\left(\mu_A^\alpha(x_i)-\mu_B^\alpha(x_i)\right)}-(1-\mu_B(x_i))e^{((1-\mu_A(x_i))^\alpha-(1-\mu_B(x_i))^\alpha)}\right]
\end{aligned}
$$

and

$$
\begin{aligned}
I_{E_\alpha}(\bar{A}\cap\bar{B}, \bar{A}\cup\bar{B}) &= \sum_{i=1}^{n}\left[1-(1-\mu_{\bar{A}\cap\bar{B}}(x_i))e^{((1-\mu_{\bar{A}\cup\bar{B}}(x_i))^\alpha-(1-\mu_{\bar{A}\cap\bar{B}}(x_i))^\alpha)}-\mu_{\bar{A}\cap\bar{B}}(x_i)e^{\left(\mu^\alpha_{\bar{A}\cup\bar{B}}(x_i)-\mu^\alpha_{\bar{A}\cap\bar{B}}(x_i)\right)}\right]\\
&= \sum_{X_1}\left[1-(1-\mu_{\bar{A}}(x_i))e^{((1-\mu\bar{B}(x_i))^\alpha-(1-\mu\bar{A}(x_i))^\alpha)}-\mu_{\bar{A}}(x_i)e^{\left(\mu^\alpha_{\bar{B}}(x_i)-\mu\bar{A}^\alpha(x_i)\right)}\right]\\
&\quad+\sum_{X_2}\left[1-(1-\mu_{\bar{B}}(x_i))e^{((1-\mu\bar{A}(x_i))^\alpha-(1-\mu\bar{B}(x_i))^\alpha)}-\mu_{\bar{B}}(x_i)e^{\left(\mu^\alpha_{\bar{A}}(x_i)-\mu\bar{B}^\alpha(x_i)\right)}\right]\\
&= \sum_{X_1}\left[1-\mu_A(x_i)e^{\left(\mu_B^\alpha(x_i)-\mu_A^\alpha(x_i)\right)}-(1-\mu_A(x_i))e^{((1-\mu_B(x_i))^\alpha-(1-\mu_A(x_i))^\alpha)}\right]\\
&\quad+\sum_{X_2}\left[1-\mu_B(x_i)e^{\left(\mu_A^\alpha(x_i)-\mu_B^\alpha(x_i)\right)}-(1-\mu_B(x_i))e^{((1-\mu_A(x_i))^\alpha-(1-\mu_B(x_i))^\alpha)}\right]\\
&= I_{E_\alpha}\left(\overline{A\cup B}, \overline{A\cap B}\right)
\end{aligned}
$$

Thus, $I_{E_\alpha}\left(\overline{A\cup B}, \overline{A\cap B}\right) = I_{E_\alpha}(\bar{A}\cap\bar{B}, \bar{A}\cup\bar{B})$

Hence, the proof of 2(c) holds.

Theorem 3 (a) $I_{E_\alpha}(A, \bar{A}) = I_{E_\alpha}(\bar{A}, A)$

(b) $I_{E_\alpha}(\bar{A}, \bar{B}) = I_{E_\alpha}(A, B)$

(c) $I_{E_\alpha}(A, \bar{B}) = I_{E_\alpha}(\bar{A}, B)$

(d) $I_{E_\alpha}(A, B) + I_{E_\alpha}(\bar{A}, B) = I_{E_\alpha}(\bar{A}, \bar{B}) + I_{E_\alpha}(A, \bar{B})$

$$\textbf{Proof}\,3(\mathrm{a})\, I_{E_\alpha}(A,\bar{A}) = \sum_{i=1}^{n}\left[1-(1-\mu_A(x_i))e^{\left((1-\mu\bar{A}(x_i))^\alpha-(1-\mu_A(x_i))^\alpha\right)}-\mu_A(x_i)e^{\left(\mu\bar{A}^\alpha(x_i)-\mu_A^\alpha(x_i)\right)}\right]$$
$$= \sum_{i=1}^{n}\left[1-(1-\mu_A(x_i))e^{\mu_A^\alpha(x_i)-(1-\mu_A(x_i))^\alpha})-\mu_A(x_i)e^{\left((1-\mu_A(x_i))^\alpha-\mu_A^\alpha(x_i)\right)}\right]$$

and

$$I_{E_\alpha}(\bar{A},A) = \sum_{i=1}^{n}\left[1-(1-\mu_{\bar{A}}(x_i))e^{\left((1-\mu_A(x_i))^\alpha-(1-\mu\bar{A}(x_i))^\alpha\right)}-\mu_{\bar{A}}(x_i)e^{\left(\mu_A^\alpha(x_i)-\mu\bar{A}^\alpha(x_i)\right)}\right]$$
$$= \sum_{i=1}^{n}\left[1-\mu_A(x_i)e^{\left((1-\mu_A(x_i))^\alpha-\mu_A^\alpha(x_i)\right)}-(1-\mu_A(x_i))e^{\mu_A^\alpha(x_i)-(1-\mu_A(x_i))^\alpha)}\right]$$
$$= I_{E_\alpha}(A,\bar{A})$$

Thus, $I_{E_\alpha}(A,\bar{A}) = I_{E_\alpha}(\bar{A},A)$
Hence, the proof of 3(a) holds.

$$3(\mathrm{b})\, I_{E_\alpha}(\bar{A},\bar{B}) = \sum_{i=1}^{n}\left[1-(1-\mu_{\bar{A}}(x_i))e^{\left((1-\mu\bar{B}(x_i))^\alpha-(1-\mu\bar{A}(x_i))^\alpha\right)}-\mu_{\bar{A}}(x_i)e^{\left(\mu\bar{B}^\alpha(x_i)-\mu\bar{A}^\alpha(x_i)\right)}\right]$$
$$= \sum_{i=1}^{n}\left[1-\mu_A(x_i)e^{\left(\mu_B^\alpha(x_i)-\mu_A^\alpha(x_i)\right)}-(1-\mu_A(x_i))e^{((1-\mu_B(x_i))^\alpha-(1-\mu_A(x_i))^\alpha)}\right]$$
$$= I_{E_\alpha}(A,B)$$

Thus, $I_{E_\alpha}(\bar{A},\bar{B}) = I_{E_\alpha}(A,B)$
Hence, the proof of 3(b) holds.

$$3(\mathrm{c})\, I_{E_\alpha}(A,\bar{B}) = \sum_{i=1}^{n}\left[1-(1-\mu_A(x_i))e^{\left((1-\mu\bar{B}(x_i))^\alpha-(1-\mu_A(x_i))^\alpha\right)}-\mu_A(x_i)e^{\left(\mu\bar{B}^\alpha(x_i)-\mu_A^\alpha(x_i)\right)}\right]$$
$$= \sum_{i=1}^{n}\left[1-(1-\mu_A(x_i))e^{\mu_B^\alpha(x_i)-(1-\mu_A(x_i))^\alpha})-\mu_A(x_i)e^{\left((1-\mu_B(x_i))^\alpha-\mu_A^\alpha(x_i)\right)}\right]$$

$$\text{Now,}\quad I_{E_\alpha}(\bar{A},B) = \sum_{i=1}^{n}\left[1-(1-\mu_{\bar{A}}(x_i))e^{\left((1-\mu_B(x_i))^\alpha-(1-\mu\bar{A}(x_i))^\alpha\right)}-\mu_{\bar{A}}(x_i)e^{\left(\mu_B^\alpha(x_i)-\mu\bar{A}^\alpha(x_i)\right)}\right]$$
$$= \sum_{i=1}^{n}\left[1-\mu_A(x_i)e^{\left((1-\mu_B(x_i))^\alpha-\mu_A^\alpha(x_i)\right)}-(1-\mu_A(x_i))e^{\mu_B^\alpha(x_i)-(1-\mu_A(x_i))^\alpha)}\right]$$
$$= I_{E_\alpha}(A,\bar{B})$$

Thus, $I_{E_\alpha}(A,\bar{B}) = I_{E_\alpha}(\bar{A},B)$
Hence, the proof of 3(c) holds.
3(d) It obviously follows from 3(b) and 3(c).
Theorem 4 Relation between $E_\alpha(A)$ and $I_{E_\alpha}(A,B)$ is given by

$$E_\alpha(A) = 1-\frac{e^{1-0.5^\alpha}}{n(e^{1-0.5^\alpha}-1)}I_{E_\alpha}\left(A,\left[\frac{1}{2}\right]\right)$$

$$\textbf{Proof } I_{E_\alpha}\left(A,\left[\frac{1}{2}\right]\right)=\sum_{i=1}^{n}\left[1-(1-\mu_A(x_i))e^{\left(0.5^\alpha-(1-\mu_A(x_i))^\alpha\right)}-\mu_A(x_i)e^{\left(0.5^\alpha-\mu_A^\alpha(x_i)\right)}\right]$$

$$=\sum_{i=1}^{n}\left[1-(1-\mu_A(x_i))\frac{e^{-(1-\mu_A(x_i))^\alpha}}{e^{-0.5^\alpha}}-\mu_A(x_i)\frac{e^{-\mu_A^\alpha(x_i)}}{e^{-0.5^\alpha}}\right]$$

$$=\sum_{i=1}^{n}\left[1-(1-\mu_A(x_i))\frac{e^{1-(1-\mu_A(x_i))^\alpha}}{e^{1-0.5^\alpha}}-\mu_A(x_i)\frac{e^{1-\mu_A^\alpha(x_i)}}{e^{1-0.5^\alpha}}\right]$$

$$=n-\frac{1}{e^{(1-0.5^\alpha)}}\sum_{i=1}^{n}\left[(1-\mu_A(x_i))e^{\left(1-(1-\mu_A(x_i))^\alpha\right)}+\mu_A(x_i)e^{\left(1-\mu_A^\alpha(x_i)\right)}+1-1\right]$$

$$=n-\frac{1}{e^{(1-0.5^\alpha)}}\sum_{i=1}^{n}\left[\mu_A(x_i)e^{\left(1-\mu_A^\alpha(x_i)\right)}+(1-\mu_A(x_i))e^{1-(1-\mu_A(x_i))^\alpha}-1\right]-\frac{n}{e^{(1-0.5^\alpha)}}$$

$$=n-\frac{1}{e^{(1-0.5^\alpha)}}n\left(e^{(1-0.5^\alpha)}-1\right)E_\alpha(A)-\frac{n}{e^{(1-0.5^\alpha)}}$$

$$=n\left(1-\frac{1}{e^{(1-0.5^\alpha)}}\right)-\frac{1}{e^{(1-0.5^\alpha)}}n\left(e^{(1-0.5^\alpha)}-1\right)E_\alpha(A)$$

$$=\frac{n(e^{(1-0.5^\alpha)}-1)}{e^{(1-0.5^\alpha)}}-\frac{n(e^{(1-0.5^\alpha)}-1)}{e^{(1-0.5^\alpha)}}E_\alpha(A)=(1-E_\alpha(A))\frac{n(e^{(1-0.5^\alpha)}-1)}{e^{(1-0.5^\alpha)}}$$

$$\Rightarrow\frac{e^{(1-0.5^\alpha)}}{n(e^{(1-0.5^\alpha)}-1)}I_{E_\alpha}\left(A,\left[\frac{1}{2}\right]\right)=(1-E_\alpha(A))$$

$$\text{Thus, } E_\alpha(A)=1-\frac{e^{(1-0.5^\alpha)}}{n(e^{(1-0.5^\alpha)}-1)}I_{E_\alpha}\left(A,\left[\frac{1}{2}\right]\right)$$

Hence, the proof of theorem 4 holds.

A comparative study

In this section, we demonstrate the efficiency of proposed fuzzy divergence measure by comparing it with some of existing fuzzy divergence measures. To do so, we present the comparative study of the proposed divergence measure with the existing fuzzy divergence measures given by Kapur [3], Prakash et al. [12], and Bajaj and Hooda [16].

Let A and B be any two fuzzy sets given A = {0.1, 0.9, 0.5}, B = {0.6, 0.7, 0.1}.

The computed values of fuzzy divergence measures $\bar{I}_\alpha(A:B)$, $I_a(A:B)$, $D_\alpha(A,B)$, $I_{E_\alpha}(A,B)$ are presented in Table 1.

Table 1 and Figure 1 depict the minimization of degree of difference of the proposed fuzzy measure. It is clear that the proposed fuzzy divergence measure is efficient than the existing fuzzy divergence measures.

Application of parametric generalized exponential measure of fuzzy divergence in strategic decision making

As we have already discussed in the introductory section, in recent years, the applications of the fuzzy divergence measure have been given in different areas: Poletti et al. [23] in bio-informatics; Bhandari et al. [21], Fan et al. [24], and Bhatia and Singh [25] in image thresholding; and Ghosh et al. [10] in automated leukocyte recognition. We

Table 1 Computed values of fuzzy divergence measures: $I_\alpha(A,B), D_\alpha(A,B), I_\alpha(A:B), I_{E_\alpha}(A,B)$

Fuzzy divergence measure	a, α = 0.1	a, α = 0.4	a, α = 0.5	a, α = 0.6	a, α = 0.9
$I_\alpha(A, B)$	0.1243	0.4743	0.5875	0.7006	1.4809
$D_\alpha(A, B)$	0.1276	0.5058	0.6273	0.7456	1.0766
$I_\alpha(A: B)$	1.1734	1.0272	0.9966	0.9681	0.8933
$I_{E_\alpha}(A, B)$	0.0977	0.2406	0.2805	0.3017	0.3416

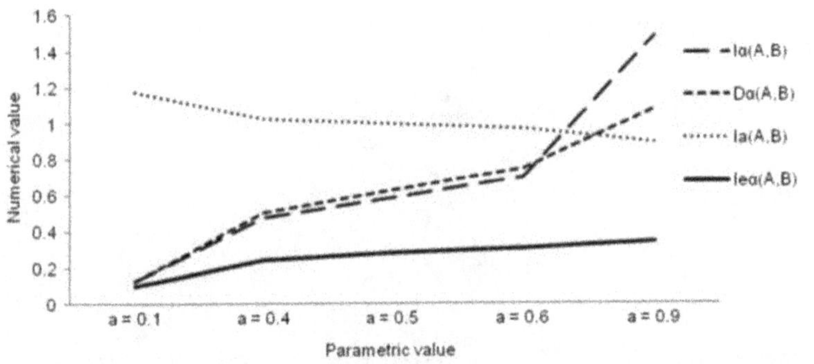

Figure 1 Comparison of numerical values of fuzzy divergence measures $I_\alpha(A,B), D_\alpha(A,B), \mathrm{I}_\alpha(A:B), \mathbf{I}_{\mathbf{E}_\alpha}(\mathbf{A},\mathbf{B})$.

provide an application of the proposed divergence measure in strategic decision-making.

Decision-making problem is the process of finding the best option from all of the feasible alternatives. It is assumed that a firm X desire to apply m strategies $S_1, S_2, S_3, \ldots, S_m$ to meet its goal. Let each strategy have different degree of effectiveness if the input associated with it is varied and let $\{I_1, I_2, I_3, \ldots, I_n\}$ be the input set or set of alternatives. Let the fuzzy set Y denotes the effectiveness of a particular strategy with uniform input. Therefore,

$$Y = \{(Y, \mu_Y(S_i))/i = 1, 2, \ldots, m\}$$

Further, let I_j be a fuzzy set denotes the degree of effectiveness of a strategy when it is implemented with input I_j.

$$I_j = \left\{\left(I_j, \mu_{I_j}(S_i)\right)/i = 1, 2, \ldots, m\right\} \text{where } j = 1, 2, \ldots, n.$$

Taking $A = Y$ and $B = I_j$ in the fuzzy divergence measure $I_{E_\alpha}(Y, I_j)$ defined in new parametric generalized exponential measure of fuzzy divergence section, and we calculate $I_{E_\alpha}(Y, I_j)$.

Then, most effective I_j is determined by $Min\{I_{E_\alpha}(Y, I_j)\}_{\substack{1\le j\le n\\ 0<\alpha\le 0.5}}$. It is assumed that I_t $(1 \le t \le n)$ determines the minimum value of $\{I_{E_\alpha}(Y, I_j)\}_{0<\alpha\le 0.5}$. With this I_t find $Max.\left\{\mu_{I_j}(S_i)\right\}_{1\le i\le m}$, let it correspond to S_p, $1 \le p \le m$.

Hence, if the strategy S_p is implemented with input budget of I_t, the firm will meet its goal in the most input-effective manner.

An illustrative example

Let $m = n = 5$ in the above model. Table 2 shows the efficiency of different strategies at uniform inputs.

Table 3 illustrates the efficiency of different strategies at particular inputs.

Table 2 Efficiency of different strategies at uniform inputs

$\mu_Y(S_1)$	$\mu_Y(S_2)$	$\mu_Y(S_3)$	$\mu_Y(S_4)$	$\mu_Y(S_5)$
0.5	0.3	0.6	0.4	0.7

Table 3 Efficiency of strategies at particular inputs

I_j	$\mu_{I_j}(S_1)$	$\mu_{I_j}(S_2)$	$\mu_{I_j}(S_3)$	$\mu_{I_j}(S_4)$	$\mu_{I_j}(S_5)$
I_1	0.3	0.4	0.5	0.8	0.6
I_2	0.6	0.3	0.9	0.2	0.4
I_3	0.7	0.5	0.8	0.9	0.6
I_4	0.8	0.7	0.6	0.5	0.3
I_5	0.5	0.8	0.4	0.7	0.5

The numerical values of divergence measure $I_{E_\alpha}(Y, I_j)\}_{\substack{1\le j\le n\\ 0<\alpha\le 0.5}}$ are presented in Table 4.

The calculated numerical values of the proposed fuzzy divergence measure indicate that alternative input I_2 is more appropriate for different values of α ($0 < \alpha \le 0.5$). An examination of the results presented in Tables 3 and 4 makes it clear that strategy S_4 is most effective. Thus, a firm will achieve its goal most effectively if the strategy S_4 is applied with an input alternative I_2.

Application of fuzzy TOPSIS and MOORA methods for strategic decision making: a comparative analysis[a]

We now present the application of TOPSIS [26] and MOORA [27] methods for strategic decision-making using the proposed fuzzy divergence measure (6).

Fuzzy TOPSIS method

Let us assume that there exists a set $I = \{I_1, I_2, I_3, \ldots, I_n\}$ of n alternative inputs and a set of m attributes (strategies) given by $S = \{S_1, S_2, S_3, \ldots, S_m\}$. The decision maker has to find the best alternative from the set I corresponding to set S of n attributes (strategies).

The computational procedure of solving the fuzzy strategic decision-making problem involves the following steps:

1. Construct a fuzzy decision matrix.

$$\begin{array}{ccccc} & S_1 & S_2 & \cdots & S_n \\ I_1 & x_{11} & x_{12} & \cdots & x_{1n} \\ I_2 & x_{21} & x_{22} & \cdots & x_{2n} \\ I_m & x_{m1} & x_{m2} & \cdots & x_{mn} \end{array}$$

$$W = [w_1, w_2, \ldots, w_n]$$

Table 4 Numerical values of divergence measure $\{I_{E_\alpha}(Y, I_j)\}_{\substack{1\le j\le n\\ 0<\alpha\le 0.5}}$

I_j	$\alpha = 0.1$	$\alpha = 0.3$	$\alpha = 0.4$	$\alpha = 0.5$
I_1	0.0456	0.0831	0.0900	0.0916
I_2	0.0449	0.0662	0.0568	0.0396
I_3	0.0779	0.1308	0.1289	0.1164
I_4	0.0748	0.1500	0.1646	0.1700
I_5	0.0764	0.1568	0.1748	0.1843

Table 5 Normalized/weighted normalized fuzzy decision-matrix

	$\mu_{I_j}(S_1)$	$\mu_{\mu I_j}(S_2)$	$\mu_{I_j}(S_3)$	$\mu_{I_j}(S_4)$	$\mu_{I_j}(S_5)$
I_1	0.2218	0.3133	0.3356	0.5010	0.5432
I_2	0.4435	0.2350	0.6040	0.1252	0.3622
I_3	0.5174	0.3916	0.5369	0.5636	0.5432
I_4	0.5914	0.5483	0.4027	0.3131	0.2716
I_5	0.3696	0.6266	0.2685	0.4383	0.4527

2. Construct the normalized fuzzy decision matrix. The normalized value n_{ij} is calculated as

$$as\ n_{ij} = x_{ij}/\sqrt{\sum_{j=1}^{m} x_{ij}^2}, j = 1, \ldots, m,\ i = 1, \ldots, n \tag{9}$$

3. Construct the weighted normalized fuzzy decision matrix, the weighted normalized value

$$v_{ij} = w_i n_{ij}, j = 1, \ldots, m,\ i = 1, \ldots, n \tag{10}$$

where weighted matrix for each strategy is as follows: W = [1,1,1,1,1] and w_i is the weight of ith attribute.

4. Determine the fuzzy positive ideal and fuzzy negative ideal solution A^+ and A^-, using the formulas

$$A^+ = \{v_1^+, v_2^+, \ldots, v_n^+\} = \left\{\max_j v_{ij}/i\in I\right\} \tag{11}$$

$$\text{and } A^- = \{v_1^-, v_2^-, \ldots, v_n^-\} = \left\{\min_j v_{ij}/i\in I\right\} \text{ respectively, where } I \text{ is associated with input set.} \tag{12}$$

5. Calculate the separation of each alternative input from positive ideal solution and negative ideal solution, respectively, using the proposed measure (6).
6. Calculate the relative closeness of each alternative to positive ideal solution using the formula

$$R_j = I_{E_\alpha}^- / \left(I_{E_\alpha}^+ + I_{E_\alpha}^-\right), j = 1, \ldots, m. \tag{13}$$

7. Rank the preference order of all alternatives according to the closeness coefficient.

Now the application of proposed measure $I_{E_\alpha}(A, B)$ with TOPSIS technique is demonstrated using the fuzzy decision matrix considered in Table 3.

Table 5 presents the normalized/weighted normalized fuzzy decision matrix corresponding to the fuzzy decision matrix given in Table 3 using the formulas (9) and (10).

Table 6 Fuzzy positive and negative ideal solution

	$\mu_{I_j}(S_1)$	$\mu_{I_j}(S_2)$	$\mu_{I_j}(S_3)$	$\mu_{I_j}(S_4)$	$\mu_{I_j}(S_5)$
A^+	0.5432	0.6040	0.5636	0.5914	0.6266
A^-	0.2218	0.1252	0.3916	0.2716	0.2685

Table 7 Distance of each alternative from positive and negative ideal solution

	$I^+_{E_\alpha}$		$I^-_{E_\alpha}$	
	$\alpha = 0.1$	$\alpha = 0.5$	$\alpha = 0.1$	$\alpha = 0.5$
I_1	0.0463	0.1568	0.0296	0.0113
I_2	0.0819	0.2835	0.0263	0.0366
I_3	0.0090	0.0173	0.0611	0.0511
I_4	0.0418	0.1338	0.0616	0.0732
I_5	0.0306	0.0986	0.0682	0.0847

Table 6 shows the fuzzy positive and negative ideal solutions A^+ and A^- using formulas (11) and (12).

The calculated numerical values of divergence/distance values of each alternative input from positive ideal solution and negative ideal solution using formula (6) are given in Table 7.

The best alternative is the one with the shortest distance to the fuzzy positive ideal solution and with the longest distance to the fuzzy negative ideal solution. The calculated values of relative closeness of each alternative to positive ideal solution using formula (13) and their corresponding ranks are shown in Table 8.

According to the closeness coefficient, the ranking of the preference order of these alternatives $I_j (j = 1, 2, 3, 4, 5)$:

$$\text{For}\, \alpha = 0.1,\, I_3 > I_5 > I_4 > I_1 > I_2.$$

$$\text{For}\, \alpha = 0.5,\, I_3 > I_5 > I_4 > I_2 > I_1.$$

Thus, here, we find that variation in values of α brings about change in ranking, but leaves the best choice unchanged. So I_3 is the most preferable alternative.

Fuzzy MOORA method

Fuzzy MOORA method for solving strategic decision-making problems is as follows. The computational procedure in fuzzy MOORA method up to step 4 is the same as discussed in fuzzy TOPSIS method above.

Step 5. Calculate the overall performance index $I_{E_\alpha}(A^+, A^-)$ for each alternative using the formula (6) and the computed values in Table 6.

Step 6. Ranking alternatives and/or selecting the most efficient one are based on the values of $I_{E_\alpha}(A^+, A^-)$.

The overall performance index $I_{E_\alpha}(A^+, A^-)$ for each alternative is calculated using formula (6) and values given in Table 6. Finally, the ranking results have been obtained using MOORA method and are presented in Table 9.

Table 8 Closeness coefficient and ranking

Alternatives (inputs)	$\alpha = 0.1$	Rank	$\alpha = 0.5$	Rank
I_1	0.3900	4	0.0672	5
I_2	0.2431	5	0.1143	4
I_3	0.8716	1	0.7471	1
I_4	0.5957	3	0.3536	3
I_5	0.6903	2	0.4621	2

Table 9 The ranking results obtained using MOORA method

	$I_{E_\alpha}(A^+, A^-)$			
	$\alpha = 0.1$	Rank	$\alpha = 0.5$	Rank
I_1	0.9816	3	0.9778	2
I_2	0.9528	5	0.9333	5
I_3	0.9951	1	0.9891	1
I_4	0.9823	2	0.9678	3
I_5	0.9775	4	0.9534	4

According to the calculation results, ranking order of alternatives is as follows:

$$\text{For}\,\alpha = 0.1, I_3 > I_4 > I_1 > I_5 > I_2.$$

$$\text{For}\,\alpha = 0.5, I_3 > I_1 > I_4 > I_5 > I_2.$$

Thus, here, we find that variation in values of α brings about change in ranking, but leaves the best choice unchanged. So I_3 is the most preferable alternative.

A comparative analysis of the proposed method and the existing methods of strategic decision-making

We now compare the proposed method of strategic decision-making with the existing method of strategic decision-making using the proposed measure (6). From the proposed method in the 'Application of parametric generalized exponential measure of fuzzy divergence in strategic decision making' section, it is clear that a firm will achieve its goal most effectively if the strategy S_4 is applied with an input alternative I_2. Thus, I_2 is best input alternative for a firm to achieve its goal. However, we above examine from the existing methods that I_3 is the most preferable input alternative. It is also noticed that the proposed method is a very short, simple, and consistent method than the existing methods of strategic decision-making having a computational procedure involving a number of steps. Thus, the proposed method of strategic decision-making is better than the existing methods.

Concluding remarks

In this paper, we have proposed and validated the generalized exponential measure of fuzzy divergence. We have established the relation between generalized exponential fuzzy entropy and the proposed fuzzy divergence measure. Particular case and some of the properties of this divergence measure are proved. The efficiency of the new divergence measure has been presented. In addition, application of the proposed divergence measure is discussed in strategic decision-making and a numerical example is given for illustration. The application of the new measure of fuzzy divergence in two existing methods of strategic decision-making is presented. A comparative analysis between the proposed method of strategic decision-making and the existing methods of strategic decision-making has also been provided. We note that our measure of fuzzy divergence comprises greater consistency and flexibility in applications because of the presence of the parameter.

Endnote

[a]The authors are thankful to an anonymous referee of this journal for bringing this point to their attention.

Authors' information

Mathematics Subject Classification (2010) 94A17, 03E72, 90B50.

Acknowledgements

The authors would like to thank the editor and two anonymous referees of this journal for their constructive suggestions. Thanks are also due to Dr. Ramphul Ohlan, Assistant Professor, Institute of Management Studies and Research, Maharshi Dayanand University, Rohtak 124 001, Haryana, India, for his comments to improve the readability of the paper.

References

1. Shannon, CE: A mathematical theory of communication. Bell. Syst. Tech. J. **27**(3), 379–423 (1948). Accessed on September 19, 2014 http://cm.bell-labs.com/cm/ms/what/shannonday/shannon1948.pdf
2. Zadeh, LA: Fuzzy sets. Inf. Control. **8**(3), 338–353 (1965)
3. Kapur, JN: Measures of Fuzzy Information. Mathematical Sciences Trust Society, New Delhi (1997).
4. De Luca, A, Termini, S: A definition of non-probabilistic entropy in the setting of fuzzy set theory. Inf. Control. **20**(4), 301–312 (1972)
5. Bhandari, D, Pal, NR: Some new information measures for fuzzy sets. Inf. Sci. **67**(3), 209–228 (1993)
6. Pal, NR, Pal, SK: Object background segmentation using new definition of entropy. IEEE Proc. **136**(4), 284–295 (1989)
7. Verma, R, Sharma, BD: On generalized exponential fuzzy entropy. World Acad. Sci. Eng. Technol. **60**, 1402–1405 (2011). Accessed on September 5, 2014 http://waset.org/publications/9417/on-generalized-exponential-fuzzy-entropy
8. Kullback, S, Leibler, RA: On information and sufficiency. Ann. Math. Stat. **22**(1), 79–86 (1951). Accessed on September 22, 2014 http://www.csee.wvu.edu/~xinl/library/papers/math/statistics/Kullback_Leibler_1951.pdf
9. Fan, J, Xie, W: Distance measures and induced fuzzy entropy. Fuzzy Sets Syst. **104**(2), 305–314 (1999)
10. Ghosh, M, Das, D, Ray, C, Chakraborty, AK: Automated leukocyte recognition using fuzzy divergence. Micron **41**, 840–846 (2010)
11. Montes, S, Couso, I, Gil, P, Bertoluzza, C: Divergence measure between fuzzy sets. Int. J. Approx. Reason. **30**, 91–105 (2002)
12. Prakash, O, Sharma, PK, Kumar, S: Two new measures of fuzzy divergence and their properties. SQU J. Sci. **11**, 69–77 (2006). Accessed on September 10, 2014 http://web.squ.edu.om/squjs/volum11/MATH041130-corrected.pdf
13. Ferreri, C: Hyper entropy and related heterogeneity divergence and information measures. Statistica **40**(2), 155–168 (1980)
14. Renyi, A: On measures of entropy and information. Proc. 4th Berkeley Symp. Math. Stat. Probab. **1**, 547–561 (1961)
15. Sharma, BD, Mittal, DP: New non-additive measures of entropy for discrete probability distributions. J. Math. Sci (Calcutta). **10**, 28–40 (1975)
16. Bajaj, RK, Hooda, DS: On some new generalized measures of fuzzy information. World Acad. Sci. Eng. Technol. **62**, 747–753 (2010)
17. Bhatia, PK, Singh, S: Three families of generalized fuzzy directed divergence. AMO **14**(3), 599–614 (2012).
18. Taneja, IJ: On Mean Divergence Measures, pp. 1–18. (2005). Accessed on September 18, 2014 http://arxiv.org/pdf/math/0501298v2.pdf
19. Zadeh, LA: Probability measures of fuzzy events. J. Math. Anal. Appl. **23**, 421–427 (1968)
20. Bouchon-Meunier, B, Rifqi, M, Bothorel, S: Towards general measures of comparison of objects. Fuzzy Sets Syst. **84**, 143–153 (1996)
21. Bhandari, D, Pal, NR, Majumder, DD: Fuzzy divergence, probability measure of fuzzy events and image thresholding. Inf. Sci. **13**, 857–867 (1992)
22. Havrda, JH, Charvat, F: Quantification methods of classification processes: concepts of structural a entropy. Kybernetika **3**, 30–35 (1967)
23. Poletti, E, Zappelli, F, Ruggeri, A, Grisan, E: A review of thresholding strategies applied to human chromosome segmentation. Comput. Methods Prog. Biomed. **108**, 679–688 (2012)
24. Fan, S, Yang, S, He, P, Nie, H: Infrared electric image thresholding using two dimensional fuzzy entropy. Energey Procedia **12**, 411–419 (2011)
25. Bhatia, PK, Singh, S: A new measure of fuzzy directed divergence and its application in image segmentation. Int. J. Intell. Sys. Appl. **4**, 81–89 (2013). Accessed on September 12, 2014. http://www.mecs-press.org/ijisa/ijisa-v5-n4/IJISA-V5-N4-8.pdf
26. Hwang, CL, Yoon, K: Multiple Attribute Decision Making: Methods and Applications. Springer, New York (1981)
27. Brauers, WKM, Zavadskas, EK: The MOORA method and its application to privatization in transition economy. Control. Cybern. **35**(2), 443–468 (2006)

Permissions

All chapters in this book were first published in JUAA, by Springer; hereby published with permission under the Creative Commons Attribution License or equivalent. Every chapter published in this book has been scrutinized by our experts. Their significance has been extensively debated. The topics covered herein carry significant findings which will fuel the growth of the discipline. They may even be implemented as practical applications or may be referred to as a beginning point for another development.

The contributors of this book come from diverse backgrounds, making this book a truly international effort. This book will bring forth new frontiers with its revolutionizing research information and detailed analysis of the nascent developments around the world.

We would like to thank all the contributing authors for lending their expertise to make the book truly unique. They have played a crucial role in the development of this book. Without their invaluable contributions this book wouldn't have been possible. They have made vital efforts to compile up to date information on the varied aspects of this subject to make this book a valuable addition to the collection of many professionals and students.

This book was conceptualized with the vision of imparting up-to-date information and advanced data in this field. To ensure the same, a matchless editorial board was set up. Every individual on the board went through rigorous rounds of assessment to prove their worth. After which they invested a large part of their time researching and compiling the most relevant data for our readers.

The editorial board has been involved in producing this book since its inception. They have spent rigorous hours researching and exploring the diverse topics which have resulted in the successful publishing of this book. They have passed on their knowledge of decades through this book. To expedite this challenging task, the publisher supported the team at every step. A small team of assistant editors was also appointed to further simplify the editing procedure and attain best results for the readers.

Apart from the editorial board, the designing team has also invested a significant amount of their time in understanding the subject and creating the most relevant covers. They scrutinized every image to scout for the most suitable representation of the subject and create an appropriate cover for the book.

The publishing team has been an ardent support to the editorial, designing and production team. Their endless efforts to recruit the best for this project, has resulted in the accomplishment of this book. They are a veteran in the field of academics and their pool of knowledge is as vast as their experience in printing. Their expertise and guidance has proved useful at every step. Their uncompromising quality standards have made this book an exceptional effort. Their encouragement from time to time has been an inspiration for everyone.

The publisher and the editorial board hope that this book will prove to be a valuable piece of knowledge for researchers, students, practitioners and scholars across the globe.

List of Contributors

Wei-Yin Chen
Department of Chemical Engineering, University of Mississippi, Anderson Hall, University, MS 38677-9740, USA

Hamed Ahmadzade
Department of Statistics, Faculty of Mathematical Sciences, Ferdowsi University of Mashhad, Mashhad 91775-1159, Iran

Mohammad Amini
Department of Statistics, Faculty of Mathematical Sciences, Ferdowsi University of Mashhad, Mashhad 91775-1159, Iran

Seyed Mahmoud Taheri
Department of Engineering Science, College of Engineering, University of Tehran, Tehran 11365-4563, Iran

Abolghasem Bozorgnia
Department of Statistics, Faculty of Mathematical Sciences, Ferdowsi University of Mashhad, Mashhad 91775-1159, Iran

Rajkumar Verma
Department of Mathematics, Jaypee Institute of Information Technology (Deemed University), Noida-201307, Uttar Pradesh, India

Bhu Dev Sharma
Department of Mathematics, Jaypee Institute of Information Technology (Deemed University), Noida-201307, Uttar Pradesh, India

Rita Girão-Silva
Department of Electrical and Computer Engineering, University of Coimbra, Pólo II, Pinhal de Marrocos, Coimbra 3030-290, Portugal
Institute of Computers and Systems Engineering of Coimbra (INESC-Coimbra), R. Antero de Quental, 199, Coimbra 3000-033, Portugal

José Craveirinha
Department of Electrical and Computer Engineering, University of Coimbra, Pólo II, Pinhal de Marrocos, Coimbra 3030-290, Portugal
Institute of Computers and Systems Engineering of Coimbra (INESC-Coimbra), R. Antero de Quental, 199, Coimbra 3000-033, Portugal

João Clímaco
Institute of Computers and Systems Engineering of Coimbra (INESC-Coimbra), R. Antero de Quental, 199, Coimbra 3000-033, Portugal

Asim Pal
Department of Mathematics, Visva-Bharati, Santiniketan, 731235, India

Banibrata Mondal
Department of Mathematics, Visva-Bharati, Santiniketan, 731235, India

Namrata Bhattacharyya
Department of Mathematics, Visva-Bharati, Santiniketan, 731235, India

Swapan Raha
Department of Mathematics, Visva-Bharati, Santiniketan, 731235, India

Amalesh Kumar Manna
Department of Applied Mathematics with Oceanology and Computer Programming, Vidyasagar University, Midnapore, West Bengal 721102, India

Jayanta Kumar Dey
Department of Mathematics, Mahishadal Raj College, Mahishadal,West Bengal, 721628, India

Shyamal Kumar Mondal
Department of Applied Mathematics with Oceanology and Computer Programming, Vidyasagar University, Midnapore, West Bengal 721102, India

Yongchao Hou
Department of Mathematical Sciences, Chaohu University, Anhui 238000, China

Gurupada Maity
Department of Applied Mathematics with Oceanology and Computer Programming, Vidyasagar University, Midnapore, West Bengal 721102, India

Sankar Kumar Roy
Department of Applied Mathematics with Oceanology and Computer Programming, Vidyasagar University, Midnapore, West Bengal 721102, India

Vijay Prakash Tomar
Department of Mathematics, Deenbandhu Chhotu Ram University of Science and Technology, Murthal 131039, Haryana, India

Anshu Ohlan
Department of Mathematics, Deenbandhu Chhotu Ram University of Science and Technology, Murthal 131039, Haryana, India

www.ingramcontent.com/pod-product-compliance
Lightning Source LLC
Chambersburg PA
CBHW080703200326
41458CB00013B/4940
* 9 7 8 1 6 8 2 8 5 0 8 4 8 *